STUDENT'S SOLUTIONS MANUAL

Milton Loyer
Penn State University

to accompany

Elementary Statistics
Eighth Edition

Mario F. Triola

Boston San Francisco New York
London Toronto Sydney Tokyo Singapore Madrid
Mexico City Munich Paris Cape Town Hong Kong Montreal

ISBN   0-201-70465-X

2 3 4 5 6 7 8 9 10 PHTH 03 02 01

TABLE OF CONTENTS

Chapter 1: Introduction to Statistics..............................1

Chapter 2: Describing, Exploring, and Comparing Data................8

Chapter 3: Probability..37

Chapter 4: Probability Distributions..............................57

Chapter 5: Normal Probability Distributions.......................72

Chapter 6: Estimates and Sample Sizes............................105

Chapter 7: Hypothesis Testing....................................122

Chapter 8: Inferences from Two Samples...........................147

Chapter 9: Correlation and Regression............................179

Chapter 10: Multinomial Experiments and Contingency Tables.........208

Chapter 11: Analysis of Variance.................................222

Chapter 12: Statistical Process Control..........................237

Chapter 13: Nonparametric Statistics.............................248

PREFACE

This manual contains the solutions to the odd-numbered exercises for each section of the textbook Elementary Statistics, Eighth Edition, by Mario Triola, and the solutions for all end-of-chapter review and cumulative review exercises of that text. In the worked problems, intermediate steps are provided for the calculations. When appropriate, additional hints and comments are included and prefaced by NOTE.

Many statistical problems are best solved using particular formats. Recognizing and following these patterns promotes understanding and develops the capacity to apply the concepts to other problems. This manual identifies and employs such formats whenever practicable.

For best results, read the text carefully before attempting the exercises, and attempt the exercises before consulting the solutions. This manual has been prepared to provide a check and extra insights for exercises that have already been completed and to provide guidance for solving exercises that have already been attempted but have not been successfully completed.

I would like to thank Mario Triola for writing an excellent elementary statistics book and for inviting me to prepare this solutions manual.

Chapter 1

Introduction to Statistics

1-2 The Nature of Data

1. Statistic, since 20.7 refers to the selected sample.

3. Parameter, since 30 refers to the entire population.

5. Discrete, since the number of absent students must be an integer.

7. Discrete, since the number owning answering machines must be an integer.

9. Ratio, since differences are meaningful and zero height has a natural meaning.

11. Interval, since differences are meaningful but ratios are not. Refer to exercise #19.

13. Interval, since differences are meaningful but ratios are not. Years are not data at the ratio level of measurement because the year zero has been arbitrarily assigned so that the year 0 does not indicate the absence of time. The year 1900, for example, does not represent twice as much time as 950 -- and the ratio would be different using the Chinese or Jewish numerical representations for those years. Since the time difference between 1900 and 1920 is the same as the time difference between 1920 and 1940, however, years are data at the interval level of measurement.

15. Ordinal, since the ratings give relative position in a hierarchy.

17. Ratio, since differences are meaningful and zero ounces has a natural meaning.

19. Temperature ratios are not meaningful because a temperature of 0° does not represent the absence of temperature in the same sense that $0 represents the absence of money. The zero temperature in the example (whether Fahrenheit or Centigrade) was determined by a criterion other than "the absence of temperature."

1-3 Uses and Abuses of Statistics

1. Because the 186,000 respondents were self-selected and not randomly chosen, they are not necessarily representative of the general population and provide no usable information about the general population. In addition, the respondents were self-selected from a particular portion of the general population -- persons watching "Nightline" and able to spend the time and money to respond.

3. a. 500 + (.05)(500) = 500 + 25 = 525
   b. 525 - (.05)(500) = 525 - 26.25 = 498.75. No, because the 5% decrease is based on a larger amount than was the previous year's 5% increase.

5. That healthier babies are born to mothers who eat lobsters doesn't mean that eating lobster caused the babies to be healthier. Mothers who eat lobster are probably more affluent than the general population and would tend to eat better, be more knowledgeable about proper pre-natal care, have better health care, etc.

7. Motorcyclists that died in crashes in which helmets may have saved their lives could not be present to testify.

9. There are several possible answers. (1) Since tallness is perceived to be a favorable attribute, people tend to overstate their heights; at the very least, people would tend to round to the next highest inch and not to the nearest inch. (2) Many people do not really accurately know their height. (3) Because Americans tend to express height in feet and inches, errors might occur either in converting heights to all inches or in misstatements like 52" for 5'2". (4) Because many cultures express height in centimeters, some people might not know or be able to readily calculate their heights in inches.

11. No. Since the second 5% price cut would be based on a lower price, two consecutive 5% price cuts yield a smaller price reduction than a single 10% price cut. Mathematically, the two consecutive 5% cuts yield a reduction of .05x + .05(x-.05x) = .0975x, or a 9.75% price cut.

13. Assuming that each of the 20 individual subjects is ultimately counted as a success or not (i.e., that there are no "dropouts" or "partial successes"), the success rates in fraction form must be one of 0/20, 1/20, 2/20,..., 19/20, 20/20. In percentages, these rates are multiples of 5 (0%, 5%, 10%,..., 95%, 100%), and values such as 53% and 58% are not mathematical possibilities.

15. a. Since 100% is the totality of whatever is being measured, removing 100% of some quantity means that none of it is left.
    b. Reducing plaque by over 300% would mean removing three times as much plaque as is there, and then removing even more!

1-4 Design of Experiments

1. Observational study, since specific characteristics are measured on unmodified subjects.

3. Experiment, since the effect of an applied treatment is measured.

5. Random, since each 212 area code telephone number has an equal chance of being selected. But this is a really a complex situation, as indicated by the following NOTES.
NOTE 1: This ignores the fact that some residences may have more than one phone number. A residence with two different phone numbers (e.g., one for the parents and one for the teenagers) has twice the chance of being selected as does a residence with a single phone number.
NOTE 2: The scenario stated the organization sought to poll "residents" with the 212 area code. If the organization polls all residents at each selected number, this is cluster sampling. If the organization polls one resident at each selected number, the sample is not a random sample of "residents" because a resident living alone and having his own phone number has a higher chance of being selected than a resident living with others (e.g., in a family) and sharing a common phone number.
NOTE 3. The poll will not include residents in the 212 area code who do not have such phone numbers. This is not a problem if the intended population is phone customers (e.g., for a poll of satisfaction with phone service), but it is if the intended population is general residents (e.g., for a poll of satisfaction with garbage service).

7. Convenience, since the sample was simply those who happen to pass by.

9. Stratified, since the population of interest (assumed to be all car owners) was divided into 5 subpopulations from which the actual sampling was done.

11. Cluster, since the population of interest (assumed to be all students at The College of Newport) was divided into classes which were randomly selected in order to interview all the students in each selected class.
NOTE: Ideally the division into classes should place each student into one and only one class (e.g., if every student must take exactly one PE class each semester, select the PE classes at random. In practice such divisions are often made in ways that place some students in none of the classes (e.g., by selecting from all 2 pm W-W-F classes) or in more than one of the classes (e.g., by selecting from all the classes offered in the college). With careful handling, imperfect divisions do not significantly affect the results.

13. Stratified, since the population of interest (assumed to be all workers) was divided into 3 subpopulations from which the actual sampling was done.

15. Systematic, since every fifth element in the population (assumed to be all drivers passing the checkpoint during its operation) was sampled.

17. There are several possible answers. (1) Write each full-time student's name on a slip of paper, place the slips in a box, mix them thoroughly, and select 200 of them. (2) Assign each full-time a number (e.g., alphabetically), and use a table of random digits (or a calculator or a computer) to generate 200 random numbers with the appropriate numbers of digits.

19. Obtain from each college bookstore a list of the textbooks currently beings used, and compile a single master list (i.e., without duplications). Number the textbooks on the master list, and use a table of random digits (or a calculator or a computer) to generate 200 numbers with the appropriate numbers of digits.

21. a. Open questions elicit the respondent's true feelings without putting words or ideas into his mind. In addition, open questions might produce responses the pollster failed to consider. Unfortunately open questions sometimes produce responses that are rambling, unintelligible or not relevant.
    b. Closed questions help to focus the respondent and prevent misinterpretation of the question. Sometimes, however, closed questions reflect only the wording and opinions of the pollster and do not allow respondents to express legitimate alternatives.
    c. Closed questions are easier to analyze because the pollster can control the number of possible responses to each question and can word the responses to establish relationships between the responses and with other questions.

23. Confounding occurs when the researcher is not able to determine which factor (often one planned and one unplanned) produced an observed effect. If a restaurant tries adding an evening buffet for one week and it is the same week a nearby theater happens to show a real blockbuster that attracts unusual crowds to the neighborhood, the restaurant can not know whether its increased business is due to the new buffet or the extra traffic created by the theater.

25. a. Possibly; no. Stratified random sampling can employ either the same sample size for each stratum or different sample sizes for the various strata. It results in a random sample only when the sample size for each stratum is proportional to the size of the stratum -- i.e., the same proportion (and not the same number) of each stratum is selected for the sample. If the strata are all the same size, then use the same sample size for each; if one stratum is half the size of the others, then its sample size should be half the other sample sizes. If one stratum is half the size of the others and the same sample size is used for each of the strata, then an element in the smaller stratum has a larger chance of being selected than an element in a larger strata -- and that violates the definition of a random sample that requires that each element has the same chance of being selected. Stratified sampling can never result in a simple random sample. It guarantees that the total sample will always include elements from each of the strata, and that total samples without any elements from one of the strata can not occur -- and that violates the definition of simple random sampling that requires that each total sample has the same chance of being selected.
    b. Possibly; no. When each element in the population is in one and only cluster, cluster sampling always results in a random sample. The chance that any element is selected is the chance that its cluster is selected; since each cluster has the same chance of being selected, each element has the same chance of being selected -- and that satisfies the definition of a random sample. Cluster sampling can never result in a simple random sample. It guarantees that total samples with elements from each of the clusters can not occur -- and that violates the definition of simple random sampling that requires that each total sample has the same chance of being selected.

Review Exercises

1. a. Discrete, since the number of shares held must be an integer.
   NOTE: Even if partial shares are allowed (e.g., 5½ shares), the number of shares must be some fractional value and not any value on a continuum -- e.g., a person could not own π shares.
   b. Ratio, since differences between values are consistent and there is a natural zero.
   c. Stratified, since the set of interest (all stockholders) was divided into subpopulations (by states) from which the actual sampling was done.
   d. Statistic, since the value is determined from a sample and not the entire population.
   e. There is no unique correct answer, but the following are reasonable possibilities. (1) The proportion of stockholders holding above that certain number of shares (which would vary from company to company) that would make them "influential." The proportion of stockholders holding below that certain number of shares (which would vary from company to company) that would make them "insignificant." (3) The numbers of shares (and hence the degree of influence) held by the largest stockholders.
   f. There are several possible valid answers. (1) The results would be from a self-selected group (i.e., those who chose to respond) and not necessarily a representative group. (2) If the questionnaire did not include information on the numbers of shares owned, the views of small stockholders (who are probably less knowledgeable about business and stocks) could not be distinguished from those of large stockholders (whose views should carry more weight).

2. a. Systematic, since the selections are made at regular intervals.
   b. Convenience, since those selected were the ones who happened to attend.
   c. Cluster, since the stockholders were organized into groups (by stockbroker) and all the stockholders in the selected groups were chosen.
   d. Random, since each stockholder has the same chance of being selected.
   e. Stratified, since the stockholders were divided into subpopulations from which the actual sampling was done.

3. Let N be the total number of full-time students and n be the desired sample size.
   a. Random. Obtain a list of all N full-time students, number the students from 1 to N, select n random numbers from 1 to N, and poll each student whose number on the list is one of the random numbers selected.
   b. Systematic. Obtain a list of all N full-time students, number the students from 1 to N, let m be the largest integer less than the fraction N/n, select a random number between 1 and m, begin with the student whose number is the random number selected, and poll that student and every mth student thereafter.
   c. Convenience. Select a location (e.g., the intersection of major campus walkways) by which most of the students usually pass, and poll the first n full-time students that pass by.
   d. Stratified. Obtain a list of all N full-time students and the gender of each, divide the list by gender, and randomly select and poll n/2 students from each gender.
   e. Cluster. Obtain a list of all the classes meeting at a popular time (e.g., 10 am Monday), estimate how many of the classes would be necessary to include n students, select that many of the classes at random, and poll all of the students in each selected class.

4. a. Blinding occurs when those involved in an experiment (either as subjects or evaluators) do not know whether they are dealing with a treatment or a placebo. It might be used in this experiment by (a) not telling the subjects whether they are receiving Sleepeze or the placebo and/or (b) not telling any post-experiment interviewers or evaluators which subjects received Sleepeze and which ones received the placebo. Double-blinding occurs when neither the subjects nor the evaluators know whether they are dealing with a treatment or a placebo.
   b. The data reported will probably involve subjective assessments (e.g., "On a scale of 1 to 10, how well did it work?") that may be subconsciously influenced by whether the subject was known to have received Sleepeze or the placebo.
   c. In a completely randomized block design, subjects are assigned to the groups (in this case to receive Sleepeze or the placebo) at random.
   d. In a rigorously controlled block design, subjects are assigned to the groups (in this case to receive Sleepeze or the placebo) in such a way that the groups are similar with respect to extraneous variables that might affect the outcome. In this experiment it may be important to make certain each group has approximately the same age distribution, degree of insomnia, number of males, number users of alcohol and/or tobacco, etc.
   e. Replication involves repeating the experiment on a sample of subjects large enough to ensure that atypical responses of a few subjects will not give a distorted view of the true situation.

5. The sample is essentially a convenience sample that might not be representative of the student body. In particular, students likely to drop out may exhibit certain common characteristics (e.g., sleeping in and/or cutting classes) that would make them under-represented in the sample because they would be less likely to pass by the polling location.

6. a. Ratio, since differences are meaningful and zero milligrams of tar has a natural meaning.
   b. Ordinal, since the ratings give relative position in a hierarchy.
   c. Nominal, since the classifications only identify categories and not relative positions on a scale.
   d. Ordinal, since the scores give relative position in a hierarchy but differences are not meaningful -- i.e., the difference in intelligence between IQ's of 40 and 50 is not the same as the difference in intelligence between IQ's of 100 and 110.
   e. Ratio, since differences are meaningful and zero points scored has a natural meaning.

Cumulative Review Exercises

NOTE: Here and throughout the text intermediate mathematical steps will be shown as an aide to those who may be having difficulty with the calculations. In practice, most of the work can be done continuously on calculators and the intermediate values are unnecessary. Even when the calculations can not be done continuously, DO NOT WRITE AN INTERMEDIATE VALUE ON YOUR PAPER AND THEN RE-ENTER IT IN THE CALCULATOR. That can practice can introduce round-off errors and copying errors. Store any intermediate values in the calculator so that you can recall them with infinite accuracy and without copying errors.

1. $\dfrac{1.23 + 4.56 + 7.89}{5} + \dfrac{13.68}{5} = 4.56$

2. $\sqrt{\dfrac{(5-7)^2 + (12-7)^2 + (4-7)^2}{3-1}} = \sqrt{\dfrac{(-2)^2 + (5)^2 + (-3)^2}{2}} = \sqrt{\dfrac{4 + 25 + 9}{2}} = \sqrt{\dfrac{38}{2}} = \sqrt{19}$
$= 4.359$

3. $\dfrac{1.96^2 \cdot (0.4)(0.6)}{0.025^2} = \dfrac{3.8416 \cdot (.24)}{.000625} = 1475.174$

4. $\dfrac{98.20 - 98.60}{0.62/\sqrt{106}} = \dfrac{-.40}{.0602} = -6.642$

5. $\dfrac{25!}{16!9!} = \dfrac{25 \cdot 24 \cdot 23 \cdot 22 \cdot 21 \cdot 20 \cdot 19 \cdot 18 \cdot 17 \cdot 16!}{16!9!} = \dfrac{25 \cdot 24 \cdot 23 \cdot 22 \cdot 21 \cdot 20 \cdot 19 \cdot 18 \cdot 17}{9 \cdot 8 \cdot 7 \cdot 6 \cdot 5 \cdot 4 \cdot 3 \cdot 2 \cdot 1} = 2042975$

NOTE: This exercise should be worked using the factorial key on the calculator. The above intermediate steps illustrate the mathematical logic involved but do not represent an efficient approach to the problem.

6. $\sqrt{\dfrac{10(513.27) - 71.5^2}{10(9)}} = \sqrt{\dfrac{5132.7 - 5112.25}{90}} = \sqrt{\dfrac{20.45}{90}} = \sqrt{.2272} = .477$

7. $\dfrac{8(151,879) - (516.5)(2176)}{\sqrt{8(34,525.75) - 516.5^2}\sqrt{8(728,520) - 2176^2}} = \dfrac{1215032 - 1123904}{\sqrt{9433.75}\sqrt{1093184}} = \dfrac{91128}{101552.0} = .897$

8. $\dfrac{(183 - 137.09)^2}{8.5} + \dfrac{(30 - 41.68)^2}{25.3} = \dfrac{(45.91)^2}{8.5} + \dfrac{(-11.68)^2}{25.3}$

$= \dfrac{2107.7281}{8.5} + \dfrac{136.4224}{25.3}$

$= 15.375 + 3.273$

$= 18.648$

9. $0.95^{150} = 4.56\text{E-}04 = 4.56 \cdot 10^{-4} = .000456$; moving the decimal point left 4 places

NOTE: Calculators and computers vary in their representation of such numbers. This manual assumes they will be given in scientific notation as a two-decimal number between 1.00 and 9.99 inclusive followed by an indication (usually E for *exponent* of the multiplying power of ten) of how to adjust the decimal point to obtain a number in the usual notation (rounded off to three significant digits).

10. $25^8 = 1.53\text{E+}11 = 1.53 \cdot 10^{11} = 153,000,000,000$; moving the decimal point right 11 places

11. $52^6 = 1.98\text{E+}10 = 1.98 \cdot 10^{10} = 19,800,000,000$; moving the decimal point right 10 places

12. $.25^5 = 9.77\text{E-}04 = 9.77 \cdot 10^{-4} = .000977$; moving the decimal point left 4 places

Chapter 2

Describing, Exploring, and Comparing Data

2-2 Summarizing Data with Frequency Tables

1. Subtracting two consecutive lower class limits indicates that the class width is 60 - 55 = 5. Since there is a gap of 1.0 between the upper class limit of one class and the lower class limit of the next, class boundaries are determined by increasing or decreasing the appropriate class limits by (1.0)/2 = 0.5. The class boundaries and class midpoints are given in the table below.

height	class boundaries	class midpoint	frequency
55 - 59	54.5 - 59.5	57	1
60 - 64	59.5 - 64.5	62	3
65 - 69	64.5 - 69.5	67	49
70 - 74	69.5 - 74.5	72	46
75 - 79	74.5 - 79.5	77	1
			100

NOTE: Although they often contain extra decimal points and may involve consideration of how the data were obtained, class boundaries are the key to tabular and pictorial data summaries. Once the class boundaries are obtained, everything else falls into place. Here the first class width is readily seen to be 59.5 - 54.5 = 5.0 and the first midpoint is (54.5 + 59.5)/2 = 57. In this manual, class boundaries will typically be calculated first and then used to determine other values. In addition, the sum of the frequencies is an informative number used in many subsequent calculations and will be shown as an integral part of each table.

3. Since the gap between classes as presented is .01, the appropriate class limits are increased or decreased by (.01)/2 = .005 to obtain the class boundaries and the following table.

GPA	class boundaries	class midpoint	frequency
0.00 - 0.49	-0.005 - 0.495	0.245	72
0.50 - 0.99	0.495 - 0.995	0.745	23
1.00 - 1.49	0.995 - 1.495	1.245	47
1.50 - 1.99	1.495 - 1.995	1.745	135
2.00 - 2.49	1.995 - 2.495	2.245	288
2.50 - 2.99	2.495 - 2.995	2.745	276
3.00 - 3.49	2.995 - 3.495	3.245	202
3.50 - 3.99	3.495 - 3.995	3.745	97
			1140

The class width is 0.495 - (-0.005) = .50; the first midpoint is (-0.005 + 0.495)/2 = 0.245.

5. The relative frequency for each class is found by dividing its frequency by 100, the sum of the frequencies. NOTE: As before, the sum is included as an integral part of the table. For relative frequencies, this should always be 1.000 (i.e., 100%) and serves as a check for the calculations.

height	relative frequency
55 - 59	.01
60 - 64	.03
65 - 69	.49
70 - 74	.46
74 - 79	.01
	1.00

7. The relative frequency for each class is found by dividing its frequency by 1140, the sum of the frequencies. NOTE: In #5, the relative frequencies were expressed as decimals; here they are expressed as percents The choice is arbitrary.

GPA	relative frequency
0.00 - 0.49	6.32%
0.50 - 0.99	2.02%
1.00 - 1.49	4.12%
1.50 - 1.99	11.84%
2.00 - 2.49	25.26%
2.50 - 2.99	24.21%
3.00 - 3.49	17.72%
3.50 - 3.99	8.51%
	100.00%

9. The cumulative frequencies are determined by repeated addition of successive frequencies to obtain the combined number in each class and all previous classes. NOTE: Consistent with the emphasis that has been placed on class boundaries, we choose to use upper class boundaries in the "less than" column. Conceptually, heights occur on a continuum and the integer values reported are assumed to be the nearest whole number representation of the precise measure of height. An exact height of 59.7, for example, would be reported as 60 and fall in the second class. The values in the first class, therefore, are better described as being "less than 59.5" (using the upper class boundary) than as being "less than 60." This distinction becomes crucial in the construction of pictorial representations in the next section. In addition, the fact that the final cumulative frequency must equal the total number (i.e, the sum of the frequency column) serves as a check for calculations. The sum of cumulative frequencies, however, has absolutely no meaning and is not included.

height	cumulative frequency
less than 59.5	1
less than 64.5	4
less than 69.5	53
less than 74.5	99
less than 79.5	100

11. The cumulative frequencies are determined by repeated addition of successive frequencies to obtain the combined number in each class and all previous classes. NOTE: Consistent with the emphasis that has been placed on class boundaries, we choose to use upper class boundaries in the "less than" column.

GPA	cumulative frequency
less than 0.495	72
less than 0.995	95
less than 1.495	142
less than 1.995	277
less than 2.495	565
less than 2.995	841
less than 3.495	1043
less than 3.995	1140

13. Assuming that "start the first class at 0.7900 lb" refers to the first lower class limit produces the frequency table below and violates none of the guidelines for constructing frequency tables.

weight (lbs)	frequency
.7900 - .7949	1
.7950 - .7999	0
.8000 - .8049	1
.8050 - .8099	3
.8100 - .8149	4
.8150 - .8199	17
.8200 - .8249	6
.8250 - .8299	4
	36

NOTE: The class boundaries above are .78995, .79495, .79995, etc. Using 0.7900 as the first lower class boundary produces boundaries of .7900, .7950, .8000, etc. This is not acceptable, as these are possible data values. This introduces subjectivity about where to place a value that falls on the boundary and violates the guideline that each of the values must belong to only one class.

14. Assuming that "start the first class at 0.7750 lb" refers to the first lower class limit produces the frequency table below and violates the guideline that frequency tables should have between 5 and 20 classes.

weight (lbs)	frequency
.7750 - .7799	4
.7800 - .7849	13
.7850 - .7899	15
.7900 - .7949	4
	36

NOTE: That this frequency table has only 4 categories, which is usually not sufficient to give a picture of the nature of the distribution, is allowable in this context -- since the class limits employed work well with the other cola data and permit meaningful comparisons across the data sets.

15. Assuming that "start the first class at 0.8100 lb" refers to the first lower class limit produces the frequency table below.

weight (lbs)	frequency
.8100 - .8149	1
.8150 - .8199	6
.8200 - .8249	16
.8250 - .8299	8
.8300 - .8349	3
.8350 - .8399	1
.8400 - .8449	1
	36

While similar to the frequency table in exercise #13, this table differs in two ways. (1) In exercise #13 [Coke], there were 5 classes below the modal class and 2 above; in exercise #15 [Pepsi], there are 2 classes below the modal class and 4 above. (2) The weights in exercise #13 appear to be less than those in exercise #15.

16. Assuming that "start the first class at 0.7700 lb" refers to the first lower class limit produces the frequency table below.

weight (lbs)	frequency
.7700 - .7749	1
.7750 - .7799	6
.7800 - .7849	14
.7850 - .7899	13
.7900 - .7949	2
	36

While similar to the frequency table in exercise #15, this table differs in two ways. (1) In exercise #15 [regular Pepsi], there were 2 classes below the modal class and 4 above; in exercise #16 [diet Pepsi], there are 2 classes below the modal class and 4 above. In both cases, however, there are more values above the modal class than below it. (2) The weights in exercise #15 appear to be greater than those in exercise #16.

17. For 11 classes to cover data ranging from a beginning lower class limit of 0 to a maximum value of 514, the class width must be at least (514 - 0)/11 = 46.7. A convenient class width would be 50, which produces the frequency table given below.

weight (lbs)	frequency
00 - 49	6
50 - 99	10
100 - 149	10
150 - 199	7
200 - 249	8
250 - 299	2
300 - 349	4
350 - 399	3
400 - 449	3
450 - 499	0
500 - 549	1
	36

19. Assuming that "start the first class at 200 lb" refers to the first lower class limit produces the frequency table below.

weight (lbs)	frequency
200 - 219	12
220 - 239	9
240 - 259	18
260 - 279	84
280 - 299	52
	175

Yes. Since the lowest recorded weight before collapse is over 200 and most of the weights are over 260, it appears the cans will withstand a pressure that varies between 158 and 165.

20. Assuming that "start the first class at 200 lb" refers to the first lower class limit produces the following frequency table.

weight (lbs)	frequency
200 - 219	6
220 - 239	5
240 - 259	12
260 - 279	36
280 - 299	87
300 - 319	28
	174

Yes. Most of the thicker cans support a weight of 280 before collapse, and they appear to be stronger. Since the thinner cans already meet the criterion given in exercise #19, however, the added strength of the thicker cans may not be worth the added cost.

21. Assuming that "start the first class at 200 lb" refers to the first lower class limit produces the frequency table below.

weight (lbs)	frequency
200 - 219	6
220 - 239	5
240 - 259	12
260 - 279	36
280 - 299	87
300 - 319	28
320 - 339	0
340 - 359	0
360 - 379	0
380 - 399	0
400 - 419	0
420 - 439	0
440 - 459	0
460 - 479	0
480 - 499	0
500 - 519	1
	175

In general, an outlier can add several rows to a frequency table. Even though most of the added rows have frequency zero, the table tends to suggest that these are possible valid values -- thus distorting the reader's mental image of the distribution.

23. The two frequency tables are given below.

a.
height	frequency
66 - 67	4
68 - 69	3
70 - 71	10
72 - 73	10
74 - 75	0
76 - 77	1
	28

b.
height	frequency
66 - 67	6
68 - 69	4
70 - 71	4
72 - 73	4
74 - 75	4
76 - 77	6
	28

Data set (b) appears to be the phony data for two reasons. (1) The frequencies in set (b) follow a regular pattern unlikely to be achieved by chance, while the frequencies in set (a) follow the type of irregular pattern expected by chance. (2) The pattern in (b) [heights fairly uniformly distributed with more at the extremes than near the middle] disagrees with the generally accepted pattern in (a) [many heights near the middle values and fewer at the extremes].

2-3 Pictures of Data

1. 42, the height of the bar centered at 0.0

3. 2, the one Monday represented by the bar centered at 1.0 and the one Monday represented by the bar centered at 1.4

5. See the figure below. The bars extend from class boundary to class boundary. Each axis is labeled numerically <u>and</u> with the name of the quantity represented. Barring an interval longer than any previously recorded, a minimum stay of 109.5 minutes assures seeing an eruption. A minimum stay of 99.5 minutes includes 199 of the 200 (i.e. 99.5%) recorded intervals and would be inadequate only if longest interval occurred *and* the tour arrived within the first ten minutes of that interval -- or about (1/200) x (10/110) = 0.00045 = .045% of the time, about once in every 2200 tours.

7. See the figure below. The bars extend from class boundary to class boundary. Each axis is labeled numerically <u>and</u> with the name of the quantity represented. Although the posted limit is 30 mph, it appears that the police ticket only those traveling at least 42 mph.

9. See the figure below. The frequencies are plotted above the class midpoints, and "extra" midpoints are added so that both polygons begin and end with a frequency of zero.

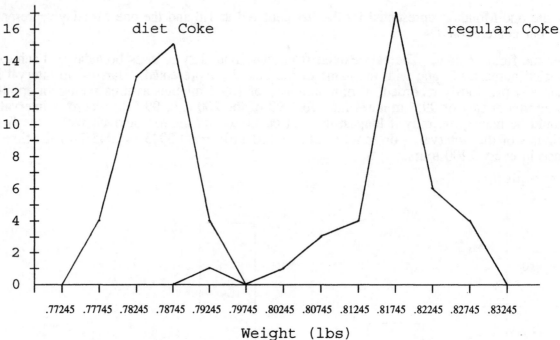

The two distributions have similar shapes, but the diet Coke weights are less than those of regular Coke.

11. See the figure below. The frequencies are plotted above the class midpoints, and "extra" midpoints are added so that both polygons begin and end with a frequency of zero.

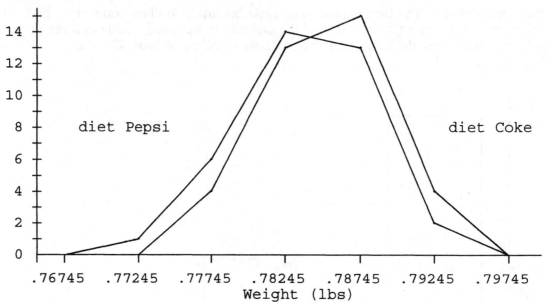

The two distributions are very similar, but the diet Pepsi weights are slightly smaller than those of diet Coke.

13. The original numbers are listed by the row in which they appear in the stem-and-leaf plot.

stem	leaves	original numbers
60	0117	600, 601, 601, 607
61	02889	610, 612, 618, 618, 619
62	13577	621, 623, 625, 627, 627
63		
64	0099	640, 640, 649, 649

15. The dotplot is constructed using the original scores as follows. Each space represents 1 unit.

```
      .                   .  .       .
 . .     .  .     : .  . : .       .        .
--T---------T--------:-T-.-.-:-:-T---------T---------:T--
 600       610       620       630       640       650
```

17. The expanded stem-and-leaf plot below on the left is one possibility. NOTE: The text claims that stem-and-leaf plots enable us to "see the distribution of data and yet keep all the information in the original list." Following the suggestion to round the nearest inch not only loses information but also uses subjectivity to round values exactly half way between. Since always rounding such values "up" creates a slight bias, many texts suggest rounding toward the even digit -- so that 33.5 becomes 34, but 36.5 becomes 36. The technique below of using superscripts to indicate the occasional decimals is both mathematically clear and visually uncluttered.

stem	leaves
3	6 7
4	0 0 1 3 3^5
4	6 6 7 8 8 9
5	0 2 2^5 3 3 4
5	7^3 7^5 8 9 9 9
6	0 0^5 1 1 1^5 2 3 3 3 3^5 4 4 4
6	5 5 6^5 7 7^5 8^5
7	0 0^5 2 2 2 2 3 3^5
7	5 6^5

19. See the figure below, with bars arranged in order of magnitude. Networking appears to be the most effective job-seeking approach.

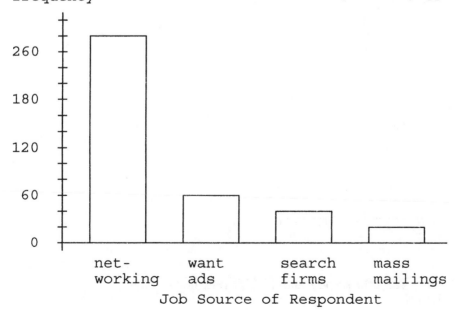

21. See the figure below. The sum of the frequencies is 50; the relative frequencies are 23/50 = 46%, 9/50 = 18%, 12/50 = 24%, and 6/50 = 12%. The corresponding central angles are $(.46)360° = 165.6°$, $(.18)360° = 64.8°$, $(.24)360° = 86.4°$, and $(.12)360° = 43.2°$. NOTE: To be complete, the figure needs to be titled with the name of the quantity being measured.

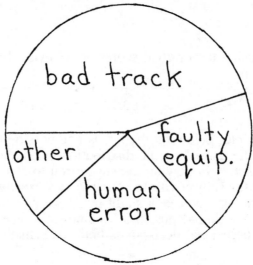

Causes of Train Derailments

23. The scatter diagram is given below. The figure should have a title, and each axis should be labeled both numerically and with the name of the variable. An "x" marks a single occurrence, while numbers indicate multiple occurrences at a point. Cigarettes high in tar also tend to be high in CO. The points cluster about a straight line from (0,0) to (18,18), indicating that the mg of CO tends to be about equal to the mg of tar.

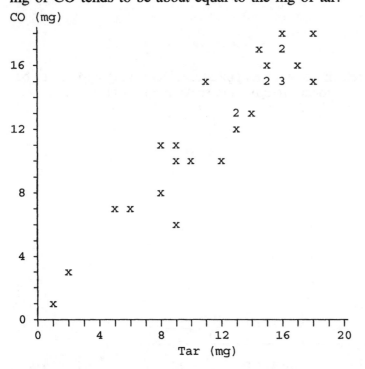

TAR AND CO CONTENT
OF SELECTED 100mm
FILTERED AMERICAN
CIGARETTES

25. According to the figure, 422,000 started and 10,000 returned. $10,000/422,000 = 2.37\%$

27. The figure indicates the number of men had just dropped to 37,000 on November 9 when the temperature was 16°F (-9°C), and had just dropped to 24,000 on November 14 when the temperature was -6°F (-21°C). The number who died during that time, therefore, was 37,000 - 24,000 = 13,000.

29. NOTE: Exercises #20 an #21 of section 2-2 dealt with frequency table representations of this data and specified a class width of 20. Using a class width with an odd number of units of measure allows the class midpoint to have the same number of decimal places as the original data and often produces more appealing visual representations. Here we use a class width of 25 -- with class midpoints of 200, 225, etc. The histogram bars extend from class boundary to class boundary (i.e., from 187.5 to 212.5 for the first class), but for convenience the labels have been placed at the class midpoints.

a. frequency

b. frequency

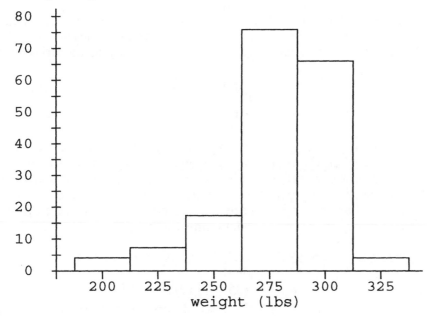

c. The basic shape of the histogram does not change, except that a distant piece has been "broken off."   NOTE: The redrawn histogram in part (b) should not be an exact copy of the one in part (a) with the distant bar erased.   Since removing the distant bar reduces the effective width of the figure significantly, the rule that the height should be approximately 3/4 of the width requires either making the remaining bars wider (and keeping the figure's height) or making them shorter (and keeping the figure's reduced width).   To do otherwise produces a figure too tall for its width -- and one that tends to visually overstate the differences between classes.

31. a. The final form of the back-to-back stem-and-leaf plot is given below, an example of adapting a standard visual form in order to better communicate the data.   While such decisions are arbitrary, we choose to display "outward" from the central stem but to keep the actors' ages in increasing order from left to right.

```
    actor's age|  |actress' age
              2|1466678
   122235677899|3|00113344445557789
00122334556788|4|111249
          13566|5|0
           0012|6|011
              6|7|4
               |8|0
```

b. Female Oscar winners tend to be younger than male Oscar winners.   If one assumes that acting ability doesn't peak differently for females and males, the data may reveal a difference in the standards by which females and males are judged.

2-4   Measures of Central Tendency

NOTE: As it is common in mathematics and statistics to use symbols instead of words to represent quantities that are used often and/or that may appear in equations, this manual employs symbols for the measures of central tendency as follows:

mean = $\bar{x}$ mode = M

median = $\tilde{x}$ midrange = m.r.

Also, this manual follows the author's guideline of presenting means, medians and ranges accurate to one more decimal place than found in the original data.   The mode, the only measure which must be one of the original pieces of data, is presented with the same accuracy as the original data.

1. Arranged in order, the 12 scores are: 8 11 13 14 14 14 15 16 17 18 25 27
   a. $\bar{x} = (\Sigma x)/n = (192)/12 = 16.0$ c. M = 14
   b. $\tilde{x} = (14 + 15)/2 = 14.5$ d. m.r. = (8 + 27)/2 = 17.5

NOTE: The median is the middle score <u>when the scores are arranged in order</u>, and the midrange is halfway between the first and last score <u>when the scores are arranged in order</u>.   It is therefore usually helpful to begin by placing the scores in order.   This will not affect the mean, and it may also aid in identifying the mode.   In addition, no measure of central tendency can have a value lower than the smallest score or higher than the largest score -- remembering this helps to protect against gross errors, which most commonly occur when calculating the mean.

3. Arranged in order, the 12 scores are: 35 46 55 65 74 83 88 93 99 107 108 119
   a. $\bar{x} = (\Sigma x)/n = (972)/12 = 81.0$ c. M = [none]
   b. $\tilde{x} = (83 + 88)/2 = 85.5$ d. m.r. = (35 + 119)/2 = 77.0

   Yes; these results are acceptable and consistent with the goal of service in 90 seconds or less.

5. Arranged in order, the scores are as follows.
   JV: 6.5 6.6 6.7 6.8 7.1 7.3 7.4 7.7 7.7 7.7
   Pr: 4.2 5.4 5.8 6.2 6.7 7.7 7.7 8.5 9.3 10.0

Jefferson Valley	Providence
n = 10	n = 10
$\bar{x} = (\Sigma x)/n = (71.5)/10 = 7.15$	$\bar{x} = (\Sigma x)/n = (71.5)/10 = 7.15$
$\tilde{x} = (7.1 + 7.3)/2 = 7.20$	$\tilde{x} = (6.7 + 7.7)/2 = 7.20$
M = 7.7	M = 7.7
m.r. = (6.5 + 7.7)/2 = 7.10	m.r. = (4.2 + 10.0)/2 = 7.10

Comparing only measures of central tendency, one might suspect the two sets are identical. The Jefferson Valley times, however, are considerably less variable.

NOTE: This is the reason most banks have gone to the single waiting line. While it doesn't make service faster, it makes service times more equitable by eliminating the "luck of the draw" -- i.e., ending up by pure chance in a fast or slow line and having unusually short or long waits.

7. Arranged in order, the scores are as follows.
   Coke: .8150 .8163 .8181 .8192 .8211 .8247
   Pepsi: .8156 .8170 .8211 .8216 .8258 .8302

Coke	Pepsi
n = 6	n = 6
$\bar{x} = (\Sigma x)/n = (4.9144)/6 = .81907$	$\bar{x} = (\Sigma x)/n = (4.9313)/6 = .82188$
$\tilde{x} = (.8181 + .8192)/2 = .81865$	$\tilde{x} = (.8211 + .8216)/2 = .82135$
M = [none]	M = [none]
m.r. = (.8150 + .8247)/2 = .81985	m.r. = (.8156 + .8302)/2 = .82290

Pepsi appears to weigh slightly more than Coke. Since the cans are sold by volume and not by weight, this is not a question of ethics -- merely a reflection of the fact that the two drink cans of the same volume have different ingredients.

9. The x values below are the class midpoints from the given frequency table.

x	f	x·f	
44.5	8	356.0	
54.5	44	2398.0	$\bar{x} = (\Sigma x \cdot f)/n$
64.5	23	1483.5	$\quad = (14870.0)/200$
74.5	6	447.0	$\quad = 74.35$
84.5	107	9041.5	
94.5	11	1039.5	
104.5	1	104.5	
	200	14870.0	

NOTE: The mean time was calculated to be 74.35 minutes. According to the rule given in the text, this value should be rounded to one decimal place. The text describes how many decimal places to present in an answer, but not the actual rounding process. When the figure to be rounded is exactly half-way between two values (i.e., the digit in the position to be discarded is a 5, and there are no further digits because the calculations have "come out even"), there is no universally accepted rounding rule. Some authors say to always round up such a value; others correctly note that always rounding up introduces a consistent bias, and that the value should actually be rounded up half the time and rounded down half the time. And so some authors suggest rounding toward the even value (e.g., .65 becomes .6 and .75 becomes .8), while others simply suggest flipping a coin. In this manual, answers exactly half-way between will be reported without rounding (i.e., stated to one more decimal than usual).

11. The x values below are the class midpoints from the given frequency table.

x	f	x·f
43.5	25	1087.5
47.5	14	665.0
51.5	7	360.5
55.5	3	166.5
59.5	1	59.5
	50	2339.0

$$\bar{x} = (\Sigma x \cdot f)/n$$
$$= (2339.0)/50$$
$$= 46.8$$

The mean speed of 46.8 mi/hr of those ticketed by the police is more than 1.5 times the posted speed limit of 30 mi/hr. NOTE: This indicates nothing about the mean speed of *all* drivers, a figure which may or may not be higher than the posted limit.

13. The following values were obtained for new textbooks, where x_i indicates the ith score from the ordered list.

author's college
n = 35
$\bar{x} = (\Sigma x)/n = 2016.85/35 = 57.624$
$\tilde{x} = x_{18} = 59.35$

UMASS
n = 40
$\bar{x} = (\Sigma x)/n = 2607.70/40 = 65.1175$
$\tilde{x} = (x_{20}+x_{21})/2 = (70.00 + 72.70)/2 = 71.35$

Assuming a random sample was taken from each institution, textbooks appear to cost more at the University of Massachusetts than at the author's college. This is probably more a reflection of the courses offered at each institution and/or the textbook selection practices of faculty than of bookstore pricing policies.

15. The following values were obtained for Boston rainfall, where x_i indicates the ith score from the ordered list.

Thursday
n = 52
$\bar{x} = (\Sigma x)/n = 3.57/52 = .069$
$\tilde{x} = (x_{26}+x_{27})/2 = (.00 + .00)/2 = .000$

Sunday
n = 52
$\bar{x} = (\Sigma x)/n = 3.52/52 = .068$
$\tilde{x} = (x_{26}+x_{27})/2 = (.00 + .00)/2 = .000$

If "it rains more on weekends" refers to the amount of rain, the data do not support the claim. The amount of rainfall appears to be virtually the same for Thursday and Sunday. If "it rains more on weekends" refers to the frequency of rain (regardless of the amount), then the proportions of days on which there was rain would have to be compared.

17. Let $\bar{x}_h$ stand for the harmonic mean: $\bar{x}_h = n/[\Sigma(1/x)]$
$$= 2/[1/40 + 1/60]$$
$$= 2/[.0417]$$
$$= 48.0$$

19. R.M.S. $= \sqrt{\Sigma x^2/n}$
$$= \sqrt{[(110)^2 + (0)^2 + (-60)^2 + (12)^2]/4}$$
$$= \sqrt{15844/4}$$
$$= \sqrt{3961}$$
$$= 62.9$$

21. Let the original data, arranged in order, be: $x_1, x_2, \ldots, x_n$
   a. Arranged in order, the new data would be: $x_1+k, x_2+k, \ldots, x_n+k$
   In general, adding (or subtracting) a constant k from each score will add (or subtract) k from each measure center.

   new $\bar{x} = [\Sigma(x+k)]/n = [\Sigma x + nk]/n = (\Sigma x)/n + (nk)/n = \bar{x}+k$
   new $\tilde{x} = \tilde{x}+k$ [from the ordered list]
   new $M = M+k$ [from the ordered list]
   new m.r. $= [(x_1+k) + (x_n+k)]/2 = [x_1 + x_n + 2k]/2 = (x_1 + x_n)/2 + (2k)/2 = $ m.r.$+k$

   b. Arranged in order, the new data would be: $k \cdot x_1, k \cdot x_2, \ldots, k \cdot x_n$
   In general, multiplying (or dividing) each score by a constant k will multiply (or divide) each measure of center by k.
   new $\bar{x} = [\Sigma(k \cdot x)]/n = [k \cdot (\Sigma x)]/n = k \cdot [(\Sigma x)/n] = k \cdot \bar{x}$
   new $\tilde{x} = k \cdot \tilde{x}$ [from the ordered list]
   new $M = k \cdot M$ [from the ordered list]
   new m.r. $= (k \cdot x_1 + k \cdot x_n)/2 = k \cdot (x_1 + x_n)/2 = k \cdot$ m.r.

   NOTE: Do not assume the principle in (a) and (b) extends automatically to other mathematical operations. A short example will show, for example, that the mean of the squares of a set of data is not necessarily the square of their original mean.

23. a. Arranged in order, the original 54 scores are:
26	29	34	40	46	48	60	62	64	65	76	79	80	86	90	94
105	114	116	120	125	132	140	140	144	148	150	150	154	166	166	180
182	202	202	204	204	212	220	220	236	262	270	316	332	344	348	356
360	365	416	436	446	514										

   $\bar{x} = (\Sigma x)/n = (9876)/54 = 182.9$

   b. Trimming the highest and lowest 10% (or $5.4 = 5$ scores), the remaining 44 scores are:
48	60	62	64	65	76	79	80	86	90	94	105	114	116	120	125
132	140	140	144	148	150	150	154	166	166	180	182	202	202	204	204
212	220	220	236	262	270	316	332	344	348	356	360				

   $\bar{x} = (\Sigma x)/n = (7524)/44 = 171.0$

   c. Trimming the highest and lowest 20% (or $10.8 = 11$ scores), the remaining 32 scores are:
79	80	86	90	94	105	114	116	120	125	132	140	140	144	148	150
150	154	166	166	180	182	202	202	204	204	212	220	220	236	262	270

   $\bar{x} = (\Sigma x)/n = (5093)/32 = 159.2$

   In this case, the mean gets smaller as more scores are trimmed. In general, means can increase, decrease, or stay the same as more scores are trimmed. The mean decreased here because the higher scores were farther from the original mean than were the lower scores.

2-5 Measures of Variation

NOTE: Although not given in the text, the symbol R will be used for the range throughout this manual. Remember that the range is the difference between the highest and the lowest scores, and not necessarily the difference between the last and the first values as they are listed. Since calculating the range involves only the subtraction of 2 original pieces of data, that measure of variation will be reported with the same accuracy as the original data.

1.

x	x $-\bar{x}$	$(x-\bar{x})^2$	x^2
8	-8	64	64
11	-5	25	121
13	-3	9	169
14	-2	4	196
14	-2	4	196
14	-2	4	196
15	-1	1	225
16	0	0	256
17	1	1	289
18	2	4	324
25	9	81	625
27	11	121	729
192	0	318	3390

$\bar{x} = (\Sigma x)/n = 192/12 = 16.0$

$R = 27 - 8 = 19$

by formula 2-4,
$$s^2 = \Sigma(x-\bar{x})^2/(n-1)$$
$$= 318/11$$
$$= 28.9091$$
$$= 28.9$$

by formula 2-5,
$$s^2 = [n(\Sigma x^2) - (\Sigma x)^2]/[n(n-1)]$$
$$= [12(3390) - (192)^2]/[12(11)]$$
$$= [3816]/[132]$$
$$= 28.9091$$
$$= 28.9$$

$s = \sqrt{28.9091} = 5.4$

NOTE: When finding the square root of the variance to obtain the standard deviation, use <u>all</u> the decimal places of the variance, and not the rounded value reported as the answer. The best way to do this is either to keep the value on the calculator display or to place it in the memory. Do <u>not</u> copy down all the decimal places and then re-enter them to find the square root, as that could introduce round-off and/or copying errors.

When using formula 2-4, constructing a table having the first three columns shown above helps to organize the calculations and makes errors less likely. In addition, verify that $\Sigma(x-\bar{x}) = 0$ before proceeding -- if such is not the case, there is an error and further calculation is fruitless. For completeness, and as a check, both formulas 2-4 and 2-5 were used above. In general, only formula 2-5 will be used throughout the remainder of this manual for the following reasons:
  (1) When the mean does not "come out even," formula 2-4 involves round-off error and/or many messy decimal calculations.
  (2) The quantities Σx and Σx^2 needed for formula 2-5 can be found directly and conveniently on the calculator from the original data without having to construct a table like the one above.

3. preliminary values: n = 121, Σx = 972, Σx^2 = 86424
   R = 119 - 35 = 84
   $s^2 = [n(\Sigma x^2) - (\Sigma x)^2]/[n(n-1)]$
   $\quad = [12(86424) - (972)^2]/[12(11)]$
   $\quad = (92304)/132$
   $\quad = 699.3$
   s = 26.4

NOTE: The quantity $[n(\Sigma x^2) - (\Sigma x)^2]$ cannot be less than zero. A negative value indicates that there is an error and that further calculation is fruitless. In addition, remember to find the value for s by taking the square root of the precise value of s^2 showing on the calculator display <u>before</u> it is rounded to one more decimal place than the original data.

5. <u>Jefferson Valley</u>   <u>Providence</u>

$n = 10$, $\Sigma x = 71.5$, $\Sigma x^2 = 513.27$ $n = 10$, $\Sigma x = 71.5$, $\Sigma x^2 = 541.09$

$R = 7.7 - 6.5 = 1.2$ $R = 10.0 - 4.2 = 5.8$

$s^2 = [n(\Sigma x^2) - (\Sigma x)^2]/[n(n-1)]$ $s^2 = [n(\Sigma x^2) - (\Sigma x)^2]/[n(n-1)]$

    $= [10(513.27) - (71.5)^2]/[10(9)]$ $= [10(541.09) - (71.5)^2]/[10(9)]$

    $= 20.45/90$ $= 298.65/90$

    $= 0.23$ $= 3.32$

$s = 0.48$ $s = 1.82$

Exercise #5 of section 2-4 indicated that the mean waiting time was 7.15 minutes at each bank. The Jefferson Valley waiting times, however, are considerably less variable. The range measures the difference between the extremes. The longest and shortest waits at Jefferson Valley differ by a little over 1 minute ($R=1.2$), while the longest and shortest waits at Providence differ by almost 6 minutes ($R=5.8$). The standard deviation measures the typical difference from the mean. A Jefferson Valley customer usually receives service within about ½ minute ($s=0.48$) of 7.15 minutes, while a Providence customer usually receives service within about 2 minutes ($s=1.82$) of the mean.

7. <u>Coke</u>   <u>Pepsi</u>

$n = 6$, $\Sigma x = 4.9144$, $\Sigma x^2 = 4.02528224$ $n = 6$, $\Sigma x = 4.9313$, $\Sigma x^2 = 4.05310181$

$R = .8247 - .8150 = .0097$ $R = .8302 - .8156 = .0146$

$s^2 = [n(\Sigma x^2) - (\Sigma x)^2]/[n(n-1)]$ $s^2 = [n(\Sigma x^2) - (\Sigma x)^2]/[n(n-1)]$

    $= [6(4.02528224) - (4.9144)^2]/[6(5)]$ $= [6(4.05310181) - (4.9313)^2]/[6(5)]$

    $= .00036608/30$ $= .00089117/30$

    $= .000012202$ $= .000029706$

$s = .00349$ $s = .00545$

There appears to be slightly more variation among the Pepsi weights than among those for Coke. In other words, Coke seems to be doing slightly better when it comes to producing a uniform product.

NOTE: Following the usual round-off rule of giving answers with one more decimal than the original data produces variances of .00001 and .00003, which have only one significant digit. This is an unusual situation occurring because values less than 1.0 become smaller when they are squared. Since the original data had 4 significant digits, we provide 5 significant digits for the variance.

9.

x	f	$f \cdot x$	$f \cdot x^2$
44.5	8	356.0	15842.00
54.5	44	2398.0	130691.00
64.5	23	1483.5	95685.75
74.5	6	447.0	33301.50
84.5	107	9041.5	764006.75
94.5	11	1039.5	98232.75
104.5	1	104.5	10920.25
	200	14870.0	1148680.00

$s^2 = [n(\Sigma f \cdot x^2) - (\Sigma f \cdot x)^2]/[n(n-1)]$

    $= [200(1148680.00) - (1487.0)^2]/[200(199)]$

    $= (8619100.00)/29800$

    $= 216.56$

$s = 14.7$

11.

x	f	f·x	f·x^2
43.5	25	1087.5	47305.25
47.5	14	665.0	31587.50
51.5	7	360.5	18565.75
55.5	3	166.5	9240.75
59.5	1	59.5	3540.25
	50	2339.0	110239.50

$s^2 = [n(\Sigma f \cdot x^2) - (\Sigma f \cdot x)^2]/[n(n-1)]$
$= [50(110239.50) - (2339.0)^2]/[50(49)]$
$= (41054.00)/2450$
$= 16.76$
$s = 4.1$

13. author's college
n = 35, Σx = 2016.85, Σx^2 = 123549.5808
$s^2 = [n(\Sigma x^2) - (\Sigma x)^2]/[n(n-1)]$
$= [35(123549.5808) - (2016.85)^2]/[35(34)]$
$= 256551.4055/1190$
$= 215.5894$
$s = 14.683$

UMASS
n = 40, Σx = 2604.70, Σx^2 = 190378.5631
$s^2 = [n(\Sigma x^2) - (\Sigma x)^2]/[n(n-1)]$
$= [40(190378.5631) - (2604.70)^2]/[40(39)]$
$= 830680.4340/1560$
$= 532.4875$
$s = 23.076$

There appears to be more variation among prices of new textbooks at UMASS than at the author's college.

15. Thursday
n = 52, Σx = 3.57, Σx^2 = 1.6699
$s^2 = [n(\Sigma x^2) - (\Sigma x)^2]/[n(n-1)]$
$= [52(1.6699) - (3.57)^2]/[52(51)]$
$= 74.0899/2652$
$= .02793$
$s = .167$

Sunday
n = 52, Σx = 3.52, Σx^2 = 2.2790
$s^2 = [n(\Sigma x^2) - (\Sigma x)^2]/[n(n-1)]$
$= [52(2.2790) - (3.52)^2]/[52(51)]$
$= 106.1176/2652$
$= .04001$
$s = .200$

The values are close, although there may be slightly more variation among the amounts for Sundays than among the amounts for Thursdays.

17. Assuming that the graduates ranged in age from 17 to 19, the Range Rule of Thumb suggests
s $\approx$ range/4 = (19-17)/4 = 2/4 = 0.5.

19. Given $\bar{x}$=75 and s=12, the Range Rule of Thumb suggests
minimum "usual" value = $\bar{x}$ - 2s = 75 - 2(12) = 75 - 24 = 51
maximum "usual" value = $\bar{x}$ + 2s = 75 + 2(12) = 75 + 24 = 99
Yes, in this context a score of 50 would be considered unusually low.

21. a. The limits 61.1 and 66.1 are 1 standard deviation from the mean. The Empirical Rule for Data with a Bell-shaped Distribution states that about 68% of the heights should fall within those limits.

b. The limits 56.1 and 76.1 are 3 standard deviations from the mean. The Empirical Rule for Data with a Bell-shaped Distribution states that about 99.7% of the heights should fall within those limits.

23. The limits 58.6 and 68.6 are 2 standard deviations from the mean. Chebyshev's Theorem states that there must be at least $1 - 1/k^2$ of the scores within k standard deviations of the mean. Here k = 2, and so the proportion of the heights within those limits is at least $1 - 1/2^2 = 1 - 1/4 = 3/4 = 75\%$.

25. A standard deviation of s = 0 is possible only when $s^2 = 0$, and $s^2 = \Sigma(x-\overline{x})^2/(n-1) = 0$ only when $\Sigma(x-\overline{x})^2 = 0$. Since each $(x-\overline{x})^2$ is non-negative, $\Sigma(x-\overline{x})^2 = 0$ only when every $(x-\overline{x})^2 = 0$ -- i.e., only when every x is equal to $\overline{x}$. In simple terms, no variation occurs only when all the scores are identical.

27. The Everlast brand is the better choice. In general, a smaller standard deviation of lifetimes indicates more consistency from battery to battery -- signalling a more dependable production process and a more dependable final product. Assuming a bell-shaped distribution of lifetimes, for example, that empirical rule states that about 68% of the lifetimes will fall within one standard deviation of the mean. Here, those limits would be
    for Everlast: 50 $\pm$ 2 or 48 months to 52 months
    for Endurance: 50 $\pm$ 6 or 44 months to 56 months
While a person might be lucky and purchase a long-lasting Endurance battery, an Everlast battery is much more likely to last for the advertised 48 months.

29.
section 1	section 2
n = 11, Σx = 201, Σx^2 = 4001	n = 11, Σx = 119, Σx^2 = 1741
R = 20 - 1 = 19	R = 19 - 2 = 17
$s^2 = [n(\Sigma x^2) - (\Sigma x)^2]/[n(n-1)]$	$s^2 = [n(\Sigma x^2) - (\Sigma x)^2]/[n(n-1)]$
$= [11(4001) - (201)^2]/[11(10)]$	$= [11(1741) - (119)^2]/[11(10)]$
= 3610/110	= 1990/110
= 32.82	= 45.36
s = 5.7	s = 6.7

The range values give the impression that section 1 had more variability than section 2. The range can be misleading because it is based only on the extreme scores. In this case, the lowest score in section 1 was so distinctly different from the others that to include it in any measure trying to give a summary about the section as a whole would skew the results. For the mean, where the value is only one of 11 used in the calculation, the effect is minimal; for the range, where the value is one of only 2 used in the calculation, the effect is dramatic. The standard deviation values give the impression that section 2 had slightly more variability.
NOTE: In this case, section 2 seems considerably more variable (or diverse), and even the standard deviation by itself fails to accurately distinguish between the sections.

31. preliminary values: n = 11, Σx = 56, Σx^2 = 396
    $\overline{x} = (\Sigma x)/n = 56/11 = 5.09$
    $s^2 = [n(\Sigma x^2) - (\Sigma x)^2]/[n(n-1)]$
    $= [11(396) - (56)^2]/[11(10)]$
    $= (1120)/110$
    = 11.09
    s = 3.33

    a. The coefficient of variation is $(s/\overline{x}) \cdot 100\% = (3.33/5.09) \cdot 100\% = 65.4\%$.

    b. The signal-to-noise ratio is $\overline{x}/s = 5.09/3.33 = 1.529$.

33. The largest standard deviation (i.e., the most spread) occurs when all the values are evenly distributed between the two extremes. Here that would be: 1, 1, 1, 9, 9, 9.
    preliminary values: n = 6, Σx = 30, Σx^2 = 246
    $\overline{x} = (\Sigma x)/n = 30/6 = 5.0$
    $s^2 = [n(\Sigma x^2) - (\Sigma x)^2]/[n(n-1)]$
    $= [6(246) - (30)^2]/[6(5)]$
    $= (576)/30$
    = 19.20
    s = 4.4

35. For greater accuracy and understanding, we use 3 decimal places and avoid shortcut formulas.

a. the original population

x	x-μ	(x-μ)^2
1	-1	1
2	0	0
3	1	1
6	0	2

$\mu = (\Sigma x)/N = 6/3 = 2$

$\sigma^2 = \Sigma(x-\mu)^2/N = 2/3 = .667$

$\sigma = .816$

b. the nine samples: using $s^2 = \Sigma(x-\bar{x})^2/(n-1)$ [for each sample, n = 2]

sample	$\bar{x}$	s^2	s
1,1	1.0	0	0
1,2	1.5	0.5	0.707
1,3	2.0	2.0	1.414
2,1	1.5	0.5	0.707
2,2	2.0	0	0
2,3	2.5	0.5	0.707
3,1	2.0	2.0	1.414
3,2	2.5	0.5	0.707
3,3	3.0	0	0
	18.0	6.0	5.656

mean of the 9 calculated variances

$(\Sigma s^2)/9 = 6.0/9 = 2/3 = .667$

c. the nine samples: using $\sigma^2 = \Sigma(x-\mu)^2/N$ [for each sample, N = 2]

sample	μ	σ^2
1,1	1.0	0
1,2	1.5	0.25
1,3	2.0	1.00
2,1	1.5	0.25
2,2	2.0	0
2,3	2.5	0.25
3,1	2.0	1.00
3,2	2.5	0.25
3,3	3.0	0
	18.0	3.00

mean of the 9 calculated variances

$(\Sigma \sigma^2)/9 = 3.0/9 = 1/3 = .333$

d. The approach in (b) of dividing by n-1 when calculating the sample variance gives a better estimate of the population variance. On the average, the approach in (b) gave the correct population variance of 2/3 = .667. The approach in (c) of dividing by n underestimated the correct population variance. When computing sample variances, divide by n-1 and not by n.

e. No. An unbiased estimator is one that gives the correct answer on the average. Since the average value of s^2 in part (b) was .667, which was the correct value calculated for σ^2 in part (a), s^2 is an unbiased estimator of σ^2. Since the average value of s in part (b) is $(\Sigma s)/9$ = 5.656/9 = .628, which is not the correct value of .816 calculated for σ in part (a), s is not an unbiased estimator of σ.
NOTE: Since the average value of $\bar{x}$ in part (b) is $(\Sigma \bar{x})/9$ = 18.0/9 = 2.0, which is the correct value calculated for μ in part (a), $\bar{x}$ is an unbiased estimator of μ.

2-6 Measures of Position

1. a. $x - \mu = 130 - 100 = 30$
   b. $30/\sigma = 30/15 = 2.00$
   c. $z = (x - \mu)/\sigma = (130 - 100)/15 = 30/15 = 2.00$
   d. The z score in part (c) and the number of standard deviations by which a score differs from the mean in part (b) are identical, being different names for the same concept.

3. In general, $z = (x - \mu)/\sigma$.
   a. $z_{85} = (85 - 69.0)/2.8 = 5.71$
   b. $z_{64} = (64 - 69.0)/2.8 = -1.79$
   c. $z_{69.72} = (69.72 - 69.0)/2.8 = 0.26$

5. $z = (x - \mu)/\sigma$
   $z_{70} = (70 - 63.6)/2.5 = 2.56$
   Yes, that height is considered unusual since $2.56 > 2.00$.

7. $z = (x - \mu)/\sigma$
   $z_{101} = (101 - 98.20)/0.62 = 4.52$
   Yes, that temperature is unusually high, since $4.52 > 2.00$. It suggests that either the person is healthy but has a very unusual temperature for a healthy person, or that person is sick (i.e., has an elevated temperature attributable to some cause).

9. In general $z = (x - \overline{x})/s$
   history: $z_{75} = (75 - 80)/12 = -0.42$
   psychology: $z_{27} = (27 - 30)/8 = -0.375$
   The psychology score has the better relative position since $-0.375 > -0.42$.

11. preliminary values: $n = 36$, $\Sigma x = 29.4056$, $\Sigma x^2 = 24.02111984$
    $\overline{x} = (\Sigma x)/n = 29.4056/36 = .81682$
    $s^2 = [n(\Sigma x^2) - (\Sigma x)^2]/[n(n-1)]$
    $\quad = [36(24.02111984) - (29.4056)^2]/[36(35)]$
    $\quad = .07100288/1260$
    $\quad = .000056351$
    $s = .007507$
    $z = (x - \overline{x})/s$
    $z_{.7901} = (.7901 - .81682)/.007507 = -3.56$
    Yes; since $-3.56 < -2.00$, .7901 is an unusual weight for regular Coke

13. Let $b = \text{\# of scores below } x$;
    $n = \text{total number of scores.}$
    In general, the percentile of score x is $(b/n) \cdot 100$.
    The percentile for a weight of .8264 is $(33/36) \cdot 100 = 92$.

15. Let $b = \text{\# of scores below } x$;
    $n = \text{total number of scores.}$
    In general, the percentile of score x is $(b/n) \cdot 100$.
    The percentile for a weight of .8192 is $(22/36) \cdot 100 = 61$.

17. To find P_{80}, $L = (80/100) \cdot 36 = 28.8$ rounded up to 29.
    Since the 29th score is .8229, $P_{80} = .8229$.

19. To find $D_6 = P_{60}$, $L = (60/100) \cdot 36 = 21.6$ rounded up to 22.
    Since the 22nd score is .8189, $D_6 = .8189$.

21. To find $Q_3 = P_{75}$, $L = (75/100) \cdot 36 = 27$ -- a whole number.
    The mean of the 27th and 28th scores, $Q_3 = (.8207 + .8211)/2 = .8209$

23. To find $D_1 = P_{10}$, $L = (10/100) \cdot 36 = 3.6$ rounded up to 4.
    Since the 4th score is .8073, $D_1 = .8073$.

NOTE: For exercises 25-36, refer to the ordered cross-numbered chart below. A cross-numbered chart gives position in the list. The column head gives the tens digit and the row lead gives the ones digit, so that **270** is #43 in the ordered list of the 54 weights.

	0	1	2	3	4	5
0		65	120	166	220	365
1	26	76	125	166	236	416
2	29	79	132	180	262	436
3	34	80	140	182	**270**	446
4	40	86	140	202	316	514
5	45	90	144	202	332	
6	48	94	148	204	344	
7	60	105	150	204	348	
8	62	114	150	212	356	
9	64	116	154	220	360	

25. Let b = # of scores below x
    n = total number of scores
    In general, the percentile of score x is $(b/n) \cdot 100$.
    The percentile of score 144 is $(24/54) \cdot 100 = 44$.

27. Let b = # of scores below x
    n = total number of scores
    In general, the percentile of score x is $(b/n) \cdot 100$.
    The percentile of score 316 is $(43/54) \cdot 100 = 80$.

29. To find P_{85}, $L = (85/100) \cdot 54 = 45.9$ rounded up to 46.
    Since the 46th score is 344, $P_{85} = 344$.

31. To find $Q_1 = P_{25}$, $L = (25/100) \cdot 54 = 13.5$ rounded up to 14.
    Since the 14th score is 86, $Q_1 = 86$.

33. To find $D_9 = P_{90}$, $L = (90/100) \cdot 54 = 48.6$ rounded up to 49.
    Since the 49th score is 360, $D_9 = 360$.

35. To find $\tilde{x} = P_{50}$, $L = (50/100) \cdot 54 = 27$ -- a whole number.
    The mean of the 27th and 28th scores, $P_{50} = (150 + 150)/2 = 150.0$.

37. In general, z scores are not affected by the particular unit of measurement that is used. The relative position of a score (whether it is above or below the mean, its rank in an ordered list of the scores, etc.) is not affected by the unit of measurement, and relative position is what a z score communicates. Mathematically, the same units (feet, centimeters, dollars, etc.) appear in both the numerator and denominator of $(x-\mu)/\sigma$ and cancel out to leave the z score unit-free. In fact the z score is also called the *standard score* for that very reason -- it is a standardized value that is independent of the unit of measure employed.

NOTE: In more technical language, changing from one unit of measure to another (feet to centimeters, °F to °C, dollars to pesos, etc.) is a linear transformation -- i.e., if x is the score in one unit, then (for some appropriate values of a and b) $y = ax + b$ is the score in the other unit. In such cases it can be shown (see exercises 2.4 #21 and 2.5 #30) that $\mu_y = a\mu_x + b$ and $\sigma_y = a\sigma_x$. The new z score is the same the old one, since
$$z_y = (y - \mu_y)/\sigma_y = [(ax + b) - (a\mu_x + b)]/a\sigma_x = [ax - a\mu_x]/a\sigma_x = (x - \mu_x)/\sigma_x = z_x$$

39. a. uniform
   b. bell-shaped
   c. In general, the shape of a distribution of not affected if all values are converted to z scores. The shape of a distribution does not depend on the unit of measurement -- e.g., if the heights of a group of people form a bell-shaped distribution, they will do so whether measured in inches or centimeters. For a histogram, for example, using a different unit of measurement simply requires a re-labeling of the horizontal axis but does not change the shape of the figure. Since z scores are merely a re-labeling to standard units, converting all the scores to z scores will not change the shape of a distribution.

41. a. The interquartile range is $Q_3 - Q_1$.
      For $Q_3 = P_{75}$, $L = (75/100) \cdot 36 = 27$ -- a whole number.
        The mean of the 27th and 28th scores, $Q_3 = (.8181 + .8189)/2 = .8185$.
      For $Q_1 = P_{25}$, $L = (25/100) \cdot 36 = 9$ -- a whole number.
        The mean of the 9th and 10th scores, $Q_1 = (.8143 + .8150)/2 = .81465$.
      The interquartile range is $.8185 - .81465 = .00385$
   b. The midquartile is $(Q_1 + Q_3)/2 = (.81465 + .8185)/2 = .8166$
   c. The 10-90 percentile range is $P_{90} - P_{10}$.
      For P_{90}, $L = (90/100) \cdot 36 = 32.4$ rounded up to 33.
        Since the 33rd score is .8251, $P_{90} = .8251$.
      For P_{10}, $L = (10/100) \cdot 36 = 3.6$ rounded up to 4.
        Since the 4th score is .8073, $P_{10} = .8073$.
      The 10-90 percentile range is $.8251 - .8073 = .0178$.
   d. Yes, $Q_2 = P_{50}$ by definition. They are always equal.
   e. For $Q_2 = P_{50}$, $L = (50/100) \cdot 36 = 18$ -- a whole number.
      The mean of the 18th and 19th scores, $Q_2 = (.8170 + .8172)/2 = .8171$.
      No; in this case $.8171 = Q_2 \neq (Q_1 + Q_3)/2 = .8166$, which demonstrates that the median does not necessarily equal the midquartile.

2-7 Exploratory Data Analysis

NOTE: The exercises in this section may be done much more easily when ordered lists of the values are available. On the next page appear ordered lists for the data of the odd-numbered exercises. The left-most column gives the ordered ID numbers 1-50. Data sets with more than 50 values have more than one column -- the second column being ordered values 51-100, etc.

1. Consider the 29 employee ages.
    For $Q_1 = P_{25}$, $L = (25/100) \cdot 25 = 6.25$ rounded up to 7.
    For $\tilde{x} = Q_2 = P_{50}$, $L = (50/100) \cdot 25 = 12.5$ rounded up to 13.
    For $Q_3 = P_{75}$, $L = (75/100) \cdot 25 = 18.75$ rounded up to 19.

    min $= x_1 = 34$
    $Q_1 = x_7 = 65$
    $Q_2 = x_{13} = 75$
    $Q_3 = x_{19} = 84$
    max $= x_{25} = 88$

The employees seem to be considerably older than most American workers.

3. Consider the 94 pulse rates.
    For $Q_1 = P_{25}$, $L = (25/100) \cdot 94 = 23.5$ rounded up to 24.
    For $\tilde{x} = Q_2 = P_{50}$, $L = (50/100) \cdot 94 = 47$ -- an integer, use 47.5.
    For $Q_3 = P_{75}$, $L = (75/100) \cdot 94 = 70.5$ rounded up to 71.

    min $= x_1 = 8$
    $Q_1 = x_{24} = 63$
    $Q_2 = x_{47.5} = (70 + 70)/2 = 70$
    $Q_3 = x_{71} = 78$
    max $= x_{94} = 100$

The values 8 and 15 are far from the other values and appear to be errors.

NOTE for exercise #3 and all future uses of this data set: The values 8 and 15 are obvious errors that will be eliminated from all subsequent analyses. Such obvious errors occur in many real life data sets. Often correct values can be deduced and the data adjusted accordingly. If, for example, all the other values were multiples of 4, one could infer that students monitored their pulse for 15 seconds and multiplied by 4 to obtain a per minute rate -- and that some students forgot to multiply by 4. It appears that the instructor gathered data by having each student monitor his own pulse rate -- and that there was not careful instruction or a re-take of suspicious values. It also appears that 6 students couldn't find their pulse -- and those places were just left blank. Although also questionable, the values in the 40's will be included in all subsequent analyses of the pulse data, but the numbers are suspect and should not be taken as accurate renderings of student pulse rates.

5. Consider the 175 axial loads of the 0.0111 in. thick cans.
    For $Q_1 = P_{25}$, $L = (25/100) \cdot 175 = 43.75$ rounded up to 44.
    For $\tilde{x} = Q_2 = P_{50}$, $L = (50/100) \cdot 175 + 87.5$ rounded up to 88.
    For $Q_3 = P_{75}$, $L = (75/100) \cdot 175 = 131.25$ rounded up to 132.

    min $= x_1 = 205$
    $Q_1 = x_{44} = 275$
    $Q_2 = x_{88} = 285$
    $Q_3 = x_{132} = 295$
    max $= x_{175} = 504$

The value 504 appears to be either an error or an anomaly.

The following ordered lists are used for the odd-numbered exercises.

	#1 age	#3 pul	#3 pul	#5 can	#5 can	#5 can	#5 can	#7 mal	#7 fem	#9 author	#9 umass
01	34	8	71	205	277	287	301	31	21	28.35	18.95
02	39	15	72	210	277	287	301	32	24	33.35	24.00
03	41	40	72	210	279	287	302	32	26	35.00	26.70
04	45	47	72	211	279	287	302	32	26	38.45	27.95
05	53	48	72	215	279	288	303	33	26	39.15	28.00
06	63	52	72	216	279	288	303	35	17	41.95	34.95
07	65	52	72	222	279	288	303	36	28	42.65	38.70
08	72	54	73	225	279	288	303	37	30	44.00	41.60
09	73	55	75	227	279	288	304	37	30	44.65	41.95
10	73	55	75	230	280	288	304	38	31	50.15	45.60
11	74	58	75	231	280	289	305	39	31	50.65	48.70
12	75	60	76	243	280	289	305	39	33	51.35	50.70
13	75	60	76	244	281	289	306	40	33	53.35	52.00
14	76	60	77	246	281	290	306	40	34	53.35	54.70
15	77	60	77	247	281	290	306	41	34	54.35	57.35
16	79	60	78	247	281	290	307	42	34	57.55	60.00
17	79	60	78	247	281	290	308	42	34	58.65	60.00
18	82	60	78	250	281	290	309	43	35	59.35	67.00
19	84	60	78	253	282	291	310	43	35	60.00	67.70
20	85	61	78	255	282	291	311	44	35	60.00	70.00
21	86	63	78	255	282	291	313	45	37	61.00	72.70
22	86	63	80	256	282	291	314	45	37	61.00	73.00
23	87	63	80	257	283	292	315	46	38	61.00	75.35
24	87	63	80	260	283	292	317	47	39	62.00	76.00
25	88	63	80	260	283	292	504	48	41	63.60	77.00
26		64	80	262	283	292		48	41	64.55	78.00
27		64	80	262	283	293		51	41	65.65	79.20
28		64	80	262	283	293		53	42	69.35	80.55
29		65	81	263	284	293		55	44	75.00	82.00
30		65	82	265	284	294		56	49	75.65	83.35
31		66	83	266	284	294		56	50	75.95	84.30
32		66	83	268	284	295		60	60	77.80	86.00
33		67	84	268	284	295		60	61	80.00	86.70
34		67	85	269	284	295		61	61	80.00	88.00
35		67	86	269	284	295		62	74	88.00	89.35
36		67	88	270	284	296		76	80		92.00
37		67	88	270	285	296					92.70
38		68	88	271	285	296					94.00
39		68	90	272	285	297					95.00
40		68	90	272	285	297					102.95
41		69	92	273	285	297					
42		69	92	274	285	298					
43		69	97	274	285	298					
44		69	100	275	286	298					
45		70		275	286	299					
46		70		275	286	299					
47		70		275	286	300					
48		70		276	287	300					
49		70		276	287	300					
50		71		276	287	300					

7. Consider the 36 actor and 36 actress values.

For $Q_1 = P_{25}$, $L = (25/100) \cdot 36 = 9$ -- an integer, use 9.5.
For $\tilde{x} = Q_2 = P_{50}$, $L = (50/100) \cdot 36 = 18$ -- an integer, use 18.5.
For $Q_3 = P_{75}$, $L = (75/100) \cdot 36 = 27$ -- an integer, use 27.5.

For the actors
min $= x_1 = 31$
$Q_1 = x_{9.5} = (37 + 38)/2 = 37.5$
$Q_2 = x_{18.5} = (43 + 43)/2 = 43$
$Q_3 = x_{27.5} = (51 + 53)/2 = 52$
max $= x_{36} = 76$

For the actresses
min $= x_1 = 21$
$Q_1 = x_{9.5} = (30 + 31)/2 = 30.5$
$Q_2 = x_{18.5} = (35 + 35)/2 = 35$
$Q_3 = x_{27.5} = (41 + 42)/2 = 41.5$
max $= x_{36} = 80$

The ages for the actresses cover a wider range and cluster around a lower value than do the ages of the actors.

9. Consider the 35 new textbook prices for the author's college.

For $Q_1 = P_{25}$, $L = (25/100) \cdot 35 = 8.75$ rounded up to 9.
For $\tilde{x} = Q_2 = P_{50}$, $L = (50/100) \cdot 35 = 17.5$ rounded up to 18.
For $Q_3 = P_{75}$, $L = (75/100) \cdot 35 = 26.25$ rounded up to 27.

min $= x_1 = 28.35$
$Q_1 = x_9 = 44.65$
$Q_2 = x_{18} = 59.35$
$Q_3 = x_{27} = 65.65$
max $= x_{34} = 88.00$

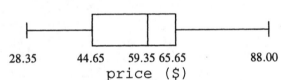

Consider the 40 new textbook prices for UMASS.

For $Q_1 = P_{25}$, $L = (25/100) \cdot 40 = 10$ -- an integer, use 10.5.
For $\tilde{x} = Q_2 = P_{50}$, $L = (50/100) \cdot 40 = 20$ -- an integer, use 20.5.
For $Q_3 = P_{75}$, $L = (75/100) \cdot 40 = 30$ -- an integer, use 30.5.

min $= x_1 = 18.95$
$Q_1 = x_{10.5} = (45.60 + 48.70)/2 = 47.15$
$Q_2 = x_{20.5} = (70.00 + 72.70)/2 = 71.35$
$Q_3 = x_{30.5} = (83.35 + 84.30)/2 = 83.825$
max $= x_{40} = 102.95$

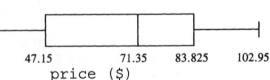

New textbooks at the University of Massachusetts appear to cost more than those at the author's school.

11. The following given values are also calculated in detail in exercise #5.

$Q_1 = 275$
$Q_2 = 285$
$Q_3 = 295$

a. IQR $= Q_3 - Q_1 = 295 - 275 = 20$
b. Since $1.5(\text{IQR}) = 1.5(20) = 30$, the modified boxplot line extends to
the smallest value above $Q_1 - 30 = 275 - 30 = 245$, which is 246
the largest value below $Q_3 + 30 = 295 + 30 = 325$, which is 317

c. Since $3.0(IQR) = 3.0(30) = 60$, mild outliers are x values for which
$$Q_1-60 \leq x < Q_1-30 \qquad \text{or} \qquad Q_3+30 < x \leq Q_3+60$$
$$215 \leq x < 245 \qquad\qquad\qquad 325 < x \leq 355$$
Those values are as follows.
lower end: 215,216,222,225,227,230,231,243,244
upper end: (none)

d. Extreme outliers in this exercise are x values for which $x < 215$ or $x > 355$.
Those values are as follows.
lower end: 205,210,210,211
upper end: 504

Review Exercises

1. The scores arranged in order are:
42 43 46 46 47 48 49 49 50 51 51 51 51 51 52 52 54 54 54 54 55
55 55 55 56 56 56 57 57 57 57 58 60 61 61 61 62 64 64 65 68 69
preliminary values: $n = 42$, $\Sigma x = 2304$, $\Sigma x^2 = 128{,}014$

a. $\bar{x} = (\Sigma x)/n = (2304)/42 = 54.9$
b. $\tilde{x} = (55 + 55)/2 = 55.0$
c. $M = 51$
d. $m.r. = (42 + 69)/2 = 55.5$
e. $R = 69 - 42 = 27$
f. $s = 6.3$ (from part g)
g. $s^2 = [n(\Sigma x^2) - (\Sigma x)^2]/[n(n-1)]$
$\qquad = [42(128{,}014) - (2304)^2]/[42(41)]$
$\qquad = (68{,}172)/1722 = 39.6$
h. For $Q_1 = P_{25}$, $L = (25/100) \cdot 42 = 10.5$ rounded up to 11. And so $Q_1 = x_{11} = 51$.
i. For P_{30}, $L = (30/100) \cdot 42 = 12.6$ rounded up to 13. And so $P_{30} = x_{13} = 51$.
j. For $D_7 = P_{70}$, $L = (70/100) \cdot 42 = 29.4$ rounded up to 30. And so $D_7 = x_{30} = 57$.

2. a. $z = (x - \bar{x})/s$
$\qquad z_{42} = (42 - 54.857)/6.292 = -2.04$
b. Yes; Teddy Roosevelt's inaugural age is unusual since $-2.04 < -2.00$.
c. According to the Range Rule of Thumb, the usual values are within 2s of $\bar{x}$.
usual minimum: $\bar{x} - 2s = 54.9 - 2(6.3) = 42.3$
usual maximum: $\bar{x} + 2s = 54.9 + 2(6.3) = 67.5$
In addition to the 42 in part (b), the ages 68 an 69 are also unusual.

3.

age	frequency
40 - 44	2
45 - 49	6
50 - 54	12
55 - 59	12
60 - 64	7
65 - 69	3
	42

4. frequency

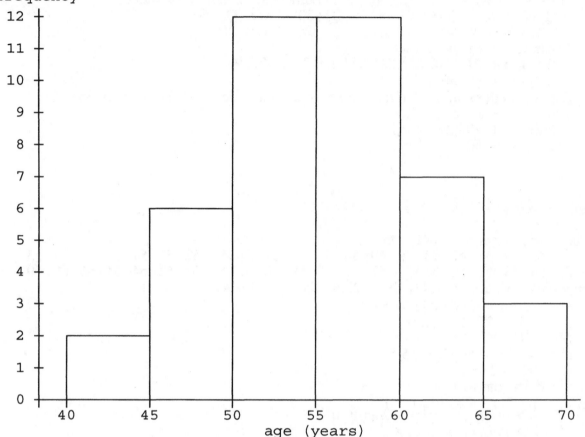

age (years)

NOTE: Unlike other data, age is usually not reported to the nearest unit. The bars in a histogram extend from class boundary to class boundary. Because of the way that ages are reported, the boundaries here are 40, 45, 50, etc -- i.e., someone 44.9 years old still reports an age of 44 and crosses into the next class only upon turning 45, not upon turning 44.5.

5. The scores are given in order in exercise #1.
   For $Q_1 = P_{25}$, $L = (25/100) \cdot 42 = 10.5$, rounded up to 11.
   For $\tilde{x} = Q_2 = P_{50}$, $L = (50/100) \cdot 42 = 21$ -- an integer, use 21.5.
   For $Q_3 = P_{75}$, $L = (75/100) \cdot 42 = 31.5$, rounded up to 32.
   The 5-number summary is as follows.
   min $= x_1 = 42$
   $Q_1 = x_{11} = 51$
   $Q_2 = x_{21.5} = (55 + 55)/2 = 55$
   $Q_3 = x_{32} = 58$
   max $= x_{42} = 69$

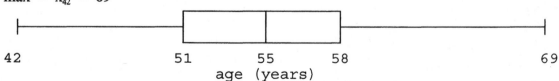

42   51   55   58   69
age (years)

6. a. Since scores 48.6 to 61.2 are within $1 \cdot s$ of $\bar{x}$, the Empirical Rule for Data with a Bell-Shaped Distribution states that about 68% of such persons fall within those limits.
   b. Since scores 42.3 to 67.5 are within $2 \cdot s$ of $\bar{x}$, the Empirical Rule for Data with a Bell-Shaped Distribution states that about 95% of such persons fall within those limits.

7. In general $z = (x - \bar{x})/s$
   management: $z_{72} = (72 - 80)/12 = -0.67$
   production: $z_{19} = (19 - 20)/5 = -0.20$
   The score on the test for production employees has the better relative position since $-0.20 > -0.67$.

8. In general, the Range Rule of Thumb states that the range typically covers about 4 standard deviations -- with the lowest and highest scores being about 2 standard deviations below and above the mean respectively. The following answers assume the ordered textbook ages range from $x_1 = 0$ years (i.e., copyrighted during the current year) to $x_n = 12$ years.
   a. The estimated mean age is $(x_1 + x_n)/2 = (0 + 12)/2 = 6$ years.
   b. The estimated standard deviation of the ages is $R/4 = (x_n - x_1)/4 = (12 - 0)/4 = 3$ years.

9. In general, adding the same value to every score will move the all scores up the number line by the same amount. This changes the location of the scores on the number line, but it does not affect the spread of the scores or the shape of the distribution -- i.e., each measure of center will change by the amount added, but the measures of variation will not be affected.
   a. Since adding 5.0 minutes to every score moves all the scores 5 units up on the number line, the mean will increase by 5.0 to $12.3 + 5.0 = 17.3$ minutes.
   b. Since adding 5.0 minutes to every score does not affect the spread of the scores, the standard deviation will not change and remain 4.0 minutes.
   c. Since adding 5.0 minutes to every score does not affect the spread of the scores, the variance will not change and will remain $(4.0)^2 = 16.0$.
   NOTE: The general principle involved here is discussed for measures of center in exercise #21a of section 2-4 and for measures of variation in exercise #30a of section 2-5.

10. Arranging the categories in decreasing order by frequency produces the following figure.

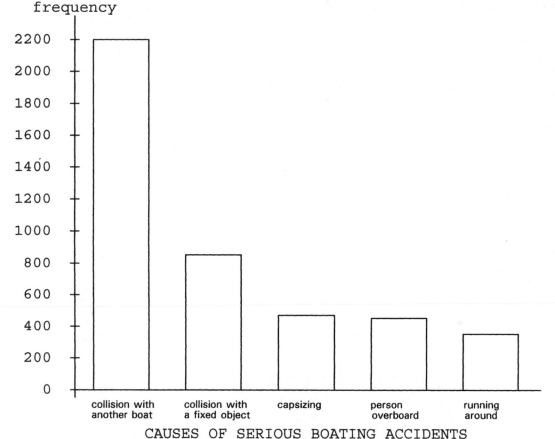

CAUSES OF SERIOUS BOATING ACCIDENTS

Cumulative Review Exercises

1. The scores arranged in order are:
   0 1 4 7 7 7 11 12 12 12 13 18 19 23 25 30
   preliminary values: $n = 16$, $\Sigma x = 201$, $\Sigma x^2 = 3625$
   a. $\bar{x} = (\Sigma x)/n = (201)/16 = 12.6$
      $\tilde{x} = (12 + 12)/2 = 12.0$
      $M = 7, 12$ (bi-modal)
      m.r. $= (0 + 30)/2 = 15.0$
   b. $R = 30 - 0 = 30$
      $s^2 = [n(\Sigma x^2) - (\Sigma x)^2]/[n(n-1)]$
         $= [16(3625) - (201)^2]/[16(15)]$
         $= (17599)/240$
         $= 73.3$
      $s = 8.6$
   c. Continuous. Even though the values were reported rounded to whole years, age can be any non-negative value on a continuum. NOTE: While age is continuous, it could be that the population from which the sample was selected was the ages rounded to whole years -- and so it might be argued that "the population from which the given years were drawn" was discrete.
   d. Ratio. Differences are consistent and there is a meaningful zero; 8 years is twice as much time as 4 years.

2. a. Mode. The median requires at least ordinal level data, and the mean and the midrange require at least interval level data.
   b. Convenience. The group was not selected by any process other than the fact that they happened to be the first names on the list.
   c. Cluster. The population was divided into units (election precincts), some of which were selected at random to be examined in their entirety.

3. No; using the mean of the state values counts each state equally, while states with more people will have a greater affect on the per capita consumption for the population in all 50 states combined. Use the state populations as weights to find the weighted mean.

Chapter 3

Probability

3-2 Fundamentals

1. Since $0 \le P(A) \le 1$ is always true, the following values less than 0 or greater than 1 cannot be probabilities.
   values less than 0: -1
   values greater than 1: 2 5/3 $\sqrt{2}$

3. Let H = getting a head when a balanced coin is tossed.
   H is one of 2 equally likely different outcomes (head or tail); use Rule 2.
   P(H) = 1/2

5. a. There are 144 + 1354 = 1498 M&M candies in the package.
      Let O = getting an orange M&M.
      O includes 144 of 1498 equally likely outcomes; use Rule 2.
      P(O) = 144/1498 = .0961
   b. There are 21 + 8 + 26 + 33 + 5 + 7 = 100 M&M candies in the data set.
      Let O = getting an orange M&M.
      O includes 8 of 100 equally likely outcomes; use Rule 2.
      P(O) = 8/100 = .08
   c. Yes; approximately, when (a) and (b) are considered individually. But considered together, the results from parts (a) and (b) suggest that the proportion of M&M's that are orange may be slightly less than 10%.

7. Let H = selecting an at bat for which McGwire hit a home run.
   H includes 70 of 506 equally likely outcomes; use Rule 2.
   P(H) = 70/508 = .138

9. The sample space of the 36 outcomes for tossing a pair of dice is given below.
   ```
   1-1  2-1  3-1  4-1  5-1  6-1
   1-2  2-2  3-2  4-2  5-2  6-2
   1-3  2-3  3-3  4-3  5-3  6-3
   1-4  2-4  3-4  4-4  5-4  6-4
   1-5  2-5  3-5  4-5  5-5  6-5
   1-6  2-6  3-6  4-6  5-6  6-6
   ```
   NOTE: Don't be confused by the fact that the dice look identical. Like identical twins, they are still different entities -- call them die A and die B. Outcome 1-6 (getting a 1 on die A and a 6 on die B), for example, is different from 6-1 (getting a 6 on die A and a 1 on die B).
   a. Let T = getting a total of 2
      T includes only 1 (the 1-1) of 36 equally likely outcomes; use Rule 2.
      P(T) = 1/36 = .028
   b. Yes; since P(T) < .05, T is considered an unusual event.

11. a. Let C = answering a question correctly with a random guess.
       C includes only 1 of 10 equally likely outcomes; use Rule 2.
       P(C) = 1/10 = .1
    b. No; since P(C) > .05, C is not considered an unusual event.

13. a. Let L = an American Airline flight arrives late.
     L occurred 400 - 344 = 56 times in a sample of 400; use Rule 1.
     P(L) ≈ 56/400 = .14
    b. No; since P(L) > .05, L is not considered an unusual event.

    NOTE: On time arrival rate by itself does not describe an airline's record. By lengthening the projected flight times, the airline could achieve a higher on time arrival rate but would have planes creating problems by arriving "too early" for their unrealistic (and inaccurate) scheduled arrival times.

15. NOTE: This problem can be approached two ways. Finding P(1 of your 2 numbers matches the winning number), the usual approach, requires using the Addition Rule of section 3-3. Finding P(the winning number matches one of your 2 numbers), which is an equivalent event, requires only the methods of this section.
    a. The winning number can be selected in 25,827,165 equally likely ways.
       Let W = you winning the grand prize
       W can occur 2 ways, if either of your numbers is selected; use Rule 2.
       P(W) = 2/25,827,165 = .0000000774
    b. Yes; since P(W) < .05, W is an unusual event.

17. There were 255 + 2302 older Americans surveyed.
    Let A = an older American has Alzheimer's disease
    A occurred 255 times in a sample of 2302; use Rule 1.
    P(A) ≈ 255/2557 = .0997
    Yes; since about 1 in 10 older Americans is affected, it is a major concern.

19. There were 225 + 275 = 500 donors surveyed.
    Let O = a person has type O blood.
    A occurred 225 times in a sample of 500; use Rule 1.
    P(O) ≈ 225/500 = .45

21. a. Let B = a person's birthday is October 18.
       B is one of 365 (assumed) equally likely outcomes; use Rule 2.
       P(B) = 1/365 = .00274
    b. Let O = a person's birthday is in October.
       O includes 31 of 365 (assumed) equally likely outcomes; use Rule 2.
       P(N) = 31/365 = .0849

23. There were 132 + 880 = 1012 respondents.
    Let D = getting a respondent who gave the doorstop answer.
    D includes 132 of 1012 equally likely outcomes; use Rule 2.
    P(D) = 132/1012 = .130

25. Let A = selecting a driver in that age bracket has an accident.
    A occurred 136 times in a sample of 400; use Rule 1.
    P(A) ≈ 136/400 = .340
    Yes; since this indicates more than 1 such driver in every 3 is likely to have an accident, this should be a concern.

    NOTE: This is properly an estimate that such a person had an accident last year, and not that such a person will have an accident next year. To extend the estimate as desired requires the additional assumption that relevant factors (speed limits, weather, alcohol laws, the economy, etc.) remain the same.

27. Let V = a Viagra user experiences a headache.
V occurred 117 times in a sample of 117 + 617 = 734; use Rule 1.
P(V) ≈ 117/734 = .159

29. a. The 4 equally likely outcomes are: BB BG GB GG.
b. P(2 girls) = 1/4 = .25
c. P(1 of each sex) = 2/4 = .5

31. a. Let M = Mean wins the race. It is given that P(M) = 3/17.
P($\bar{M}$) = 1 - P(M) = 1 - 3/17 = 14/17
odds against Mean winning = P($\bar{M}$)/P(M)
$$= (14/17)/(3/17)$$
$$= 14/3$$
$$= 14 \text{ to } 3$$
b. The text's formula *(payoff odds)* = *(net profit)/(amount bet)* may be solved for *(net profit)*.
(net profit) = (payoff odds) · (amount bet)
$$= (4/1) \cdot (\$4)$$
$$= \$16$$

33. Chance fluctuations in testing situations typically prevent subjects from scoring exactly the same each time the test is given. When treated subjects show an improvement, there are two possibilities: (1) the treatment was effective and the improvement was due to the treatment or (2) the treatment was not effective and the improvement was due to chance. The accepted standard is that an event whose probability of occurrence is less than .05 is an unusual event. In this scenario, the probability that the treatment group shows an improvement even if the drug has no effect is calculated to be .04. If the treatment group does show improvement, then, either the treatment truly is effective or else an unusual event has occurred. Operating according to the accepted standard, one should conclude that the treatment was effective.

35. If the odds against A are a:b, then P(A) = b/(a+b).
Let M = Millennium wins the next race.
If the odds against M are 3:5, then a=3 and b=5 in the statement above.
P(M) = 5/(3+5) = 5/8 = .625

37. Properly identified, the 5 numbers in the summary are:
x_1 = 82, Q_1 = 91, Q_2 = 100, Q_3 = 109, x_n = 118.
Since 109 = Q_3 = P_{75}, we say that P(x > 109) = .25
NOTE: Technically, P_{75} = 109 means that at least 75% of the scores less than or equal to 109 and at least 25% of the scores are greater than or equal to 109. This means the best we can truly say is that P(x > 109) ≤ .25.

39. No matter where the two flies land it is possible to cut the orange in half to include both flies on the same half. Since this is a certainty, the probability is 1. [Compare the orange to a globe. Turn the orange so the spot where fly A lands corresponds to, say, New York City. Consider all the circles of longitude. Wherever fly B lands it is possible to slice the globe in half along some circle of longitude that places fly A and fly B on the same half.]
NOTE: If the orange is marked into two predesignated halves before the flies land, the problem is different; once fly A lands, fly B has 1/2 a chance of landing on the same half. If both flies are to land on a specified one of the two predesignated halves, the problem is different still; fly A has 1/2 a chance of landing on the specified half, and only 1/2 the time will fly B pick the same half -- the final answer would then be 1/2 of 1/2, or 1/4.

3-3 Addition Rule

1. a. No, it's possible for MTV to be the principal news source of a voter under the age of 30.
   b. No, it's possible for someone treated with an experimental drug to experience improved symptoms.
   c. Yes, if the outcome is an even number then it cannot be an odd number.

3. a. $P(\overline{A}) = 1 - P(A)$
      $= 1 - .05$
      $= .95$
   b. $P(girl) = P(\overline{boy})$
         $= 1 - P(boy)$
         $= 1 - .513$
         $= .487$

5. Make a chart like the one on the right.
   let F = getting a female
      G = getting a person coded green

		COLOR		
		Red	Green	
GENDER	Male	3	2	5
	Female	4	6	10
		7	8	15

   There are two approaches.

   * Use broad categories and allow for double-counting (i.e., the "formal addition rule").
      $P(F \text{ or } G) = P(F) + P(G) - P(F \text{ and } G)$
         $= 10/15 + 8/15 - 6/15$
         $= 12/15$
         $= .8$

   * Use individual mutually exclusive categories that involve no double-counting (i.e., the "intuitive addition rule"). NOTE: For simplicity in this problem, we use MR for "M and R" and FR for "F and R" and so on.
      $P(F \text{ or } G) = P(FR \text{ or } MG \text{ or } FG)$
         $= P(FR) + P(MG) + P(FG)$
         $= 4/15 + 2/15 + 6/15$
         $= 12/15$
         $= .8$

   NOTE: In general, using broad categories and allowing for double-counting is a "more powerful" technique that "lets the formula do the work" and requires less analysis by the solver. Except when such detailed analysis is instructive, this manual uses the first approach.

NOTE: Throughout the manual we follow the pattern in exercise #5 of using the first letter [or other natural designation] of each category to represent that category. And so in exercise #7 P(D) = P(selecting a person who died) and P(M) = P(selecting a man) and so on. If there is ever cause for ambiguity, the notation will be clearly defined. Since mathematics and statistics use considerable notation and formulas, it is important to clearly define what various letters stand for.

7. Make a chart like the one on the right.
   a. $P(W \text{ or } G) = P(W) + P(G) - P(W \text{ and } G)$
         $= 422/2223 + 45/2223 - 0/2223$
         $= 467/2223$
         $= .210$
   b. $P(W \text{ or } S) = P(W) + P(S) - P(W \text{ and } S)$
         $= 422/2223 + 706/2223 - 318/2223$
         $= 810/2223$
         $= .364$

		FATE		
		Survived	Died	
	Men	332	1360	1692
GROUP	Women	318	104	422
	Boys	29	35	64
	Girls	27	18	45
		706	1517	2223

9. Let B = a person's birthday is October 18
$$P(\bar{B}) = 1 - P(B)$$
$$= 1 - 1/365$$
$$= 364/365$$
$$= .997$$

11. Make a chart like the one on the right.

		TICKET?		
		Yes	No	
GENDER	male	39	11	50
	female	8	24	32
		47	35	82

$$P(M \text{ or } Y) = P(M) + P(Y) - P(M \text{ and } Y)$$
$$= 50/82 + 47/82 - 39/82$$
$$= 58/82$$
$$= .707$$

13. Make a chart like the one on the right.

		Rh FACTOR		
		+	−	
	A	35	5	40
GROUP	B	8	2	10
	AB	4	1	5
	O	39	6	45
		86	14	100

$$P(\bar{A}) = 1 - P(A)$$
$$= 1 - 40/100$$
$$= 60/100$$
$$= .60$$

15. Refer to exercise #13.
$$P(A \text{ or } Rh-) = P(A) + P(Rh-) - P(A \text{ and } Rh-)$$
$$= 40/100 + 14/100 - 5/100$$
$$= 49/100$$
$$= .49$$

17. Refer to exercise #13.
$$P(\overline{Rh+}) = 1 - P(Rh+)$$
$$= 1 - 86/100$$
$$= 14/100$$
$$= .14$$

19. Refer to exercise #13.
$$P(AB \text{ or } Rh+) = P(AB) + P(Rh+) - P(AB \text{ and } Rh+)$$
$$= 5/100 + 86/100 - 4/100$$
$$= 87/100$$
$$= .87$$

21. Make a chart like the one on the right.
Let A = a person is age 18-21
N = a person does not respond

		RESPOND?		
		Yes	No	
AGE	18-21	73	11	84
	22-29	255	20	275
		328	31	359

$$P(A \text{ or } N) = P(A) + P(N) - P(A \text{ and } N)$$
$$= 84/359 + 31/359 - 11/359$$
$$= 104/359$$
$$= .290$$

23. Make a chart like the one on the right.

		HARASSMENT?		
		Yes	No	
GENDER	male	240	380	620
	female	180	200	380
		420	580	1000

$$P(N) = 580/1000$$
$$= .580$$

25. The general formula is: P(A or B) = P(A) + P(B) - P(A and B).
    Solving for P(A and B) yields: P(A and B) = P(A) + P(B) - P(A or B)
    a. For P(A) = 3/11 and P(B) = 4/11 and P(A or B) = 7/11, the Addition Rule produces
        P(A and B) = P(A) + P(B) - P(A or B)
                   = 3/11 + 4/11 - 7/11
                   = 0
    This means that A and B are mutually exclusive.
    b. For P(A) = 5/18 and P(B) = 11/18 and P(A or B) = 13/18, the Addition Rule produces
        P(A and B) = P(A) + P(B) - P(A or B)
                   = 5/18 + 11/18 - 13/18
                   = 3/18
    This means that A and B are not mutually exclusive.

27. If the *exclusive or* is used instead of the *inclusive or*, then the double-counted probability must
    be completely removed (i.e., must be subtracted twice) and the formula becomes
        P(A or B) = P(A) + P(B) - 2·P(A and B)

3-4 Multiplication Rule: Basics

1. a. Independent, since getting a head on the first flip does not affect the outcome on the second
      flip.
   b. Dependent, since speeding while driving to class makes it more likely to get a traffic ticket
      while driving to class.
   c. Independent, since the car's not starting does not affect the performance of the kitchen light.
      NOTE: One could argue that a car's not starting may be due to improper maintenance -- the
      same type of behavior that would be likely, for example, to leave burned out kitchen light
      bulbs unchanged. If so, then a person's car not starting suggests he may be one who doesn't
      properly maintain things -- which suggests that his kitchen light might not be working either.
      This manual ignores such highly speculative hypothetical scenarios.

3. There are 5 red + 4 yellow + 3 green = 12 lenses.
   let A = the first lens selected is yellow
       B = the second lens selected is yellow
   In general, P(A and B) = P(A)·P(B | A)
   a. P(A and B) = P(A)·P(B | A)
                 = (4/12)·(4/12)
                 = 1/9
                 = .111
   b. P(A and B) = P(A)·P(B | A)
                 = (4/12)·(3/11)
                 = 1/11
                 = .091

5. let F = selecting a female
   P(F) = 15/25, for the first selection only
   P(F₁ and F₂) = P(F₁)·P(F₂ | F₁)
               = (15/25)·(14/24)
               = 7/20
               = .35

7. let W = selecting a wrong answer
   P(W) = 4/5, for each selection
   $$P(W_1 \text{ and } W_2 \text{ and } W_3) = P(W_1) \cdot P(W_2) \cdot P(W_3)$$
   $$= (4/5) \cdot (4/5) \cdot (4/5)$$
   $$= 64/125$$
   $$= .512$$

9. a. let J = a person is born on July 4
   P(J) = 1/365, for each selection
   $$P(J_1 \text{ and } J_2) = P(J_1) \cdot P(J_2)$$
   $$= (1/365) \cdot (1/365)$$
   $$= 1/133225$$
   $$= .00000751$$

   b. let S = a person's birthday is favorable to being the same
   P(S) = 365/365 for the first person, but then the second person must match
   $$P(\text{the birthdays are the same}) = P(S_1 \text{ and } S_2)$$
   $$= P(S_1) \cdot P(S_2 \mid S_1)$$
   $$= (365/365) \cdot (1/365)$$
   $$= 1/365$$
   $$= .00274$$

11. Since the sample size is no more than 5% of the size of the population, treat the selections as being independent even though the selections are made without replacement.
    let G = a selected CD is good
    P(G) = .97, for each selection
    $$P(\text{batch accepted}) = P(\text{all good})$$
    $$= P(G_1 \text{ and } G_2 \text{ and...and } G_{12})$$
    $$= P(G_1) \cdot P(G_2) \cdot ... \cdot P(G_{12})$$
    $$= (.97)^{12}$$
    $$= .694$$

13. let G = a girl is born
    P(G) = .50, for each couple
    $$P(G_1 \text{ and } G_2 \text{ and...and } G_{10}) = P(G_1) \cdot P(G_2) \cdot ... \cdot P(G_{10})$$
    $$= (.50)^{10}$$
    $$= .000977$$
    Yes, the gender selection method appears to be effective. The choice is between two possibilities: (1) the gender selection method has no effect and a very unusual event occurred or (2) the gender selection method is effective.

15. Since the sample size is no more than 5% of the size of the population, treat the selections as being independent even though the selections are made without replacement.
    let L = a person is left-handed
    P(L) = .10, for each selection
    a. $$P(L_1 \text{ and } L_2 \text{ and } L_3) = P(L_1) \cdot P(L_2) \cdot P(L_3)$$
    $$= (.10)(.10)(.10)$$
    $$= .001$$
    b. $$P(L_1 \text{ and } L_2 \text{ and...and } L_{20}) = P(L_1) \cdot P(L_2) \cdot ... \cdot P(L_{20})$$
    $$= (.10)^{20})$$
    $$= .00000000000000000001$$
    The probability of having 20 persons randomly chosen from the general population all be left-handed is so small that it is essentially not possible. Something is wrong. Either the students are not from the general population (e.g., are part of a program to help left-handers cope in a right-handed world) or (more likely) the survey is flawed because the respondents either are not telling the truth or miscoded their responses due to unclear directions.

17. let T = the tire named is consistent with all naming the same tire
    P(T) = 4/4 for the first person, but then the others must match him
    P(all name the same tire) = $P(T_1 \text{ and } T_2 \text{ and } T_3 \text{ and } T_4)$
    $$= P(T_1) \cdot P(T_2 \mid T_1) \cdot P(T_3 \mid T_1 \text{ and } T_2) \cdot P(T_4 \mid T_1 \text{ and } T_2 \text{ and } T_3)$$
    $$= (4/4)(1/4)(1/4)(1/4)$$
    $$= 1/64$$
    $$= .0156$$

19. Since the sample size is no more than 5% of the size of the population, treat the selections as being independent even though the selections are made without replacement.
    let N = a cell phone is not defective
    P(N) = 1 - .04
    $\quad\;\; = .96$, for each selection
    $P(N_1 \text{ and } N_2 \text{ and}...\text{and } N_{25}) = P(N_1) \cdot P(N_2) \cdot ... \cdot P(N_{25})$
    $$= (.96)^{25}$$
    $$= .360$$
    No, there is not strong evidence to conclude that the new process is better. Even if the new process made no difference, the observed result is not considered unusual and could be expected to occur about 36% of the time.

21. Refer to the table at the right.
    Since the sample size is no more than 5% of the size of the population, treat the selections as being independent even though the selections are made without replacement.

		TEST RESULT Pos	Neg	
HIV?	Yes	57	3	60
	No	997	18,943	19,940
		1054	18,946	20,000

    P(Pos) = 1054/20000 = .0527, for each selection
    $P(Pos_1 \text{ and } Pos_2) = P(Pos_1) \cdot P(Pos_2)$
    $$= (.0527) \cdot (.0527)$$
    $$= .00278$$

23. Refer to the table and comments for exercise #21.
    P(Yes) = 60/20000 = .003, for each selection
    $P(Yes_1 \text{ and } Yes_2) = P(Yes_1) \cdot P(Yes_2)$
    $$= (.003) \cdot (.003)$$
    $$= .000009$$

25. let D = a birthday is different from any yet selected
    $P(D_1) = 366/366$ NOTE: With nothing to match, it <u>must</u> be different.
    $P(D_2 \mid D_1) = 365/366$
    $P(D_3 \mid D_1 \text{ and } D_2) = 364/366$

    ...
    $P(D_5 \mid D_1 \text{ and } D_2 \text{ and}...\text{and } D_4) = 362/366$

    ...
    $P(D_{25} \mid D_1 \text{ and } D_2 \text{ and}...\text{and } D_{24}) = 342/366$
    a. P(all different) = $P(D_1 \text{ and } D_2 \text{ and } D_3)$
    $$= P(D_1) \cdot P(D_2) \cdot P(D_3)$$
    $$= (366/366) \cdot (365/366) \cdot (364/366)$$
    $$= .992$$
    b. P(all different) = $P(D_1 \text{ and } D_2 \text{ and}...\text{and } D_5)$
    $$= P(D_1) \cdot P(D_2) \cdot ... \cdot P(D_5)$$
    $$= (366/366) \cdot (365/366) \cdot ... \cdot (362/366)$$
    $$= .973$$

c. P(all different) = P(D_1 and D_2 and...and D_{25})
$$= P(D_1) \cdot P(D_2) \cdot ... \cdot P(D_{25})$$
$$= (366/366) \cdot (365/366) \cdot ... \cdot (342/366)$$
$$= .432$$

NOTE: A program to perform this calculation can be constructed using a programming language, or using most spreadsheet or statistical software packages. In BASIC, for example, use

```
10 LET P=1
15 PRINT "How many birthdays?"
20 INPUT D
30  FOR K=1 TO D-1
40  LET P=P*(366-K)/366
50  NEXT K
55 PRINT "The probability they all are different is"
60 PRINT P
70 END
```

27. This is problem can be done by two different methods. In either case,
let A = getting an ace
   S = getting a spade
* consider the sample space
   The first card could be any of 52 cards; for each first card, there are 51 possible second cards. This makes a total of $52 \cdot 51 = 2652$ equally likely outcomes in the sample space. How many of them are A_1S_2?
   The aces of hearts, diamonds and clubs can be paired with any of the 13 clubs for a total of $3 \cdot 13 = 39$ favorable possibilities. The ace of spades can only be paired with any of the remaining 12 members of that suit for a total of 12 favorable possibilities. Since there are $39 + 12 = 51$ favorable possibilities among the equally likely outcomes,
   $P(A_1S_2) = 51/2652$
   $ = .0192$
* use the formulas
   Let As and Ao represent the ace of spades and the ace of any other suit respectively. Break A_1S_2 into mutually exclusive parts so the probability can be found by adding and without having to consider double-counting.
   $P(A_1S_2) = P[(As_1$ and $S_2)$ or $(Ao_1$ and $S_2)]$
   $ = P(As_1$ and $S_2) + P(Ao_1$ and $S_2)$
   $ = P(As_1) \cdot P(S_2 \mid As_1) + P(Ao_1) \cdot P(S_2 \mid Ao_1)$
   $ = (1/52)(12/51) + (3/52)(13/51)$
   $ = 12/2652 + 39/2652$
   $ = 51/2652$
   $ = .0192$

3-5 Multiplication Rule: Complements and Conditional Probability

1. If it is not true that "at least one of them is a girl," then "they are all boys."

3. If it is not true that "all students receive a grade of A," then "at least one student does not receive a grade of A."

5. let B = a child is a boy
   P(B) = .5, for each birth
   P(at least one girl) = 1 - P(all boys)
   $$= 1 - P(B_1 \text{ and } B_2 \text{ and } B_3)$$
   $$= 1 - P(B_1) \cdot P(B_2) \cdot P(B_3)$$
   $$= 1 - (.5) \cdot (.5) \cdot (.5)$$
   $$= 1 - .125$$
   $$= .875$$

7. let N = rolling a number that is not a six
   P(N) = 5/6, for each roll
   P(at least one six) = 1 - P(all numbers that are not six)
   $$= 1 - P(N_1 \text{ and } N_2 \text{ and } N_3 \text{ and } N_4 \text{ and } N_5)$$
   $$= 1 - P(N_1) \cdot P(N_2) \cdot P(N_3) \cdot P(N_4) \cdot P(N_5)$$
   $$= 1 - (5/6) \cdot (5/6) \cdot (5/6) \cdot (5/6) \cdot (5/6)$$
   $$= 1 - 3125/7776$$
   $$= 4651/7776$$
   $$= .598$$

9. let H = getting a head
   P(H) = .5, for each toss
   $$P(H_3 \mid H_1 \text{ and } H_2) = P(H_3)$$
   $$= .5$$

11. let B = a person has given birth to a baby
    W = a person is a woman
    $$P(W \mid B) = 1 \quad \text{[a certainty]}$$

13. let F = the alarm clock fails
    P(F) = .01, for each clock
    P(at least one works) = 1 - P(all fail)
    $$= 1 - P(F_1 \text{ and } F_2 \text{ and } F_3)$$
    $$= 1 - P(F_1) \cdot P(F_2) \cdot P(F_3)$$
    $$= 1 - (.01)(.01)(.01)$$
    $$= 1 - .000001$$
    $$= .999999$$

NOTE: Rounded to 3 significant digits as usual, the answer is 1.00. In cases when rounding to 3 significant digits produces a probability of 1.00, this manual gives the answer with sufficient significant digits to distinguish the answer from a certainty.

15. let N = a person is HIV negative
    P(N) = .9, for each person in the at-risk population
    P(HIV positive result) = P(at least person is HIV positive)
    $$= 1 - P(\text{all persons HIV negative})$$
    $$= 1 - P(N_1 \text{ and } N_2 \text{ and } N_3)$$
    $$= 1 - P(N_1) \cdot P(N_2) \cdot P(N_3)$$
    $$= 1 - (.9)(.9)(.9)$$
    $$= 1 - .729$$
    $$= .271$$

NOTE: This plan is very efficient. Suppose, for example, there were 3,000 people to be tested. Only in .271 = 27.1% of the groups would a retest need to be done for each of the 3 individuals. Those (.271) · (1,000) = 271 groups would generate 271 · 3 = 813 retests. The total number of tests required is then 1813 (1000 original tests + 813 retests), only 60% of the 3,000 tests that would have been required to test everyone individually.

17. Refer to the table at the right.
    This problem may be done two ways.
    * reading directly from the table
       P(M | D) = 1360/1517
                = .897
    * using the formula
       P(M | D) = P(M and D)/P(D)
                = (1360/2223)/(1517/2223)
                = .6118/.6824
                = .897

GROUP	FATE		
	Survived	Died	
Men	332	1360	1692
Women	318	104	422
Boys	29	35	64
Girls	27	18	45
	706	1517	2223

NOTE: In general, the manual will use the most obvious approach -- which most often is the first one of applying the basic definition by reading directly from the table.

19. Refer to the table and comments for exercise #17.
    P([B or G] | S) = [29 + 27]/706
                    = 56/706
                    = .0793

21. Refer to the table at the right.
    a. Each entry is 3 times the value
       in Table 3-2.
    b. P(Yes | Pos) = 1425/2100
                    = .679
    c. No, the assumed size of the population
       does not affect the probability values.

HIV?	TEST RESULT		
	Pos	Neg	
Yes	1425	75	1,500
No	675	12,825	13,500
	2100	12,900	15,000

23. For 25 randomly selected people,
    a. P(no 2 share the same birthday) = .432 [see exercise #25c in section 3-4]
    b. P(at least 2 share the same birthday) = 1 - P(no 2 share the same birthday)
                                             = 1 - .432
                                             = .568

3-6 Probabilities Through Simulations

1. Using the first four digits of each number and letting even numbers represent males and odd number represent females produces the following list of 20 numbers of girls per family of 4 children: 2 3 3 0 1 2 1 2 4 1 2 3 1 2 3 1 2 3 3 2
   Letting x be the number of girls per simulated family produces the following frequency distribution.

x	f
0	1
1	5
2	7
3	6
4	1
	20

Letting G be getting a family with 3 girls, we estimate P(G) ≈ 6/20 = .30.
NOTE: This is fairly close to the true value of 4/16 = .25 [see exercise #5]. In this case the simulation produced a good estimate.

3. Reading the digits in groups of 3, by rows and ignoring 0 and digits greater than 6, produces the following list of 20 simulated rollings of 3 dice: [the final 54 are not used]
    461 643 211 344 446 631 516 431 155 664 462 125 525 251 452 544 652 625 323 335
   The 20 corresponding total are:
    11 13 4 11 14 10 12 8 11 16 12 8 12 8 11 13 13 13 8 11
   Letting T be getting a total of 10, we estimate $P(T) \approx 1/20 = .05$
   NOTE: This is considerably lower than the true value of $27/216 = .125$ [see exercise #7]. In this case the simulation did not produce a good estimate.

5. The following Minitab commands produce the desired simulation, where 0 represents a boy and 1 a girl.

```
MTB > RANDOM 100 C1-C4;
SUBC> INTEGER 0 1.
MTB > LET C5 = C1+C2+C3+C4
MTB > TABLE C5
```

   Let x be the number of girls per family of 4. The table produced by this simulation is given below at the left, and the table obtained from the sample space of the all 16 possibilities is given at the right.

x	f		x	f
0	5		0	1
1	23		1	4
2	36		2	6
3	30		3	4
4	6		4	1
	100			16

   Letting G be getting a family of 3 girls, we estimate $P(G) \approx 30/100 = .30$. This is close to the true value $P(G) = 4/16 = .25$.

7. The following Minitab commands produce the desired simulation, where C1-C3 represent the 3 dice.

```
MTB > RANDOM 100 C1-C3;
SUBC> INTEGER 1 6.
MTB > LET C4 = C1+C2+C3
MTB > TABLE C4
```

   Let x be the total showing on the 3 dice. The table produced by this simulation is given below at the left, and the table obtained from the sample space of the all 216 possibilities is given at the right.

x	f		x	f
3	1		3	1
4	0		4	3
5	4		5	6
6	5		6	10
7	9		7	15
8	9		8	21
9	12		9	25
10	10		10	27
11	12		11	27
12	12		12	25
13	9		13	21
14	10		14	15
15	2		15	10
16	4		16	6
17	1		17	3
18	0		18	1
	100			216

Letting T be getting a total of 10, we estimate $P(T) \approx 10/100 = .10$. This is close to the true value $P(T) = 27/216 = .125$

9. The probability of originally selecting the door with the prize is 1/3. He can either not switch or switch.

* If a person makes a selection and does not switch, he should get the prize 1/3 of the time.
* If a person makes a selection and does switch under the conditions given, there are only two possibilities:

    A = he had originally selected the winning door and now switched to a losing door
    B = he had originally selected a losing door and now switched to the winning door.
      [He cannot switch from a losing door to another losing door. If he has a losing door and the door Monte Hall opens is a losing door, then the only door to switch to is the winning door.]

Since A and B are the only possibilities when he switches and P(A) = 1/3, then P(B) = 2/3. And so he should get the prize 2/3 of the time.

The better strategy is to switch.

NOTE: While the above analysis makes the simulation trivial and unnecessary, the simulation can be done as follows:

  1. n times randomly select door 1,2 or 3 to hold the prize
  2. n times randomly select door 1,2 or 3 as the contestant's original selection
  3. determine m = the number of times the doors match
  4. the contestant who doesn't switch wins m/n [should be approximately 1/3] of the time
     the contestant who switches wins (n-m)/n [should be approximately 2/3] of the time

The following Minitab commands produce the desired simulation.

```
MTB > RANDOM 100 C1-C2;
SUBC> INTEGER 1 3.
MTB > LET C3 = C1-C2
MTB > TABLE C3
```

A zero in C3 indicates that C1 and C2 match, and that the contestant's original choice was correct. For this simulation, the n=100 trials produced m=24 0's in C3. The simulated contestant who doesn't switch wins 24/100 = 24% of the time and the one who switches wins 76/100 = 76% of the time.

11. No; his reasoning is not correct. No; the proportion of girls will not increase. Each family would consist of zero to several consecutive girls, followed by a boy. The types of possible families and their association probabilities would be as follows:

P(B) = 1/2   = 16/32
P(GB) = (1/2)·(1/2) = 1/4   = 8/32
P(GGB) = (1/2)·(1/2)·(1/2) = 1/8   = 4/32
P(GGGB) = (1/2)·(1/2)·(1/2)·(1/2) = 1/16 = 2/32
P(GGGGB) = (1/2)·(1/2)·(1/2)·(1/2)·(1/2) = 1/32
etc.

Each collection of 32 families could be expected to look like this, where * = one family with 5 or more girls and a boy:

B	B	B	B	B	B	B	B
B	B	B	B	B	B	B	B
GB	GB	GB	GB	GB	GB	GB	GB
GGB	GGB	GGB	GGB	GGGB	GGGB	GGGGB	*

A gender count reveals 32 boys and 31 or more girls, still approximately an equal distribution of genders. In practice, however, some families would likely stop after having a certain number of children -- whether they had a boy or not. If that number was 5 for each family, then * = GGGGG and the expected gender count for the 32 families would be 31 boys and 31 girls, exactly an equal distribution of genders.

3-7 Counting

1. $14! = 14 \cdot 13 \cdot 12 \cdot 11 \cdot 10 \cdot 9 \cdot 8 \cdot 7 \cdot 6 \cdot 5 \cdot 4 \cdot 3 \cdot 2 \cdot 1 = 87,178,291,200$
   NOTE: It's staggering how such a small number like 14 can have such a large value for its factorial. In practice this means that if you have 14 books on a shelf you could arrange them in over 87 billion different orders!

3. $_{50}P_3 = 50!/47! = (50 \cdot 49 \cdot 48 \cdot 47!)/47! = 50 \cdot 49 \cdot 48 = 117,600$
   NOTE: This technique of "cancelling out" or "reducing" the problem by removing the factors $47! = 47 \cdot 96 \cdot \ldots \cdot 1$ from both the numerator and the denominator is preferred over actually evaluating 50!, actually evaluating 47!, and then dividing those two very large numbers. In general, a smaller factorial in the denominator can be completely divided into a larger factorial in the numerator to leave only the "excess" factors not appearing the in the denominator. This is the technique employed in this manual -- e.g., see #5 below, where the 43! is cancelled from both the numerator and the denominator. In addition, $_nP_r$ and $_nC_r$ will always be integers; calculating a non-integer value for either expression indicates an error has been made. More generally, the answer to any <u>counting</u> problem (but not a <u>probability</u> problem) must always be a whole number; a fractional number indicates that an error has been made.

5. Let W = winning the described lottery with a single selection
   The total number of possible combinations is
   $_{49}C_6 = 49!/(43!6!) = (49 \cdot 48 \cdot 47 \cdot 46 \cdot 45 \cdot 44)/6! = 13,983,816$
   Since only one combination wins,
   P(W) = 1/13,983,816

7. Let W = winning the described lottery with a single selection
   The total number of possible combinations is
   $_{44}C_6 = 44!/(38!6!) = (44 \cdot 43 \cdot 42 \cdot 41 \cdot 40 \cdot 39)/6! = 7,059,052$
   Since only one combination wins,
   P(W) = 1/7,059,052

9. let W = winning the described lottery with a single selection
   The total number of possible arrangements is
   $_{51}P_6 = 51!/45! = 51 \cdot 50 \cdot 49 \cdot 48 \cdot 47 \cdot 46 = 12,966,811,200$
   Since only one arrangement wins,
   P(W) = 1/12,966,811,200

11. a. Since a different arrangement of selected names is a different slate, use
    $_{12}P_4 = 12!/8! = 12 \cdot 11 \cdot 10 \cdot 9 = 11,880$
    b. Since a different arrangement of selected names is the same committee, use
    $_{12}C_4 = 12!/(8!4!) = (12 \cdot 11 \cdot 10 \cdot 9)/4! = 495$

13. There are 5 players. Once the 3 are chosen for the United Way event, the other 2 for the Heart Fund event are automatically determined. The problem reduces to how many ways can 3 be selected for the United Way Event. Since a different arrangement of the selected names is the same delegation, use
    $_5C_3 = 5!/(2!3!) = (5 \cdot 4)/2! = 10$

15. a. $4! = 4 \cdot 3 \cdot 2 \cdot 1 = 24$
    b. Let A = the cities are in alphabetical order
    Since only one of the routes has the cities in alphabetical order, P(A) = 1/24

17. let S = the resulting route is one with the shortest distance
    The total number of routes is 8! = 40,320
    Since 2 of the routes represent the shortest distance, P(S) = 2/40,320 = 1/20,160

19. The first note is given by *. There are 3 possibilities (R,U,D) for each of the next 15 notes.
There are $3 \cdot 3 \cdot 3 \cdot 3 \cdot 3 \cdot 3 \cdot 3 \cdot 3 \cdot 3 \cdot 3 \cdot 3 \cdot 3 \cdot 3 \cdot 3 \cdot 3 = 3^{15} = 14,348,907$ possible sequences.
NOTE: This assumes each song has at least 16 notes.

21. a. There are 2 possibilities (B,G) for each baby. The number of possible sequences is
$2 \cdot 2 \cdot 2 \cdot 2 \cdot 2 \cdot 2 \cdot 2 \cdot 2 \cdot 2 \cdot 2 \cdot 2 \cdot 2 \cdot 2 \cdot 2 \cdot 2 \cdot 2 \cdot 2 \cdot 2 \cdot 2 \cdot 2 = 2^{20} = 1,048,576$
   b. The number of ways to arrange 20 items consisting of 10 B's and 10 G's is
$20!/(10!10!) = 20 \cdot 19 \cdot 18 \cdot 17 \cdot 16 \cdot 15 \cdot 14 \cdot 13 \cdot 12 \cdot 11/10! = 184,756$
   c. Let E = getting equal numbers of boys and girls in 20 births. Based on parts (a) and (b),
P(E) = 184756/1048576 = .176
   d. The preceding results indicate that getting 10 boys and 10 girls when 20 babies are randomly
selected is not an unusual event -- for one such random selection. But the probability of
getting that result twice in a row would be $(.176) \cdot (.176) = .0310$, which would be
considered unusual since .0310 < .05. Getting 10 boys and 10 girls "consistently" should
arouse suspicion.

23. Use the factorial rule with n = 12 to determine the total number of possible arrangements.
   n! = 479,001,600
Let I = selecting the houses so that the readings are in increasing order.
Since only one of the equally likely possibilities gives the readings in increasing order,
   P(I) = 1/479,001,600 = .00000000209
Yes, her concern about the test equipment is justified. NOTE: The houses were presumably
measured in a geographical order and not randomly. If she is working her way toward the
center of a radon concentration, such a pattern would be expected. Since radon dissipates in
circulating air (home readings are typically taken in basements), an extra measurement in a
well-ventilated upstairs room should determine whether there is an equipment malfunction.

25. The total number of ways of selecting 5 eggs from among 12 is
   $_{12}C_5 = 12!/(7!5!) = (12 \cdot 11 \cdot 10 \cdot 9 \cdot 8)/5! = 792.$
   a. Let A = selecting all 3 of the cracked eggs. The total number of ways of selecting 3 eggs
from among the 3 cracked ones and 2 eggs from among the 9 good ones is
$_3C_3 \cdot _9C_2 = [3!/(0!3!)] \cdot [9!/(7!2!)] = [1] \cdot [36] = 36$
P(A) = 36/792 = .0455
   b. Let N = selecting none of the cracked eggs. The total number of ways of selecting 0 eggs
from among the 3 cracked ones and 5 eggs from among the 9 good ones is
$_3C_0 \cdot _9C_5 = [3!/(3!0!)] \cdot [9!/(4!5!)] = [1] \cdot [126] = 126$
P(N) = 126/792 = .1591
   c. Let T = selecting 2 of the cracked eggs. The total number of ways of selecting 2 eggs
from among the 3 cracked ones and 3 eggs from among the 9 good ones is
$_3C_2 \cdot _9C_3 = [3!/(1!2!)] \cdot [9!/(6!3!)] = [3] \cdot [84] = 252$
P(T) = 252/792 = .3182
NOTE 1: Since the answer in part (a) required 4 decimal places to provide 3 significant digit
accuracy, the answers in parts (b) and (c) are also given to 4 decimal places for consistency
and ease of comparison. In general, this pattern will apply throughout the manual.
NOTE 2: There is only one possibility not covered in parts (a)-(c) above -- viz., one of the
cracked eggs is selected. As a check, it is recommended that this probability be calculated. If
these 4 mutually exclusive and exhaustive probabilities do not sum to 1.000, then an error has
been made.
   Let O = selecting 1 of the cracked eggs. The total number of ways of selecting 1 egg
from among the 3 cracked ones and 4 eggs from among the 9 good ones is
$_3C_1 \cdot _9C_4 = [3!/(2!1!)] \cdot [9!/(5!4!)] = [3] \cdot [126] = 378$
P(O) = 378/792 = .4773

$P(N \text{ or } O \text{ or } T \text{ or } A) = P(N) + P(O) + P(T) + P(A)$
$$= .1591 + .4773 + .3182 + .0455$$
$$= 1.000 \quad [\text{the 4 stated values actually sum to 1.0001 due to round-off}]$$
And so everything seems in order. Performing an extra calculation that can provide a check to both the methodology and the mathematics is a good habit that is well worth the effort.

27. This problem may be approached in two different ways.

* general probability formulas of earlier sections

   Let F = selecting a card favorable for getting a flush.

   $P(F_1) = 52/52$, since the first card could be anything

   $P(F_2) = 12/51$, since the second card must be from the same suit as the first

   etc.

   $P(\text{getting a flush}) = P(F_1 \text{ and } F_2 \text{ and } F_3 \text{ and } F_4 \text{ and } F_5)$
   $$= P(F_1) \cdot P(F_2) \cdot P(F_3) \cdot P(F_4) \cdot P(F_5)$$
   $$= (52/52) \cdot (12/51) \cdot (11/50) \cdot (10/49) \cdot (9/48)$$
   $$= .00198$$

* counting techniques of this section

   The total number of possible 5 card selections is

   $_{52}C_5 = 52!/(47!5!) = 2,598,960$

   The total number of possible 5 card selections from one particular suit is

   $_{13}C_5 = 13!/(8!5!) = 1287$

   The total number of possible 5 card selections from any of the 4 suits is

   $4 \cdot 1287 = 5148$

   $P(\text{getting a flush}) = 5148/2598960 = .00198$

29. There are 26 possible first characters, and 36 possible characters for the other positions. Find the number of possible names using 1,2,3,...,8 characters and then add to get the total.

characters	possible names		
1	26	=	26
2	26·36	=	936
3	26·36·36	=	33,696
4	26·36·36·36	=	1,213,056
5	26·36·36·36·36	=	43,670,016
6	26·36·36·36·36·36	=	1,572,120,576
7	26·36·36·36·36·36·36	=	56,596,340,736
8	26·36·36·36·36·36·36·36	=	2,037,468,266,496
		total =	2,095,681,645,538

31. a. The calculator factorial key gives $50! = 3.04140932 \times 10^{64}$.

   Using the approximation, $K = (50.5) \cdot \log(50) + .39908993 - .43429448(50)$
   $$= 85.79798522 + .39908993 - 21.71472400$$
   $$= 64.48235115$$

   And then $n! = 10^K$
   $$= 10^{64.48235115}$$
   $$= 3.036345216 \times 10^{64}$$

   NOTE: The two answers differ by 5.1×10^{61} (i.e., by 51 followed by 60 zeros -- "zillions and zillions"). While that error may seem large, the numbers being dealt with are very large and that difference is only a $(5.1 \times 10^{61})/(3.04 \times 10^{64}) = 1.7\%$ error.

   b. Using the approximation, with n=300,

   $K = (n+.5)(\log n) + .39908993 - .43429448n$
   $$= (300.5)(\log 300) + .39908993 - .43429448(300)$$
   $$= 614.4$$

   $300! = 10^{614.4}$

   Since the number of digits in 10^x is the next whole number above x, 300! has 615 digits.

Review Exercises

1. Refer to the table at the right.
   let Y = a subject experienced soreness
       N = a subject did not experience soreness
       D = a subject used the drug Nicorette
       P = a subject used the placebo

   P(D) = 152/305 = .498

		TREATMENT		
		Drug	Placebo	
SORENESS?	Yes	43	35	78
	No	109	118	227
		152	153	305

2. Refer to the table and notation of exercise #1.
   P(N) = 227/305 = .744

3. Refer to the table and notation of exercise #1.
   P(D or Y) = P(D) + P(Y) - P(D and Y)
   $\qquad$ = 152/305 + 78/305 - 43/305
   $\qquad$ = 187/305
   $\qquad$ = .613

4. Refer to the table and notation of exercise #1.
   P(P or N) = P(P) + P(N) - P(P and N)
   $\qquad$ = 153/305 + 227/305 - 118/305
   $\qquad$ = 262/305
   $\qquad$ = .859

5. Refer to the table and notation of exercise #1.
   $P(Y_1 \text{ and } Y_2) = P(Y_1) \cdot P(Y_2 \mid Y_1)$
   $\qquad = (78/305) \cdot (77/304)$
   $\qquad = .0648$

6. Refer to the table and notation of exercise #1.
   $P(P_1 \text{ and } P_2) = P(P_1) \cdot P(P_2 \mid P_1)$
   $\qquad = (153/305) \cdot (152/304)$
   $\qquad = .251$

7. Refer to the table and notation of exercise #1.
   P(Y | P) = 35/153 = .229

8. Refer to the table and notation of exercise #1.
   P(Y or P) = P(Y) + P(P) - P(Y and P)
   $\qquad$ = 78/305 + 153/305 - 35/305
   $\qquad$ = 196/305
   $\qquad$ = .643

9. let G = an item is defective
   P(G) = .27, for each item
   a. $P(\overline{G}) = 1 - .27 = .73$
   b. $P(G_1 \text{ and } G_2) = P(G_1) \cdot P(G_2)$
   $\qquad = (.27) \cdot (.27)$
   $\qquad = .0729$
   c. P(at least one good) = 1 - P(all not good)
   $\qquad = 1 - P(\overline{G}_1 \text{ and } \overline{G}_2 \text{ and } \overline{G}_3)$
   $\qquad = 1 - P(\overline{G}_1) \cdot P(\overline{G}_2) \cdot P(\overline{G}_3)$
   $\qquad = 1 - (.73)^3$
   $\qquad = .611$

10. Since the sample size is no more than 5% of the size of the population, treat the selections as being independent even though the selections are made without replacement.
    let G = a CD is good
    P(G) = .98, for each selection
    P(reject batch) = P(at least one is defective)
    $$= 1 - P(\text{all are good})$$
    $$= 1 - P(G_1 \text{ and } G_2 \text{ and } G_3 \text{ and } G_4)$$
    $$= 1 - P(G_1) \cdot P(G_2) \cdot P(G_3) \cdot P(G_4)$$
    $$= 1 - (.98)^4$$
    $$= .0776$$
    NOTE: P(reject batch) = $1 - (2450/2500) \cdot (2449/2499) \cdot (2448/2498) \cdot (2447/2497)$ = .0777 if the problem is done considering the effect of the non-replacement.

11. let B = getting a boy
    P(B) = .5, for each child
    P(at least one girl) = 1 - P(all boys)
    $$= 1 - P(B_1 \text{ and } B_2 \text{ and } B_3 \text{ and } B_4 \text{ and } B_5)$$
    $$= 1 - P(B_1) \cdot P(B_2) \cdot P(B_3) \cdot P(B_4) \cdot P(B_5)$$
    $$= 1 - (.5)^5$$
    $$= .969$$

12. a. let W = the committee of 3 is the 3 wealthiest members
       The number of possible committees of 3 is $_{10}C_3 = 10!/(7!3!) = 120$.
       Since only one of those committees is the 3 wealthiest members,
       P(W) = 1/120 = .00833
    b. Since order is important, the number of possible slates of officers is
       $_{10}P_3 = 10!/7! = 720$

13. let E = a winning even number occurs
    a. P(E) = 18/38 [=9/19]
    b. odds against E = $P(\overline{E})/P(E)$
       $$= (20/38)/(18/38)$$
       $$= 20/18$$
       $$= 10/9, \text{ usually expressed as } 10{:}9$$
    c. payoff odds = (net profit):(amount bet)
       If the payoff odds are 1:1, the net profit is $5 for a winning $5 bet.

14. Since the sample size is no more than 5% of the size of the population, treat the selections as being independent even though the selections are made without replacement.
    let R = getting a Republican
    P(R) = .30, for each selection
    $P(R_1 \text{ and } R_2 \text{ and...and } R_{12})$ = $P(R_1) \cdot P(R_2) \cdot ... \cdot P(R_{12})$
    $$= (.30)^{12}$$
    $$= .000000531$$
    No, the pollster's claim is not correct.

15. let L = a selected 27 year old male lives for one more year
    P(L) = .9982, for each such person
    $P(L_1 \text{ and } L_2 \text{ and...and } L_{12})$ = $P(L_1) \cdot P(L_2) \cdot ... \cdot P(L_{12})$
    $$= (.9982)^{12}$$
    $$= .979$$

16. let G = a selected switch is good
P(G) = 44/48, for the first selection only
a. $P(G_1$ and G_2 and G_3 and $G_4) = P(G_1) \cdot P(G_2) \cdot P(G_3) \cdot P(G_4)$
$= (44/48) \cdot (43/47) \cdot (42/46) \cdot (41/45)$
$= 3258024/4669920$
$= .69766$
Since this is the only way to arrange 4G's,
P(all 4 are good) = .698
b. $P(G_1$ and G_2 and G_3 and $\bar{G}_4) = P(G_1) \cdot P(G_2) \cdot P(G_3) \cdot P(\bar{G}_4)$
$= (44/48) \cdot (43/47) \cdot (42/46) \cdot (4/45)$
$= 317856/4669920$
$= .06806$
Since 3G's and $1\bar{G}$ can be arranged in 4!/(3!1!) = 4 ways (of which this is only one),
P(exactly 3 are good) = $4 \cdot (.06806)$ = .272
c. $P(G_1$ and G_2 and $\bar{G}_3$ and $\bar{G}_4) = P(G_1) \cdot P(G_2) \cdot P(\bar{G}_3) \cdot P(\bar{G}_4)$
$= (44/48) \cdot (43/47) \cdot (4/46) \cdot (3/45)$
$= 22704/4669920$
$= .00486$
Since 2G's and $2\bar{G}$'s can be arranged in 4!/(2!2!) = 6 ways (of which this is only one),
P(exactly 2 are good) = $6 \cdot (.00486)$ = .0292

Cumulative Review Exercises

1. scores arranged in order: 2 4 7 8 10 10 13 13 14 15 17 29
summary statistics: n = 12, Σx = 142, Σx^2 = 2222
a. $\bar{x} = (\Sigma x)/n = 142/12 = 11.8$
b. $\tilde{x} = (10 + 13)/2 = 11.5$
c. $s^2 = [n \cdot (\Sigma x^2) - (\Sigma x)^2]/[n(n-1)]$
$= [12(2222) - (142)^2]/[(12)(11)]$
$= 49.24$
s = 7.0
d. s^2 = 49.2 [from part (c)]
e. minimum usual age = $\bar{x}$ - 2s = 11.8 - 2(7.0) = -2.2, truncated to 0
maximum usual age = $\bar{x}$ + 2s = 11.8 + 2(7.0) = 25.8
No, a new jet [i.e., x = 0] is within the above limits and would not be considered unusual.
f. ratio, since differences are meaningful and there is a natural meaning for 0
g. let C = getting the correct answer
Since there are 4 given choices and only 1 of them is correct, P(C) = 1/4
h. let A = getting an age of at least 10
Since there are 12 ages and 8 of them are at least 10, P(A) = 8/12 = .667
i. let L = getting an age less than 10
P(L) = 4/12, for the first selection only
$P(L_1$ and $L_2) = P(L_1) \cdot P(L_2 \mid L_1)$
$= (4/12) \cdot (3/11)$
$= .0909$
j. let L = getting an age less than 10
P(L) = 4/12, for each selection
$P(L_1$ and $L_2) = P(L_1) \cdot P(L_2)$
$= (4/12) \cdot (4/12)$
$= .1111$

2. Let x = the height selected.
   The values identified on the boxplot are as follows:
   x_1 [minimum] = 56.1
   P_{25} = 62.2
   P_{50} = 63.6
   P_{75} = 65.0
   x_n [maximum] = 71.1

NOTE: Assume that the number of heights is so large that (1) repeated sampling does not affect the probabilities for subsequent selections and (2) it can be said that P_a has a% of the heights below it and (100-a)% of the heights above it.

a. The exact values of $\bar{x}$ and s cannot be determined from the information given. The Range Rule of Thumb [the highest and lowest values are approximately 2 standard deviations above and below the mean], however, indicates that s $\approx$ $(x_n - x_1)/4$ and $\bar{x}$ $\approx$ $(x_1 + x_n)/2$. In this case we estimate $\bar{x}$ $\approx$ (56.1 + 71.1)/2 = 63.6

   NOTE: The boxplot, which gives $\tilde{x}$ = P_{50} = 63.6, is symmetric. This suggests that the original distribution is approximately so. In a symmetric distribution the mean and median are equal. This also suggests $\bar{x}$ $\approx$ 63.6.

b. P(56.1 < x < 62.2) = $P(x_1 < x < P_{25})$
                       = .25 - 0
                       = .25

c. P(x < 62.2 or x > 63.6) = P(x < 62.2) + P(x > 63.6)
                           = $P(x < P_{25})$ + $P(x > P_{50})$
                           = .25 + .50
                           = .75

   NOTE: No height is less than 62.2 <u>and</u> greater than 63.6 -- i.e., the addition rule for mutually exclusive events can be used.

d. Let B = selecting a height between 62.2 and 63.6.
   P(B) = $P(P_{25} < x < P_{50})$
        = .50 - .25
        = .25
   $P(B_1 \text{ and } B_2)$ = $P(B_1) \cdot P(B_2 \mid B_1)$
                = (.25)·(.25)
                = .0625

e. let S = a selected woman is shorter than the mean
       T = a selected woman is taller than the mean
       E = the event that a group of 5 women consists of 3 T's and 2 S's
   The total number of arrangements of T's and S's is 2·2·2·2·2 = 32
   The number of arrangements of 3 T's and 2 S's is 5!/(3!2!) = 10
   P(E) = 10/32 = .3125

   NOTE: The exact value of the mean cannot be determined from the information given. From part (a) we estimate $\bar{x}$ $\approx$ 63.6, which is also the median. If this is true then *in this particular problem* $P(x < \bar{x})$ = P(S) = .50 and $P(x > \bar{x})$ P(T) = .50. This means that all arrangements of T's and S's are equally likely and
   P(E) = (# of ways E can occur)/(total number of possible outcomes)

Chapter 4

Probability Distributions

4-2 Random Variables

1. a. Continuous, since weight can be any value on a continuum.
   b. Discrete, since the cost must be an integer number cents.
   c. Continuous, since time can be any value on a continuum.
   d. Continuous, since volume can be any value on a continuum.
   e. Discrete, since the number of cans must be an integer.

NOTE: When working with probability distributions and formulas in the exercises that follow, always keep these important facts in mind.
   * If one of the conditions for a probability distribution does not hold, the formulas do not apply and produce numbers that have no meaning.
   * $\Sigma x \cdot P(x)$ gives the mean of the x values and must be a number between the highest and lowest x values.
   * $\Sigma x^2 \cdot P(x)$ gives the mean of the x^2 values and must be a number between the highest and lowest x^2 values.
   * $\Sigma P(x)$ must always equal 1.000.
   * Σx and Σx^2 have no meaning and should not be calculated.
   * The quantity $[\Sigma x^2 \cdot P(x) - \mu^2]$ cannot possibly be negative; if it is, then there is a mistake.
   * Always be careful to use the <u>unrounded</u> mean in the calculation of the variance and to take the square root of the <u>unrounded</u> variance to find the standard deviation.

3. This is a probability distribution since $\Sigma P(x)=1$ is true and $0 \leq P(x) \leq 1$ is true for each x.

x	P(x)	x·P(x)	x^2	x^2·P(x)	
0	.125	0	0	0	$\mu = \Sigma x \cdot P(x)$
1	.375	.375	1	.375	$= 1.500$, rounded to 1.5
2	.375	.750	4	1.500	$\sigma^2 = \Sigma x^2 \cdot P(x) - \mu^2$
3	.125	.375	9	1.125	$= 3.000 - (1.500)^2$
	1.000	1.500		3.000	$= .750$
					$\sigma = .866$, rounded to 0.9

5. This is a probability distribution since $\Sigma P(x)=1$ is true and $0 \leq P(x) \leq 1$ is true for each x.

x	P(x)	x·P(x)	x^2	x^2·P(x)	
0	.0625	0	0	0	$\mu = \Sigma x \cdot P(x)$
1	.2500	.2500	1	.2500	$= 2.0000$, rounded to 2.0
2	.3750	.7500	4	1.5000	$\sigma^2 = \Sigma x^2 \cdot P(x) - \mu^2$
3	.2500	.7500	9	2.2500	$= 5.0000 - (2.0000)^2$
4	.0625	.2500	16	1.0000	$= 1.0000$
	1.0000	2.0000		5.0000	$\sigma = 1.0000$, rounded to 1.0

7. This is a probability distribution since $\Sigma P(x)=1$ is true and $0 \le P(x) \le 1$ is true for each x.

x	P(x)	x·P(x)	x²	x²·P(x)	
0	.512	0	0	0	$\mu = \Sigma x \cdot P(x)$
1	.301	.301	1	.301	$= .730$, rounded to 0.7
2	.132	.264	4	.528	$\sigma^2 = \Sigma x^2 \cdot P(x) - \mu^2$
3	.055	.165	9	.495	$= 1.324 - (.73)^2$
	1.000	.730		1.324	$= .791$
					$\sigma = .889$, rounded to 0.9

9. This is a probability distribution since $\Sigma P(x)=1$ is true and $0 \le P(x) \le 1$ is true for each x.

x	P(x)	x·P(x)	x²	x²·P(x)	
0	.36	0	0	0	$\mu = \Sigma x \cdot P(x)$
1	.48	.48	1	.48	$= .80$, rounded to 0.8
2	.16	.32	4	.64	$\sigma^2 = \Sigma x^2 \cdot P(x) - \mu^2$
	1.00	.80		1.12	$= 1.12 - (.80)^2$
					$= .48$
					$\sigma = .693$, rounded to 0.7

11. This is a probability distribution since $\Sigma P(x)=1$ is true and $0 \le P(x) \le 1$ is true for each x.

x	P(x)	x·P(x)	x²	x²·P(x)	
0	.289	0	0	0	$\mu = \Sigma x \cdot P(x)$
1	.407	.407	1	.407	$= 1.101$, rounded to 1.1
2	.229	.458	4	.916	$\sigma^2 = \Sigma x^2 \cdot P(x) - \mu^2$
3	.065	.195	9	.585	$= 2.077 - (1.101)^2$
4	.009	.036	16	.144	$= .865$
5	.001	.005	25	.025	$\sigma = .939$, rounded to 0.9
	1.000	1.101		2.077	

13. a.

x	P(x)	x·P(x)	
175	1/38	175/38	$E = \Sigma x \cdot P(x)$
-5	37/38	-185/38	$= -10/38$
	38/38	-10/38	$= \$-0.263$ [i.e., a loss of 26.3¢]

The expected loss is 26.3¢ for a $5 bet.

b.

x	P(x)	x·P(x)	
5	18/38	90/38	$E = \Sigma x \cdot P(x)$
-5	20/38	-100/38	$= -10/38$
	38/38	-10/38	$= \$-0.263$ [i.e., a loss of 26.3¢]

The expected loss is 26.3¢ for a $5 bet.

c. Since the bets in parts (a) and (b) each have (exactly the same) negative expectation, the best option is not to bet at all.

NOTE: Problems whose probabilities do not "come out even" as decimals can be very sensitive to rounding errors. Since the P(x) values are used in several subsequent calculations, express them as exact fractions instead of rounded decimals. If it is not convenient to continue with fractions throughout the entire problem, then use sufficient decimal places (typically one more than "usual") in the x·P(x) and x²·P(x) columns to guard against cumulative rounding errors.

15. a. Mike "wins" $100,000 - $250 = $99,750 if he dies.
Mike "loses" $250 if he lives.

b.

x	P(x)	x·P(x)	
99750	.0015	149.625	$E = \Sigma x \cdot P(x)$
-250	.9985	-249.625	$= -100.000$ (i.e., a loss of $100)
	1.0000	-100.000	

c. Since the company is making a $100 profit at this price, it would break even if it sold the policy for $100 less -- i.e. for $150. NOTE: This oversimplified analysis ignores the cost of doing business. If the costs (printing, salaries, etc.) associated with offering the policy is $25, for example, then the company's profit is only $75 and selling the policy for $75 less (i.e., for $175) would represent the break even point for the company.

17. Use the Range Rule of Thumb to identify unusual results.
The minimum usual value is $\mu - 2\sigma = 11.4 - 2\cdot(.75) = 9.9$.
Yes; since "fewer than 8" is less than 9.9, that would be considered an unusual result.

19. a. Since the probabilities must sum to 1.000, the missing value (found by subtracting all the other values from 1.000) is P(x=2) = .161.
b. Use probabilities to identify unusual results.
P(at least 3 sevens) = $P(x \geq 3)$
　　　　　　　　　　 = P(x=3 or x=4 or x=5)
　　　　　　　　　　 = P(x=3) + P(x=4) + P(x=5)
　　　　　　　　　　 = .032 + .003 + .000
　　　　　　　　　　 = .035
Yes; since .035 < .05, "at least 3 sevens" would be considered an unusual result.

21. Use probabilities to identify unusual results.
P(10 or more girls) = $P(x \geq 10)$
　　　　　　　　　　 = P(x=10 or x=11 or x=12 or x=13 or x=14)
　　　　　　　　　　 = P(x=10) + P(x=11) + P(x=12) + P(x=13) + P(x=14)
　　　　　　　　　　 = .061 + .022 + .006 + .001 + .000
　　　　　　　　　　 = .090
No; since .090 > .05, "10 or more girls" would not be considered an unusual result for 14 births. The gender selection technique may be effective, but this is not evidence to conclude that it is. While results like those observed might be expected if the technique is effective, they could still reasonably (i.e., with probability of 5% or more) occur if the technique is not effective.

23. The 16 equally like outcomes in the sample space are given at the right. If x represents the number of girls, counting the numbers of favorable outcomes indicates

outcome	x	outcome	x
BBBB	0	GBBB	1
BBBG	1	GBBG	2
BBGB	1	GBGB	2
BGBB	1	GGBB	2
BGGB	2	GGGB	3
BGBG	2	GGBG	3
BBGG	2	GBGG	3
BGGG	3	GGGG	4

```
P(x = 0) = 1/16 = .0625
P(x = 1) = 4/16 = .2500
P(x = 2) = 6/16 = .3750
P(x = 3) = 4/16 = .2500
P(x = 4) = 1/16 = .0625
```

x	P(x)	x·P(x)	x^2	x^2·P(x)
0	.0625	0	0	0
1	.2500	.2500	1	.2500
2	.3750	.7500	4	1.5000
3	.2500	.7500	9	2.2500
4	.0625	.2500	16	1.0000
	1.0000	2.0000		5.0000

$\mu = \Sigma x\cdot P(x)$
　 = 2.0000, rounded to 2.0
$\sigma^2 = \Sigma x^2\cdot P(x) - \mu^2$
　 = 5.0000 - (2.0000)^2
　 = 1.0000
σ = 1.0000, rounded to 1.0

25. In order for P(x) to be a probability distribution, it must be true that
    (1) $\Sigma P(x) = 1$
    (2) $0 \leq P(x) \leq 1$ for every permissible value of x.
    a. P(x) is a probability distribution because for x = 1,2,3,4
       (1) $\Sigma P(x) = \Sigma(1/4)$
       $= 1/4 + 1/4 + 1/4 + 1/4$
       $= 1$
       (2) $0 \leq P(x) \leq 1$ for each x, since $P(x) = 1/4$ for each x
    b. P(x) is a probability distribution because for x = 0,1,2
       (1) $\Sigma P(x) = \Sigma 1/[2 \cdot (2-x)! x!]$
       $= 1/[2 \cdot 2! 0!] + 1/[2 \cdot 1! 1!] + 1/[2 \cdot 0! 2!]$
       $= 1/4 + 1/2 + 1/4$
       $= 1$
       (2) $0 \leq P(x) \leq 1$ for each x, since the P(x) values are given above as 1/4, 1/2, 1/4
          for x = 0,1,2
    c. P(x) is not a probability distribution because for x = 1,2,3,...
       (1) $\Sigma P(x) = \Sigma[1/(2x)]$
       $= 1/2 + 1/4 + 1/6 + 1/8 + ...$
       $= .5000 + .2500 + .1666 + .1250 + ...$
       $= 1.0416 + ...$
       > 1
       (2) [Condition (2) happens to be satisfied, but that is irrelevant since condition (1) is not
          satisfied and both conditions must be met to have a probability distribution.]
    d. P(x) is a probability distribution because for x = 1,2,3...
       (1) $\Sigma P(x) = \Sigma(.5)^x$
       $= (.5) \cdot \Sigma(.5)^{x-1}$
       $= (.5) \cdot [1 + .5 + .5^2 + .5^3 + ...]$
       $= (.5) \cdot [1/(1-.5)]$
       $= (.5) \cdot [1/.5]$
       $= 1$
       (2) $0 \leq P(x) \leq 1$ for each x, since (i) P(x) > 0 for each x and (ii) P(x) = .5 for x = 1
          and (iii) P(x) keeps decreasing as x grows larger
       NOTE: Part (d) uses the fact from algebra that
       $1 + r + r^2 + r^3 + r^4 + ... = 1/(1-r)$ for any r such that $-1 < r < 1$
       proof: (1)   let $S = 1 + r + r^2 + r^3 + ...$
              (2)   then $rS = \underline{\quad r + r^2 + r^3 + r^4 + ...}$
          subtracting (1)-(2) $S-rS = 1$ [the terms with r's subtract out...]
                             $S(1-r) = 1$
                                 $S = 1/(1-r)$

27. a. For random selections with replacement, P(x) = .1 for each x = 0,1,2,3,4,5,6,7,8,9.

x	P(x)	x·P(x)	x^2	x^2·P(x)	
0	.1	0	0	0	$\mu = \Sigma x \cdot P(x)$
1	.1	.1	1	.1	$= 4.5$
2	.1	.2	4	.4	$\sigma^2 = \Sigma x^2 \cdot P(x) - \mu^2$
3	.1	.3	9	.9	$= 28.5 - (4.5)^2$
4	.1	.4	16	1.6	$= 8.25$
5	.1	.5	25	2.5	$\sigma = 2.8723$, rounded to 2.9
6	.1	.6	36	3.6	
7	.1	.7	49	4.9	
8	.1	.8	64	6.4	
9	.1	.9	81	8.1	
	1	4.5		28.5	

b. The z score for each x is $z = (x-\mu)/\sigma = (x-4.5)/2.8723$

z	P(z)	z·P(z)	z²	z²·P(z)
-1.5667	.1	-.15667	2.45455	.245455
-1.2185	.1	-.12185	1.48485	.148485
-0.8704	.1	-.08704	.75758	.075758
-0.5222	.1	-.05222	.27272	.027272
-0.1741	.1	-.01741	.03030	.003030
0.1741	.1	.01741	.03030	.003030
0.5222	.1	.05222	.27272	.027272
0.8704	.1	.08704	.75758	.075758
1.2185	.1	.12185	1.48485	.148485
1.5667	.1	.15667	2.45455	.245455
	1.0	0		1.000000

$\mu = \Sigma z \cdot P(z)$
$= 0$
$\sigma^2 = \Sigma z^2 \cdot P(z) - \mu^2$
$= 1 - (0)^2$
$= 1$
$\sigma = 1$

c. Yes; for the z scores of any population of size n,
$\Sigma z = \Sigma(x-\mu)/\sigma = (1/\sigma) \cdot \Sigma(x-\mu) = (1/\sigma) \cdot [\Sigma x - \Sigma\mu] = (1/\sigma) \cdot [n\mu - n\mu] = 0$
$\Sigma z^2 = \Sigma[(x-\mu)/\sigma]^2 = (1/\sigma^2) \cdot \Sigma(x-\mu)^2 = (1/\sigma^2) \cdot n\sigma^2 = n$
$\mu = \Sigma z/n = 0/n = 0$
$\sigma^2 = \Sigma(z-\mu)^2/n = \Sigma z^2/n = n/n = 1$

29. Since each die has 6 faces, there will be $6 \cdot 6 = 36$ possible outcomes. If each of the 12 sums 1,2,3,...,12 is to be equally likely, each sum must occur exactly 3 times. As a logical starting point, suppose that one of the dice is normal. If one die contains the usual digits 1,2,3,4,5,6, the other die
    (1) must have three 0's to pair with the 1 to get three sums of 1
    (2) must have three 6's to pair with the 6 to get three sums of 12
The 36 possible outcome generated by such dice would be

1-0	2-0	3-0	4-0	5-0	6-0
1-0	2-0	3-0	4-0	5-0	6-0
1-0	2-0	3-0	4-0	5-0	6-0
1-6	2-6	3-6	4-6	5-6	6-6
1-6	2-6	3-6	4-6	5-6	6-6
1-6	2-6	3-6	4-6	5-6	6-6

Inspection of these sums indicates that each of the values 1,2,3,...12 appears exactly 3 times so that $P(x) = 3/36 = 1/12$ for x=1,2,3,...,12. The solution, therefore, is to mark one die normally 1,2,3,4,5,6 and mark the other die 0,0,0,6,6,6.

4-3 Binomial Experiments

NOTE: The four requirements for a binomial experiment are
  #1 There are a fixed number of trials.
  #2 The trials are independent.
  #3 Each trial has two possible named outcomes.
  #4 The probabilities remain constant for each trial.

1. Yes; all four requirements are met [assuming each multiple choice question has the same number of choices, only one of which is correct].

3. No; requirement #3 is not met. There are more than 2 possible outcomes.

5. No; requirement #3 is not met. There are more than 2 possible outcomes.

7. No; requirement #3 is not met. There are more than 2 possible outcomes.

9. let W = guessing the wrong answer
     C = guessing the correct answer
   P(W) = 4/5, for each question
   P(C) = 1/5, for each question
   a. $P(WWC) = P(W_1 \text{ and } W_2 \text{ and } C_3)$
   $\qquad\qquad\quad = P(W_1) \cdot P(W_2) \cdot P(C_3)$
   $\qquad\qquad\quad = (4/5) \cdot (4/5) \cdot (1/5)$
   $\qquad\qquad\quad = 16/125$
   $\qquad\qquad\quad = .128$
   b. There are 3 possible arrangements: WWC, WCW, CWW
      Following the pattern in part (a)
      $P(WWC) = P(W_1 \text{ and } W_2 \text{ and } C_3) = P(W_1) \cdot P(W_2) \cdot P(C_3) = (4/5) \cdot (4/5) \cdot (1/5) = 16/125$
      $P(WCW) = P(W_1 \text{ and } C_2 \text{ and } W_3) = P(W_1) \cdot P(C_2) \cdot P(W_3) = (4/5) \cdot (1/5) \cdot (4/5) = 16/125$
      $P(CWW) = P(C_1 \text{ and } W_2 \text{ and } W_3) = P(C_1) \cdot P(W_2) \cdot P(W_3) = (1/5) \cdot (4/5) \cdot (4/5) = 16/125$
   c. P(exactly one correct answer) = P(WWC or WCW or CWW)
      $\qquad\qquad\qquad\qquad\qquad\quad = P(WWC) + P(WCW) + P(CWW)$
      $\qquad\qquad\qquad\qquad\qquad\quad = 16/125 + 16/125 + 16/125$
      $\qquad\qquad\qquad\qquad\qquad\quad = 48/125$
      $\qquad\qquad\qquad\qquad\qquad\quad = .384$

11. from table A-1 in the .01 column and the 7-1 row, .066

13. from table A-1 in the .60 column and the 8-3 row, .124

15. from table A-1 in the .01 column and the 11-4 row, .000$^+$

NOTE: To use the binomial formula, one must identify 3 quantities: n,x,p. Table A-1, for example, requires only these 3 values to supply a probability. Since what the text calls "q" always equals 1-p, it can be so designated without introducing unnecessary notation [just as no special notation is utilized for the quantity n-x, even though it appears twice in the binomial formula]. This has the additional advantage of ensuring that the probabilities p and 1-p sum to 1.00 and protecting against an error in the separate calculation and/or identification of "q." In addition, reversing the order of (n-x)! and x! in the denominator of the $_nC_x$ coefficient term seems appropriate. That agrees with the $n!/(n_1!n_2!)$ logic of the "permutation rule when some objects are alike" for n_1 objects of one type and n_2 objects of another type, and that places the "x" and "n-x" in the same order in both the denominator of the coefficient term and the exponents. Such a natural ordering also leads to fewer errors. Accordingly, this manual expresses the binomial formula as

$$P(x) = [n!/x!(n-x)!] \cdot p^x \cdot (1-p)^{n-x}$$

17. $P(x) = [n!/x!(n-x)!] \cdot p^x \cdot (1-p)^{n-x}$
    $P(x=3) = [5!/3!2!] \cdot (.25)^3 \cdot (.75)^2$
    $\qquad\quad = [10] \cdot (.25)^3 \cdot (.75)^2$
    $\qquad\quad = [10] \cdot (.0156) \cdot (.5625)$
    $\qquad\quad = .0879$

IMPORTANT NOTE: The intermediate values of 10, .0156 and .5625 are given to help those with incorrect answers to identify the portion of the problem in which the mistake was made. This practice will be followed in all problems (i.e., not just binomial problems) throughout the manual. In practice, all calculations can be done in one step on the calculator. You may choose to (or be asked to) write down such intermediate values for your own (or the instructor's) benefit, but <u>never round off in the middle of a problem</u>. <u>Do not write the values down on paper and then re-enter</u>

them in the calculator -- use the memory to let the calculator remember with complete accuracy any intermediate values that will be used in subsequent calculations. In addition, always make certain that the quantity [n!/x!(n-x)!] is a whole number and that the final answer is between 0 and 1.

19. $P(x) = [n!/x!(n-x)!] \cdot p^x \cdot (1-p)^{n-x}$
    $P(x=4) = [10!/4!6!] \cdot (1/3)^4 \cdot (2/3)^6$
    $\quad\quad\quad = [210] \cdot (1/81) \cdot (64/729)$
    $\quad\quad\quad = .228$

21. $P(x \geq 4) = P(x=4 \text{ or } x=5)$
    $\quad\quad\quad\quad = P(x=4) + P(x=5)$
    $\quad\quad\quad\quad = .1956 + .0459$
    $\quad\quad\quad\quad = .2415$

23. $P(x > 1) = 1 - P(x \leq 1)$
    $\quad\quad\quad\quad = 1 - P(x=0 \text{ or } x=1)$
    $\quad\quad\quad\quad = 1 - [P(x=0) + P(x=1)]$
    $\quad\quad\quad\quad = 1 - [.0206 + .1209]$
    $\quad\quad\quad\quad = 1 - [.1415]$
    $\quad\quad\quad\quad = .8585$

    NOTE: This also could have been directly as $P(x > 1) = P(x=2 \text{ or } x=3 \text{ or } x=4 \text{ or } x=5)$. In general the manual will choose the most efficient technique for solving problems that may be approached in more than one way.

25. let x = the number of taxpayers that are audited
    binomial problem: n = 5 and p = .01, use Table A-1
    a. $P(x=3) = .000^+$
    b. $P(x \geq 3) = P(x=3) + P(x=4) + P(x=5)$
    $\quad\quad\quad\quad = .000^+ + .000^+ + .000^+$
    $\quad\quad\quad\quad = .000^+$
    c. Since $.000^+ < .05$, "three or more are audited" is an unusual result and we conclude that factors other than chance are at work. It appears that either customers of the Hemingway Company are being targeted by the IRS for audits or that customers with the types of returns that are often audited are selecting the Hemingway Company to do their returns. NOTE: A classic scenario of this type is the hospital that has higher than average patient mortality. This can be due either to the care being below average or to the accepting of a higher than average proportion of more serious cases.

27. let x = the number of wrong answers
    binomial problem: n = 10 and p = .15, use the binomial formula
    a. $P(x) = [n!/x!(n-x)!] \cdot p^x \cdot (1-p)^{n-x}$
    $\quad P(x=1) = [10!/1!9!] \cdot (.15)^1 \cdot (.85)^9$
    $\quad\quad\quad\quad = [10] \cdot (.1500) \cdot (.2316)$
    $\quad\quad\quad\quad = .347$
    b. $P(x=0) = [10!/0!10!] \cdot (.15)^0 \cdot (.85)^{10}$
    $\quad\quad\quad\quad = [1] \cdot (1) \cdot (.1969)$
    $\quad\quad\quad\quad = .197$
    $\quad P(x \leq 1) = P(x=0) + P(x=1)$
    $\quad\quad\quad\quad = .197 + .347$
    $\quad\quad\quad\quad = .544$
    c. No; since $.544 > .05$, "at most one wrong number" is not an unusual occurrence when the error rate is 15%.

29. let x = the number of men that are color blind
    binomial problem: n = 6 and p = .09, use the binomial formula
    $P(x) = [n!/x!(n-x)!] \cdot p^x \cdot (1-p)^{n-x}$
    $P(x=2) = [6!/2!4!] \cdot (.09)^2 \cdot (.91)^4$
    $ = [15] \cdot (.0081) \cdot (.6857)$
    $ = .0833$

31. let x = the number of special program students who graduated
    binomial problem: n = 10 and p = .94, use the binomial formula
    $P(x) = [n!/x!(n-x)!] \cdot p^x \cdot (1-p)^{n-x}$
    a. $P(x \geq 9) = P(x=9) + P(x=10)$
    $ = [10!/9!1!] \cdot (.94)^9 \cdot (.06)^1 + [10!/10!0!] \cdot (.94)^{10} \cdot (.06)^0$
    $ = [10] \cdot (.5730) \cdot (.06) + [1] \cdot (.5386) \cdot (1)$
    $ = .344 + .539$
    $ = .883$
    b. $P(x=8) = [10!/8!2!] \cdot (.94)^8 \cdot (.06)^2$
    $ = [36] \cdot (.6096) \cdot (.0036)$
    $ = .079$
    $P(x \leq 7) = 1 - P(x > 7)$
    $ = 1 - [P(x=8) + P(x=9) + P(x=10)]$
    $ = 1 - [.079 + .344 + .539]$
    $ = 1 - [.962]$
    $ = .038$
    Yes; since $P(x \leq 7) = .038 < .05$, getting only 7 that graduated would be an unusual result.
    NOTE: Recall that "x successes among n trials is unusually low if P(x or fewer) < .05."

33. let x = the number of correct answers
    binomial problem: n = 10 and p = .20, use Table A-1
    $P(60\% \text{ or higher}) = P(x \geq 6)$
    $\phantom{P(60\% \text{ or higher})} = P(x=6) + P(x=7) + P(x=8) + P(x=9) + P(x=10)$
    $\phantom{P(60\% \text{ or higher})} = .006 + .001 + 0^+ + 0^+ + 0^+$
    $\phantom{P(60\% \text{ or higher})} = .007$
    No, the probability is not high enough to make random guessing a reasonable substitute for studying.

35. let x = the number of defective bulbs
    binomial problem: n = 24 and p = .04, use the binomial formula
    $P(x) = [n!/x!(n-x)!] \cdot p^x \cdot (1-p)^{n-x}$
    $P(\text{accept shipment}) = P(x \leq 1)$
    $\phantom{P(\text{accept shipment})} = P(x=0) + P(x=1)$
    $\phantom{P(\text{accept shipment})} = [24!/0!24!] \cdot (.04)^0 \cdot (.96)^{24} + [24!/1!23!] \cdot (.04)^1 \cdot (.96)^{23}$
    $\phantom{P(\text{accept shipment})} = [1] \cdot (1) \cdot (.3754) + [24] \cdot (.04) \cdot (.3911)$
    $\phantom{P(\text{accept shipment})} = .375 + .375$
    $\phantom{P(\text{accept shipment})} = .750$

37. let x = the number of components tested to find 1st defect
    geometric problem: p = .2, use the geometric formula
    $P(x) = p \cdot (1-p)^{x-1}$
    $P(x=7) = (.2) \cdot (.8)^6$
    $ = (.2) \cdot (.2621)$
    $ = .0524$

39. Extending the pattern to cover 6 types of outcomes, where $\Sigma x = n$ and $\Sigma p = 1$,
$$P(x_1,x_2,x_3,x_4,x_5,x_6) = [n!/(x_1!x_2!x_3!x_4!x_5!x_6!)] \cdot p_1^{x_1} \cdot p_2^{x_2} \cdot p_3^{x_3} \cdot p_4^{x_4} \cdot p_5^{x_5} \cdot p_6^{x_6}$$
$n = 20$
$p_1 = p_2 = p_3 = p_4 = p_5 = p_6 = 1/6$
$x_1 = 5, x_2 = 4, x_3 = 3, x_4 = 2, x_5 = 3, x_6 = 3$
Use the multinomial formula.
$$\begin{aligned}P(x_1,x_2,x_3,x_4,x_5,x_6) &= [n!/(x_1!x_2!x_3!x_4!x_5!x_6!)] \cdot p_1^{x_1} \cdot p_2^{x_2} \cdot p_3^{x_3} \cdot p_4^{x_4} \cdot p_5^{x_5} \cdot p_6^{x_6}\\
&= [20!/(5!4!3!2!3!3!)] \cdot (1/6)^5 \cdot (1/6)^4 \cdot (1/6)^3 \cdot (1/6)^2 \cdot (1/6)^3 \cdot (1/6)^3\\
&= [20!/(5!4!3!2!3!3!)] \cdot (1/6)^{20}\\
&= [1.955 \cdot 10^{12}] \cdot (2.735 \cdot 10^{-16})\\
&= .000535\end{aligned}$$

4.4 Mean, Variance and Standard Deviation for the Binomial Distribution

1. $\mu = n \cdot p = (100) \cdot (.25) = 25.0$
$\sigma^2 = n \cdot p \cdot (1-p) = (100) \cdot (.25) \cdot (.75) = 18.75$
$\sigma = 4.3$
minimum usual value $= \mu - 2\sigma = 25.0 - 2 \cdot (4.3) = 16.4$
maximum usual value $= \mu + 2\sigma = 25.0 + 2 \cdot (4.3) = 33.6$

3. $\mu = n \cdot p = (237) \cdot (2/3) = 158.0$
$\sigma^2 = n \cdot p \cdot (1-p) = (237) \cdot (2/3) \cdot (1/3) = 52.67$
$\sigma = 7.3$
minimum usual value $= \mu - 2\sigma = 158.0 - 2 \cdot (7.3) = 143.4$
maximum usual value $= \mu + 2\sigma = 158.0 + 2 \cdot (7.3) = 172.6$

5. let x = the number of correct answers
binomial problem: n = 20, p = .5
a. $\mu = n \cdot p = (20) \cdot (.5) = 10.0$
$\sigma^2 = n \cdot p \cdot (1-p) = (20) \cdot (.5) \cdot (.5) = 5.00$
$\sigma = 2.2$
b. maximum usual value $= \mu + 2\sigma = 10.0 + 2 \cdot (2.2) = 14.4$
No; since 12 is less than or equal to the maximum usual value, it would not be unusual for a student to pass by getting at least 12 correct answers.

7. let x = the number of wins
binomial problem: n = 100, p = 1/38
a. $\mu = n \cdot p = (100) \cdot (1/38) = 2.6$
$\sigma^2 = n \cdot p \cdot (1-p) = (100) \cdot (1/38) \cdot (37/38) = 2.56$
$\sigma = 1.6$
b. minimum usual value $= \mu - 2\sigma = 2.6 - 2 \cdot (1.6) = -0.6$ [use 0, by practical considerations]
No; since 0 is greater than or equal to the minimum usual value, it would not be unusual not to win once in the 100 trials.

9. let x = the number of high school graduates
binomial problem: n = 1250, p = .82
a. $\mu = n \cdot p = (1250) \cdot (.82) = 1025.0$
$\sigma^2 = n \cdot p \cdot (1-p) = (1250) \cdot (.82) \cdot (.18) = 184.50$
$\sigma = 13.6$

b. maximum usual value = $\mu + 2\sigma$ = 1025.0 + 2·(13.6) = 1052.2

Yes; since 1107 is greater than the maximum usual value, it would be unusual to get that many high school graduates in a random sample of Americans. But while the 1250 people surveyed were a random sample of Dutchess County residents, they are not a random sample of all Americans -- and the 82% figure was based on all Americans. Since Americans tend to live and associate with others of their same socio-economic class, most communities are probably well above or well below the 82% figure -- with only relatively few communities near the nationwide average.

11. let x = the number testing positive for the HIV virus

binomial problem: n = 150, p = .10

a. μ = n·p = (150)·(.10) = 15.0

σ^2 = n·p·(1-p) = (150)·(.10)·(.90) = 13.50

σ = 3.7

b. minimum usual value = $\mu - 2\sigma$ = 15.0 - 2·(3.7) = 7.6

No; since 12 is greater than or equal to the minimum usual value, it would not be unusual to get that many persons with the HIV virus if the program had no effect and the infection rate were still 10%. The result does not give enough evidence to conclude that the program is effective.

13. let x = the number of customers filing complaints

binomial problem: n = 850, p = .032

a. μ = n·p = (850)·(.032) = 27.2

σ^2 = n·p·(1-p) = (850)·(.032)·(.968) = 26.33

σ = 5.1

b. minimum usual value = $\mu - 2\sigma$ = 27.2 - 2·(5.1) = 17.0

Yes; since 7 is less than the minimum usual value, it would be unusual to get that many complaints if the program had no effect and the rate were still 3.2%. The result does give enough evidence to conclude that the program is effective in lowering the rate of complaints.

15. let x = the number of commercials that are one minute long

binomial problem: n = 50, p = .06

a. μ = n·p = (50)·(.06) = 3.0

σ^2 = n·p·(1-p) = (50)·(.06)·(.94) = 2.82

σ = 1.7

b. maximum usual value = $\mu + 2\sigma$ = 3.0 + 2·(1.7) = 6.4

Yes; since 16 is greater than the maximum usual value, it would be unusual to get that many one-minute commercials if the 6% still applied. The recent result suggests that one-minute commercials are now more than 6% of the total.

17. let x = the number of baby girls

binomial problem: n = 100, p = .5

μ = n·p = (100)·(.5) = 50.0

σ^2 = n·p·(1-p) = (100)·(.5)·(.5) = 25

σ = 5.0

a. Yes; a probability histogram would reveal an approximately bell-shaped distribution. The most likely value is x=50, with the likelihoods decreasing symmetrically for x values smaller or larger then 50.

b. Since 40 and 60 correspond to $\mu \pm 2\sigma$, the empirical rule indicates that approximately 95% of the time x will fall between 40 and 60.

c. Since 35 and 65 correspond to $\mu \pm 3\sigma$, the empirical rule indicates that approximately 99.7% of the time x will fall between 35 and 65.

d. Since 40 and 60 correspond to $\mu \pm 2\sigma$, use Chebyshev's Theorem with k=2. The portion of cases within k·σ of μ is at least $1 - 1/k^2$. Here, the likelihood that the number of girls falls between 40 and 60 is at least $1 - 1/2^2$ = 1 - 1/4 = 3/4 -- i.e., at least 75%.

NOTE: Parts (b) and (d) agree, since 95% is "at least 75%." Chebyshev's Theorem applies to all distributions and must be more general, the empirical rule applies only to bell-shaped distributions and can be more specific.

19. NOTE: Because mathematical formulation of this problem involves complicated algebra and does not promote better understanding of the concepts involved, we suggest a trial-and-error approach using Table A-1. The procedure given below may not be the most efficient, but it is easy to follow and promotes better understanding of the concepts involved.

let x = the number of edible pizzas
binomial problem: n is unknown, p = .8, we want $P(x \geq 5) \geq .99$
for n=5, $P(x \geq 5) = P(x=5)$
$= .328$
for n=6, $P(x \geq 5) = P(x=5) + P(x=6)$
$= .393 + .262$
$= .655$
for n=7, $P(x \geq 5) = P(x=5) + P(x=6) + P(x=7)$
$= .275 + .367 + .210$
$= .852$
for n=8, $P(x \geq 5) = P(x=5) + P(x=6) + P(x=7) + P(x=8)$
$= .147 + .294 + .336 + .168$
$= .945$
for n=9, $P(x \geq 5) = P(x=5) + P(x=6) + P(x=7) + P(x=8) + P(x=9)$
$= .066 + .176 + .302 + .302 + .134$
$= .980$
for n=10, $P(x \geq 5) = P(x=5) + P(x=6) + P(x=7) + P(x=8) + P(x=9) + P(x=10)$
$= .026 + .088 + .201 + .302 + .268 + .107$
$= .992$

The minimum number of pizzas necessary to be at least 99% sure that there will be 5 edible pizzas available is n = 10.

4-5 The Poisson Distribution

1. $P(x) = \mu^x \cdot e^{-\mu}/x!$ with $\mu = 3$ and $x = 4$
$P(x=4) = (3)^4 \cdot e^{-3}/4!$
$= (81) \cdot (.0498/24$
$= .168$

3. $P(x) = \mu^x \cdot e^{-\mu}/x!$ with $\mu = 50$ and $x = 48$
$P(x=48) = (50)^{48} \cdot e^{-50}/48!$
$= (3.553 \cdot 10^{81}) \cdot (1.929 \cdot 10^{-22})/(1.241 \cdot 10^{61})$
$= .0552$

5. let x = the number of student arrivals per office hour
Poisson problem: $\mu = 2.3$
$P(x) = \mu^x \cdot e^{-\mu}/x!$
a. $P(x=0) = (2.3)^0 \cdot e^{-2.3}/0! = (1) \cdot (.1003)/1 = .100$
b. $P(x=1) = (2.3)^1 \cdot e^{-2.3}/1! = (2.3) \cdot (.1003)/1 = .231$
c. $P(x=2) = (2.3)^2 \cdot e^{-2.3}/2! = (5.29) \cdot (.1003)/2 = .265$
d. $P(x=3) = (2.3)^3 \cdot e^{-2.3}/3! = (12.167) \cdot (.1003)/6 = .203$
e. $P(x=6) = (2.3)^6 \cdot e^{-2.3}/6! = (148.036) \cdot (.1003)/720 = .0206$

7. let x = the number of aircraft hijackings per day
   Poisson problem: μ = 29/365 = .07945
   $P(x) = \mu^x \cdot e^{-\mu}/x!$
   $P(x=0) = (.07945)^0 \cdot e^{-.079452}/0! = (1) \cdot (.9236)/1 = .924$
   $P(x=1) = (.07945)^1 \cdot e^{-.079452}/1! = (.07945) \cdot (.9236)/1 = .073$
   $P(x=0$ or $x=1) = P(x=0) + P(x=1)$
   $= .924 + .073$
   $= .997$

   Yes; a single response team is sufficient. We expect more than one hijacking in a single day only about .003 = .3% of the time, or roughly once a year.

9. let x = the number of homicides per day
   Poisson problem: μ = 116/365 = .3178
   $P(x) = \mu^x \cdot e^{-\mu}/x!$
   a. $P(x=0) = (.3178)^0 \cdot e^{-.3178}/0! = (1) \cdot (.7277)/1 = .7277$
   b. $P(x=1) = (.3178)^1 \cdot e^{-.3178}/1! = (.3178) \cdot (.7277)/1 = .2313$
   c. $P(x=2) = (.3178)^2 \cdot e^{-.3178}/2! = (.1010) \cdot (.7277)/2 = .0368$
   d. $P(x=3) = (.3178)^3 \cdot e^{-.3178}/3! = (.0321) \cdot (.7277)/6 = .0038$
   e. $P(x=4) = (.3178)^4 \cdot e^{-.3178}/4! = (.0102) \cdot (.7277)/24 = .0003$

   The following tables compares the actual results with the calculated probabilities. They agree quite closely.

x	f	r.f.	P(x)
0	268	.7342	.7277
1	79	.2164	.2313
2	17	.0466	.0368
3	1	.0027	.0038
4+	0	0	.0004
		1.0000	1.0000

   (by subtraction, $P(x \geq 4 = .0004)$

11. binomial problem: n = 100, p = .10
    since n = 100 ≥ 100 and np = 100 · (.10) = 10 ≥ 10
    use Poisson distribution
    $P(x) = \mu^x \cdot e^{-\mu}/x!$ with μ = np = 100 · (.10) = 10
    $P(x=101) = (10)^0 \cdot e^{-10}/101!$
    $= (1) \cdot (.0000454)/101!$
    = a number so small it is zero to well over 100 decimal places

    This agrees very well with the fact that x=101 is an impossible result for the original binomial problem.

Review Exercises

1. a. A random variable is a characteristic that assumes a single value (usually a numerical value), determined by chance, for each outcome of an experiment.
   b. A probability distribution is a statement of the possible values a random variable can assume and the probability associated with each of those values. To be valid it must be true for each value x in the distribution that $0 \leq P(x) \leq 1$ and that $\Sigma P(x) = 1$.
   c. Yes, the given table is a valid probability distribution because it meets the definition and conditions in part (b) above.
   NOTE: The table on the next page summarizes the calculations for parts (d) and (e).
   d. $\mu = \Sigma x \cdot P(x) = 3.4400 = 3.4$ (rounded)
   e. $\sigma^2 = \Sigma x^2 \cdot P(x) - \mu^2 = 12.3152 - (3.4400)^2 = .4816$; $\sigma = .7$

x	P(x)	x·P(x)	x^2	$x^2 \cdot P(x)$
0	.0004	0	0	0
1	.0094	.0094	1	.0094
2	.0870	.1740	4	.3480
3	.3562	1.0686	9	3.2058
4	.5470	2.1880	16	8.7520
	1.0000	3.4400		12.3152

f. Yes; since $P(x=0) = .0004 < .05$.

2. let x = the number of TV's tuned to *ER*
   binomial problem: n = 20 and p = .34, use the binomial formula
   a. $E(x) = \mu = n \cdot p = (20) \cdot (.34) = 6.8$
   b. $\mu = n \cdot p = (20) \cdot (.34) = 6.8$
   c. $\sigma^2 = n \cdot p \cdot (1-p) = (20) \cdot (.34) \cdot (.66) = 4.488$
      $\sigma = 2.1$
   d. $P(x) = [n!/x!(n-x)!] \cdot p^x \cdot (1-p)^{n-x}$
      $P(x=5) = [20!/5!15!] \cdot (.34)^5 \cdot (.66)^{15}$
      $= [15504] \cdot (.0045435) \cdot (.0019641)$
      $= .138$
   e. maximum usual value $= \mu + 2\sigma = 6.8 + 2 \cdot (2.1) = 11$
      Yes; since 12 is greater than the maximum usual value it would be unusual to find that 12 sets are tuned to *ER*.

3. let x = the number of companies that test for drug abuse
   binomial problem: n = 10 and p = .80, use Table A-1
   a. $P(x=5) = .026$
   b. $P(x \geq 5) = P(x=5) + P(x=6) + P(x=7) + P(x=8) + P(x=9) + P(x=10)$
      $= .026 + .088 + .201 + .302 + .268 + .107$
      $= .992$
   c. $\mu_2 = n \cdot p = (10) \cdot (.80) = 8.0$
      $\sigma^2 = n \cdot p \cdot (1-p) = (10) \cdot (.80) \cdot (.20) = 1.60$
      $\sigma = 1.3$
   d. No; since $P(x=6) = .088$, which is not unusual. A particular result is unusual if P(that result or a more extreme result) < .05.

4. let x = the number of workers fired for inability to get along with others
   binomial problem: n = 5 and p = .17, use the binomial formula
   $P(x) = [n!/x!(n-x)!] \cdot p^x \cdot (1-p)^{n-x}$
   a. $P(x=4) = [5!/4!1!] \cdot (.17)^4 \cdot (.83)^1 = (5) \cdot (.000835) \cdot (.83) = .00347$
      $P(x=5) = [5!/5!0!] \cdot (.17)^5 \cdot (.83)^0 = (1) \cdot (.000142) \cdot (1) = .000142$
      $P(x \geq 4) = P(x=4) + P(x=5)$
      $= .00347 + .00014$
      $= .00361$
   b. Yes; x=4 would be an unusual result for a company with the standard rate of 17%. A particular result is unusual if P(that result or a more extreme result) < .05, and $P(x \geq 4) = .00361 < .05$.

5. let x = the number of babies born per day
   Poisson problem: $\mu = 212/365 = .5808$
   $P(x) = \mu^x \cdot e^{-\mu}/x!$
   a. $\mu = 212/365 = .5808$

b. $P(x=2) = (.5808)^2 \cdot e^{-.5908}/2!$
$= (.3374) \cdot (.5594)/2$
$= .094$

c. $P(x=3) = (.5808)^3 \cdot e^{-.5908}/3!$
$= (.1959) \cdot (.5594)/6$
$= .018$

Since $.018 < .05$ this is an unusual event -- but this translates into $(.018)(365) \approx 7$ times a year. Whether that justifies constructing a facility to accommodate 3 births on the same day is a medical decision. In general, medical decision use standards more strict than the generally accepted 5% guideline.

Cumulative Review Exercises

1. a. The table below at the left was used to calculate the mean and standard deviation.

x	f	f·x	f·x^2		x	r.f.
0	55	0	0		0	.786
1	2	2	2		1	.029
2	1	2	4		2	.014
3	1	3	9		3	.014
4	0	0	0		4	.000
5	3	15	75		5	.043
6	0	0	0		6	.000
7	2	14	98		7	.029
8	4	32	256		8	.057
9	2	18	162		9	.029
	70	86	606			1.000

$\bar{x} = [\Sigma(f \cdot x)]/[\Sigma f] = 86/70 = 1.2$
$s^2 = \{[\Sigma f] \cdot [\Sigma(f \cdot x^2)] - [\Sigma(f \cdot x)]^2\}/\{[(\Sigma f)] \cdot [(\Sigma f) - 1]\}$
$= [(70) \cdot (606) - (86)^2]/[(70) \cdot (69)]$
$= 37091/4830$
$= 7.251$
$s = 2.7$

b. The table is given above at the right and was constructed using r.f. $= f/(\Sigma f) = f/70$.

c. The first two columns of the table below constitute the requested probability distribution. The other columns were added to calculate the mean and the standard deviation.

x	P(x)	x·P(x)	x^2	x^2·P(x)
0	.1	0	0	0
1	.1	.1	1	.1
2	.1	.2	4	.4
3	.1	.3	9	.9
4	.1	.4	16	1.6
5	.1	.5	25	2.5
6	.1	.6	36	3.6
7	.1	.7	49	4.9
8	.1	.8	64	6.4
9	.1	.9	81	8.1
	1	4.5		28.5

$\mu = \Sigma x \cdot P(x)$
$= 4.5$
$\sigma^2 = \Sigma x^2 \cdot P(x) - \mu^2$
$= 28.5 - (4.5)^2$
$= 8.25$
$\sigma = 2.8723$, rounded to 2.9

d. The presence of so many 0's make it clear that the last digits do not represent a random sample. Factors other than accurate measurement (which, like social security numbers and other such data, would represent random final digits) determined the recorded distances.

2. a. The summary statistics are n = 20, Σx = 193, Σx^2 = 2027

$\bar{x}$ = $(\Sigma x)/n$ = 193/20 = 9.65

s^2 = $[n \cdot (\Sigma x^2) - (\Sigma x)^2]/[n \cdot (n-1)]$

= $[20 \cdot (2027) - (193)^2]/[20 \cdot (19)]$

= 3291/380

= 8.66

s = 2.9

b. Of the 20 sample results, 8 were 12's. Use the relative frequency to estimate the probability.

estimated P(x=12) = 8/20 = .400

correct P(x=12) = 1/36 = .0278

These results do not agree. It appears that the dice are not balanced.

c. let x = the number of 12's

binomial problem: n = 20 and p = 1/36, use the binomial formula

$P(x)$ = $[n!/x!(n-x)!] \cdot p^x \cdot (1-p)^{n-x}$

$P(x=0)$ = $[20!/0!20!] \cdot (1/36)^0 \cdot (35/36)^{20}$ = $(1) \cdot (1) \cdot (.569)$ = .569

$P(x \geq 1)$ = 1 - P(x<1)

= 1 - P(x=0)

= 1 - .569

= .431

d. The results obtained in court would be difficult to dispute. Here are some possible ideas and where they lead.

* The probability of at least one 12 is reasonable -- .431, from part (c). Maybe the probability of eight 12's isn't as low as one would guess.

binomial problem: n = 20 and p = 1/36, use the binomial formula

$P(x)$ = $[n!/x!(n-x)!] \cdot p^x \cdot (1-p)^{n-x}$

$P(x=8)$ = $[20!/8!12!] \cdot (1/36)^8 \cdot (35/36)^{12}$

= $(125.970) \cdot (3.54 \cdot 10^{-13}) \cdot (.713)$

= .0000000318

Oops! That's one calculation that the defense definitely shouldn't show the court. Even using $P(x \geq 8)$, since the rarity of an event is P(that result or a more extreme value), would not be helpful -- as the probabilities for more than 8 12's are even smaller and will not accumulate to any significant value.

* Perhaps the criterion for what constitutes and unusual event would be helpful. For a balanced pair of dice, the court has agreed that μ = 7.0 and σ = 2.4. Unusual values are those outside $\mu \pm 2 \cdot \sigma$

7.0 $\pm$ 2 $\cdot$ (2.4)

7.0 $\pm$ 4.8

2.2 to 11.8

The sample average in part (a) was $\bar{x}$ = 9.65. Since this is well within these limits, does that mean the result from the accused person's dice should not be considered unusual?

Unfortunately, not. The above limits are for <u>one</u> x value, not for the <u>mean</u> of 20 x values. Rolling the dice <u>once</u> and getting a 9 or 10 would not be considered unusual. The statement says nothing about rolling a pair of dice 20 times and getting a <u>mean</u> of 9 or 10. [The Central Limit Theorem in the following chapter does address this concern, but the results would be very damaging to the accused.]

* The only hope appears to be to complain in non-statistical language that the sample is too small with a diversionary speech such as the following: "We all know that these dice could be rolled millions and millions of times. Using a sample of size 20 to reach a conclusion about all the millions and millions of possible results within these dice is just as foolish as allowing the opinions of a haphazard sample of size 20 to speak for the millions and millions of law-abiding citizens in this great country."

Chapter 5

Normal Probability Distributions

5-2 The Standard Normal Distribution

1. The height of the rectangle is .2. Probability corresponds to area, and the area of a rectangle is (width)·(height).

$$P(x < 4) = \text{(width)·(height)}$$
$$= (4-0)·(.2)$$
$$= (4)·(.2)$$
$$= .8$$

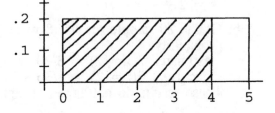

3. The height of the rectangle is .2. Probability corresponds to area, and the area of a rectangle is (width)·(height).

$$P(3 < x < 4) = \text{(width)·(height)}$$
$$= (4-3)·(.2)$$
$$= (1)·(.2)$$
$$= .2$$

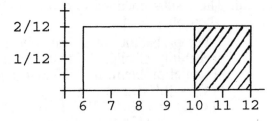

5. The height of the rectangle is 1/6. Probability corresponds to area, and the area of a rectangle is (width)·(height).

$$P(x > 10) = \text{(width)·(height)}$$
$$= (12-10)·(1/6)$$
$$= (2)·(1/6)$$
$$= .333$$

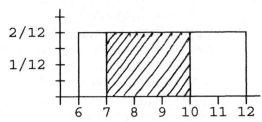

7. The height of the rectangle is 1/6. Probability corresponds to area, and the area of a rectangle is (width)·(height).

$$P(7 < x < 10) = \text{(width)·(height)}$$
$$= (10-7)·(1/6)$$
$$= (3)·(1/6)$$
$$= .5$$

NOTE: The sketch is the key to exercises 9-32. It tells whether to add two Table A-2 probabilities, to subtract two Table A-2 probabilities, to subtract a Table A-2 probability from .5000, to add a Table A-2 probability to .5000, etc. It also often provides a check against gross errors by indicating at a glance whether the final probability is less than or greater than .5000. Remember that the symmetry of the normal curve implies two important facts:

* There is always .5000 above and below the middle (i.e., at z = 0).
* P(-a < z < 0) = P(0 < z < a) for all values of "a."

9. P(0 < z < 1.50)
   = .4332

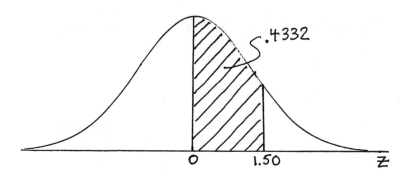

11. P(-1.96 < z < 0)
    = P(0 < z < 1.96)
    = .4750

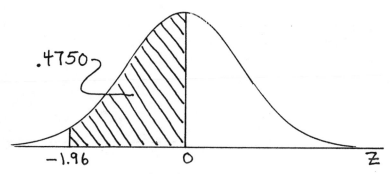

13. P(z < -1.79)
    = P(z < 0) - P(-1.79 < z < 0)
    = .5000 - .4633
    = .0367

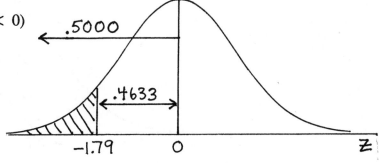

15. P(z > 2.05)
    = P(z > 0) - P(0 < z < 2.05)
    = .5000 - .4798
    = .0202

17. P(.50 < z < 1.50)
    = P(0 < z < 1.50) - P(0 < z < .50)
    = .4332 - .1915
    = .2417

19. P(-2.00 < z < -1.00)
    = P(-2.00 < z < 0) - P(-1.00 < z < 0)
    = .4772 - .3413
    = .1359

21. P(z < 1.62)
    = P(z < 0) + P(0 < z < 1.62)
    = .5000 + .4474
    = .9474

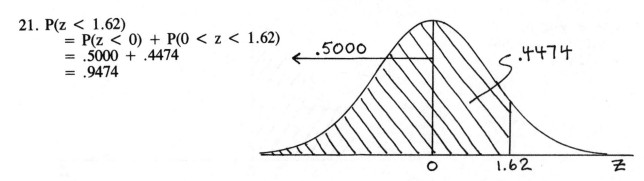

23. P(z > -.27)
    = P(-.27 < z < 0) + P(z > 0)
    = .1064 + .5000
    = .6064

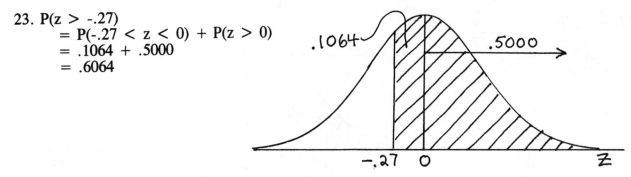

25. P(-1.08 < z < .33)
    = P(-1.08 < z < 0) + P(0 < z < .33)
    = .3599 + .1293
    = .4892

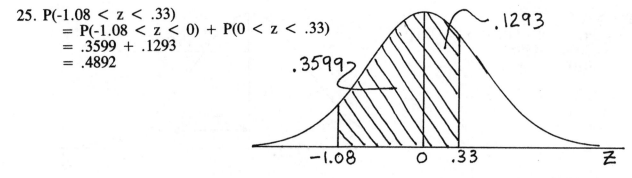

27. P(z > 0)
    = .5000

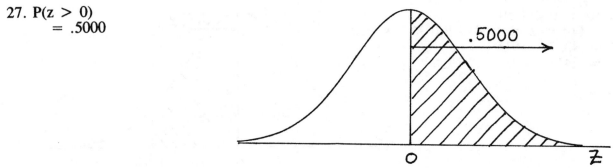

29. P(-1.96 < z < 1.96)
  = P(-1.96 < z < 0) + P(0 < z < 1.96)
  = .4750 + .4750
  = .9500

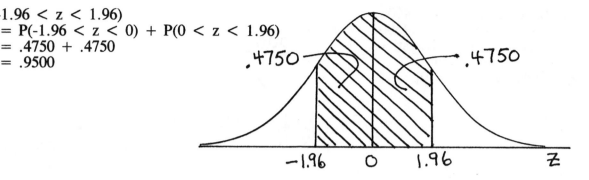

31. P(z > -2.575)
  = P(-2.575 < z < 0) + P(z > 0)
  = .4950 + .5000
  = .9950

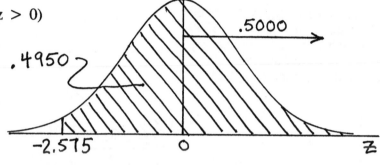

NOTE: The sketch is the key to exercises 33-40. It tells what probability is between 0 and the z score of interest (i.e. the area A to look up when reading Table A-2 "backwards." It also provides a check against gross errors by indicating at a glance whether a z score is above or below 0. Remember that the symmetry of the normal curve implies two important facts:
  * There is always .5000 above and below the middle (i.e., at z = 0).
  * P(-a < z < 0) = P(0 < z < a) for all values of "a."

33. For P_{90}, A = .4000.
  The closest entry is A = .3997, for which z = 1.28 [positive, since it is to the right of the middle, where z = 0].

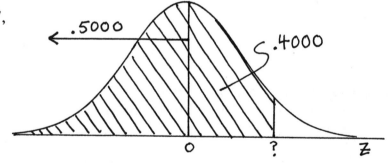

35. For Q_1, A = .2500.
  The closest entry is A = .2486, for which z = -.67 [negative, since it is to the left of the middle, where z = 0].

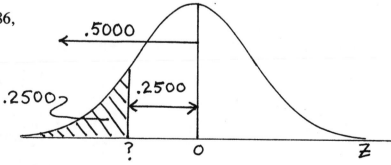

37. For the lowest 5%, A = .4500.
The closest entry is A = .4500,
for which z = -1.645 [negative,
since it is to the left of
the middle, where z = 0].

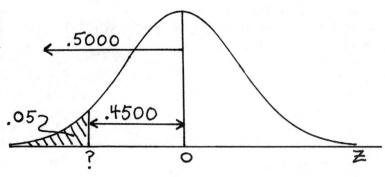

39. For the lowest 3%, A = .4700.
The closest entry is A = .4699,
for which z = -1.88 [negative,
since it is to the left of
the middle, where z = 0].

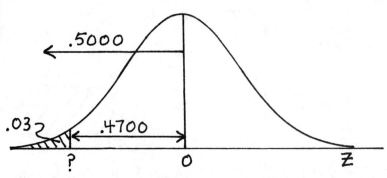

41. Rewrite each of the given statements in terms of z, recalling that z is the number of standard deviations a score is from the mean.

a. P(-1.00 < z < 1.00)
     = P(-1.00 < z < 0) + P(0 < z < 1.00)
     = .3413 + .3413
     = .6826

b. P(-1.96 < z < 1.96)
     = P(-1.96 < z < 0) + P(0 < z < 1.96)
     = .4750 + .4750
     = .9500

c. P(-3.00 < z < 3.00)
     = P(-3.00 < z < 0) + P(0 < z < 3.00)
     = .4987 + .4987
     = .9974

d. P(-1.00 < z < 2.00)
   = P(-1.00 < z < 0) + P(0 < z < 2.00)
   = .3413 + .4772
   = .8185

e. P(z < -2.00 or z > 2.00)
   = P(z < -2.00) + P(z > 2.00)
   = [P(z < 0) - P(-2.00 < z < 0)]
       + [P(z > 0) - P(0 < z < 2.00)]
   = [.5000 - .4772] + [.5000 - .4772]
   = .0228 + .0228
   = .0456

43. a. P(0 < z < a) = .3907
    A = .3907
    a = 1.23

b. P(-b < z < b) = .8664
   P(0 < z < b) = .8664/2 = .4332
   A = .4332
   b = 1.50

c. P(z > c) = .0643
   P(0 < z < c) = .5000 - .0643 = .4357
   A = .4357
   c = 1.52

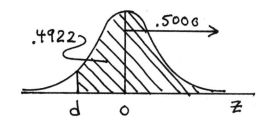

d. P(z > d) = .9922
   P(d < z < 0) = .9922 - .5000 = .4922
   A = .4922
   d = -2.42 [negative, since it is
        to the left of the middle,
        where z = 0]

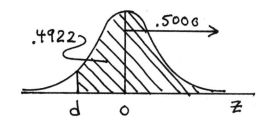

e. $P(z < e) = .4500$
   $P(e < z < 0) = .5000 - .4500 = .0500$
   $A = .0500$ [.0517 is closest entry]
   $e = -.13$ [negative, since it is
                to the left of the middle,
                where $z = 0$]

5-3 Nonstandard Normal Distributions:
Finding Probabilities

NOTE: In each nonstandard normal distribution, x scores are converted to z scores using the formula $z = (x-\mu)/\sigma$ and rounded to two decimal places. As in the previous section, drawing and labeling the sketch is the key to successful completion of the exercises.

1. $\mu = 143$
   $\sigma = 29$
   $P(143 < x < 172)$
     $= P(0 < z < 1.00)$
     $= .3413$

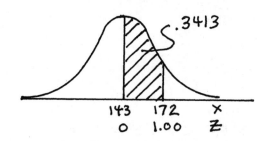

3. $\mu = 143$
   $\sigma = 29$
   $P(x > 150)$
     $= P(z > .24)$
     $= .5000 - .0948$
     $= .4052$

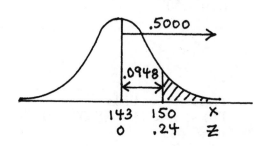

5. $\mu = 63.6$
   $\sigma = 2.5$
   $P(65.5 < x < 68.0)$
     $= P(.76 < z < 1.76)$
     $= .4608 - .2764$
     $= .1844$

7. $\mu = 63.6$
$\sigma = 2.5$
$P(64.5 < x < 72.0)$
$= P(.36 < z < 3.36)$
$= .4999 - .1406$
$= .3593$

9. $\mu = 998$
$\sigma = 202$
$P(x < 900)$
$= P(z < -.49)$
$= .5000 - .1879$
$= .3121$

11. $\mu = 3.586$
$\sigma = .074$
$P(x < 3.5)$
$= P(z < -1.16)$
$= .5000 - .3770$
$= .1230$

No; it is not true that many consumers are being cheated. Since .1230 > .05, it is not unusual for a packet to contain less than the indicated amount. As consumers do not purchase individual packets, however, they are not really being cheated. When a package of several packets is purchased, the number that are overweight more than compensates for those that are underweight, so that the overall purchase contains more than the indicated amount.

13. $\mu = 268$
$\sigma = 15$
$P(x \geq 308)$
$= P(z \geq 2.67)$
$= .5000 - .4962$
$= .0038$

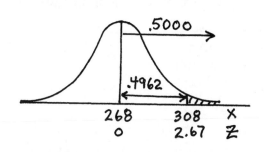

Since .0038 < .05, such a pregnancy is rare -- but certainly not impossible. Out of every one million pregnancies, for example, we expect 3,800 to last 308 days or longer.

15. $\mu = 12.19$
    $\sigma = .11$
    $P(x < 12)$
    $\quad = P(z < -1.73)$
    $\quad = .5000 - .4852$
    $\quad = .0418$

No; it is not true that many consumers are being cheated. Since .0418 < .05, getting a can with less than 12 ounces is unusual. And while a "defect" rate of 4% may seem high for a reputable product, the facts that $\mu > 12$ and that most people buy several cans over a period of time means that the consumer still comes out ahead in the long run.

17. $\mu = 69.0$
    $\sigma = 2.8$
    $P(64 < x < 78)$
    $\quad = P(-1.79 < z < 3.21)$
    $\quad = .4633 + .4999$
    $\quad = .9632$

No; it is not true that height prevents too many men from joining the Marines.

19. $\mu = 3570$
    $\sigma = 500$
    $P(x < 2200)$
    $\quad = P(z < -2.74)$
    $\quad = .5000 - .4969$
    $\quad = .0031$

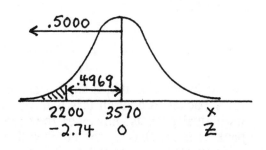

The percentage of babies in the "at-risk" category is .31%. If a hospital experiences 900 births a year, we expect about 3 of them to be "at-risk."

21. $\mu = 178.1$
    $\sigma = 40.7$
    $P(x > 300)$
    $\quad = P(z > 3.00)$
    $\quad = .5000 - .4987$
    $\quad = .0013$

This probability may not cause an individual male aged 18 to 24 to be concerned. It may be, however, that the medical community should be concerned. In a metropolitan area of 100,000 such people, for example, we expect about 130 to be "at-risk."

23. $\mu = 100$
    $\sigma = 15$

a. $P(x > 131.5)$
    $= P(z > 2.10)$
    $= .5000 - .4821$
    $= .0179$

b. $(.0179) \cdot (75,000) = 1,342.5$

25. No matter what the original problem, z scores have no units. Consider, for example, the z score for a weight of 150 from the population of women used in this section.
    $$z = \frac{150 \text{ lb} - 143 \text{ lb}}{29 \text{ lb}} = \frac{7 \text{ lb}}{29 \text{ lb}} = .24, \text{ the "lb" units having "cancelled out"}$$

27. Multiply each score by .4536 kg/lb will "shrink" each score and the distances between scores by that factor so that
    $\mu_{kg} = (.4536 \text{ kg/lb}) \cdot \mu_{lb} = (.4536 \text{ kg/lb}) \cdot (143 \text{ lb}) = 64.9 \text{ kg}$
    $\sigma_{kg} = (.4536 \text{ kg/lb}) \cdot \sigma_{lb} = (.4536 \text{ kg/lb}) \cdot (29 \text{ lb}) = 13.2 \text{ kg}$
    The distribution will still be normal, because multiplying each score will change only the scale but not the shape of the figure.

5-4 Nonstandard Normal Distributions: Finding Scores

NOTE: The relationship between standardized and raw scores is given by $z = (x-\mu)/\sigma$. That formula may be solved, as needed, for any of the quantities on the right to yield
    $$x = \mu + z\sigma \qquad \sigma = (x-\mu)/z \qquad \mu = x - z\sigma$$
As in the previous section, drawing and labeling the sketch is the key to successful completion of the exercises. Remember that the z score for the mean is zero, and that z scores to the left of the mean must be negative.

1. $\mu = 10$
   $\sigma = 2$
   for P_{90}, A = .4000 [.3997] and z = 1.28
   $x = \mu + z\sigma$
   $= 10 + (1.28)(2)$
   $= 10 + 2.6$
   $= 12.6$

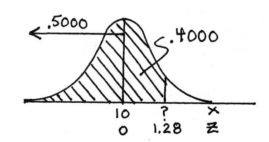

3. $\mu = 10$
   $\sigma = 2$
   for D_2, A = .3000 [.2995] and z = -.84
   $x = \mu + z\sigma$
   $= 10 + (-.84)(2)$
   $= 10 - 1.7$
   $= 8.3$

5. $\mu = 143$
   $\sigma = 29$
   for Q_3, $A = .2500$ [.2486] and $z = .67$
   $x = \mu + z\sigma$
   $\quad = 143 + (.67)(29)$
   $\quad = 143 + 19$
   $\quad = 162$ lbs

7. $\mu = 143$
   $\sigma = 29$
   for P_{15}, $A = .3500$ [.3508] and $z = -1.04$
   $x = \mu + z\sigma$
   $\quad = 143 + (-1.04)(29)$
   $\quad = 143 - 30$
   $\quad = 113$ lbs

9. $\mu = 63.6$
   $\sigma = 2.5$
   for the tallest 5%, $A = .4500$ and $z = 1.645$
   $x = \mu + z\sigma$
   $\quad = 63.6 + (1.645)(2.5)$
   $\quad = 63.6 + 4.1$
   $\quad = 67.7$ inches

"A height that is in the top 5% of all women" does not identify a specific value for women to know whether they are eligible to apply. If the population continues to grow taller, as it has over the past decades, the membership requirement will not be constant and will require monitoring the population in some way.

11. $\mu = 1017$
    $\sigma = 207$
    for the top 42%, $A = .0800$ [.0793] and $z = .20$
    $x = \mu + z\sigma$
    $\quad = 1017 + (.20)(207)$
    $\quad = 1017 + 41$
    $\quad = 1058$

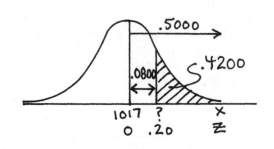

"A score in the top 42%" does not identify a specific value that students can aim for and/or determine whether they are eligible to apply. It may generate undesirable negative interaction if students perceive that they are competing against each other and not against some standard.

13. $\mu = 8.2$
    $\sigma = 1.1$
    a. $P(x < 5)$
       $= P(z < -2.91)$
       $= .5000 - .4982$
       $= .0018$

b. for the poorest 1%, A = .4900 [.4901] and z = -2.33
   $x = \mu + z\sigma$
     $= 8.2 + (-2.33)(1.1)$
     $= 8.2 - 2.6$
     $= 5.6$ years

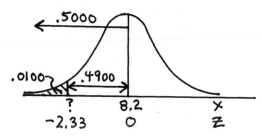

In practice, the warranty time would be rounded (down) to a "nice" amount -- probably 5.5 years or (equivalently) 66 months.

15. $\mu = 9.4$
    $\sigma = 4.2$
    a. $P(x < 15)$
       $= P(z < 1.33)$
       $= .5000 + .4082$
       $= .9082$

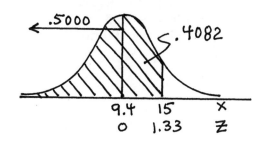

b. for P_{99}, A = .4900 [.4901] and z = 2.33
   $x = \mu + z\sigma$
     $= 9.4 + (2.33)(4.2)$
     $= 9.4 + 9.8$
     $= 19.2$ lbs

17. $\mu = 268$
$\sigma = 15$
for the lowest 4%, A = .4600 [.4599] and z = -1.75.
$x = \mu + z\sigma$
$= 268 + (-1.75)(15)$
$= 268 - 26$
$= 242$ days

19. $\mu = 178.1$
$\sigma = 40.7$
a. P(x > 250)
$= P(z > 2.01)$
$= .5000 - .4778$
$= .0222$

b. for the top 3%, A = .4700 [.4699] and z = 1.88
$x = \mu + z\sigma$
$= 178.1 + (1.88)(40.7)$
$= 178.1 + 76.5$
$= 254.6$ mg/100 mL

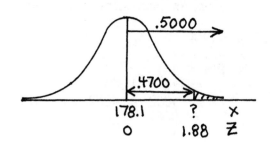

21. $\mu = 25$
$\sigma = 5$

a. Adding 50 to each grade will move the entire class up by 50 points, but it will not affect distances between the grades.
$\mu_N = \mu_O + 50 = 25 + 50 = 75$
$\sigma_N = \sigma_O = 5$

b. This is probably not fair. It is reasonable to assume that the original scores were low because the test was either too long or too hard. Adding 50 points to each score is equivalent to saying that if the test had been shorter or easier each student would have improved by the same amount. In truth, the abler students would probably benefit more from having additional time or easier questions than would their less able classmates -- i.e., differences between students would be accentuated by a fairer test, and the new standard deviation would be higher than the original one.

c. The lower limits for each grade would be as follows.
A: for P_{90}, A = .4000 [.3997] and z = 1.28
  x = μ + zσ = 25 + (1.28)(5) = 31.4
A: for P_{70}, A = .2000 [.1985] and z = .52
  x = μ + zσ = 25 + (.52)(5) = 27.6
A: for P_{30}, A = .2000 [.1985] and z = -.52
  x = μ + zσ = 25 + (-.52)(5) = 22.4
A: for P_{10}, A = .4000 [.3997] and z = -1.28
  x = μ + zσ = 25 + (-1.28)(5) = 18.6

d. Which method of curving is more fair depends on which end of the scale you fall. Adding 50 points to each grade to get μ = 75 and σ = 5 would probably produce few or no scores above 90 and below 60 (which represent 3 standard deviations above and below the mean). On a strict 90-80-70-60 grading system that means few or no A's and few or no F's -- this agrees with the comments in part (b) about such a plan not differentiating between the students as a fairer test might have done. Assuming approximately a normal distribution of scores, the scheme in part (c) produces approximately 10% A's and 10% F's -- whether such grades are deserved or not. A student near the bottom of the class would vote for curve (a) because it would likely give him a D instead of an F; a student near the top of the class would vote for curve (c) since it would likely give him an A instead of a B. The best remedy would be a fairer test, as any curve ultimately assigns grades according to performance relative to the rest of the class instead of according to an absolute standard of academic accomplishment. Ultimately (c) is the lesser of two evils because it takes into account the spread of the scores as well as the mean -- it grants A's those scoring more than 1.28 standard deviations above the mean and F's to those scoring more than 1.28 standard deviations below the mean.

23. μ = 143
 σ = 29

a. $\mu \pm 2\sigma$ = 143 $\pm$ 2·(29)
   = 143 $\pm$ 58
   = 85 to 201

b. P(-2.00 < z < 2.00)
   = .4772 + .4772
   = .9544
   = 95.44%

c. for A = .9500/2 = .4750, z = 1.96
  $\mu \pm 1.96\sigma$ = 143 $\pm$ 1.96·(29)
         = 143 $\pm$ 56.84
         = 86.16 to 199.84

5-5 The Central Limit Theorem

NOTE: When using individual scores (i.e., making a statement about one x score from the original distribution), convert x to z using the mean and standard deviation of the x's and $z = (x-\mu)/\sigma$.

When using a sample of n scores (i.e., making a statement about $\bar{x}$), convert $\bar{x}$ to z using the mean and standard deviation of the $\bar{x}$'s and $z = (\bar{x}-\mu_{\bar{x}})/\sigma_{\bar{x}}$.

IMPORTANT NOTE: After calculating $\sigma_{\bar{x}}$, <u>STORE IT</u> in the calculator to recall it with total accuracy whenever it is needed in subsequent calculations. <u>DO NOT</u> write it down on paper rounded off (even to several decimal places) and then re-enter it in the calculator whenever it is needed. This avoids both round-off errors and recopying errors.

1. a. normal distribution
   $\mu = 143$
   $\sigma = 29$
   $P(143 < x < 155)$
   $\quad = P(0 < z < .41)$
   $\quad = .1591$

   b. normal distribution, since the original distribution is so
   $\mu_{\bar{x}} = \mu = 143$
   $\sigma_{\bar{x}} = \sigma/\sqrt{n} = 29/\sqrt{36} = 4.833$
   $P(143 < \bar{x} < 155)$
   $\quad = P(0 < z < 2.48)$
   $\quad = .4934$

3. a. normal distribution
   $\mu = 143$
   $\sigma = 29$
   $P(x > 160)$
   $\quad = P(z > .59)$
   $\quad = .5000 - .2224$
   $\quad = .2776$

   b. normal distribution, since the original distribution is so
   $\mu_{\bar{x}} = \mu = 143$
   $\sigma_{\bar{x}} = \sigma/\sqrt{n} = 29/\sqrt{50} = 4.101$
   $P(\bar{x} > 160)$
   $\quad = P(z > 4.15)$
   $\quad = .5000 - .4999$
   $\quad = .0001$

5. a. normal distribution, since the original distribution is so
$\mu_{\bar{x}} = \mu = 143$
$\sigma_{\bar{x}} = \sigma/\sqrt{n} = 29/\sqrt{16} = 7.250$
$P(150 < \bar{x} < 155)$
   $= P(.97 < z < 1.66)$
   $= .4515 - .3340$
   $= .1175$

b. When the original distribution is normal, the sampling distribution of the $\bar{x}$'s is normal regardless of the sample size.

7. normal distribution, by the Central Limit Theorem
$\mu_{\bar{x}} = \mu = .500$
$\sigma_{\bar{x}} = \sigma/\sqrt{n} = .289/\sqrt{50} = .0409$
$P(.6 < \bar{x} < .7)$
   $= P(2.45 < z < 4.89)$
   $= .4999 - .4929$
   $= .0070$

Yes; since .0070 < .05, it would be unusual for a properly-functioning computer random number generator to generate 50 such numbers.

9. a. normal distribution, by the Central Limit Theorem
$\mu_{\bar{x}} = \mu = 8.2$
$\sigma_{\bar{x}} = \sigma/\sqrt{n} = 1.1/\sqrt{50} = .156$
$P(\bar{x} \le 7.8)$
   $= P(z \le -2.57)$
   $= .5000 - .4949$
   $= .0051$

b. Yes; it appears that the TV sets sold by the Portland Electronics store are of less than average quality.

11. a. normal distribution
$\mu = 32.473$
$\sigma = 5.601$
$P(x < 29.000)$
   $= P(z < -.62)$
   $= .5000 - .2324$
   $= .2676$

b. NOTE: Completion of the problem assumes that "a decade of ten years" means 10 years chosen at random and used to form an artificial "decade." Since weather may have long range cycles and patterns (so that the rainfall in any one year is not independent of the rainfall in adjacent years), a chronological decade could not be assumed to represent a random sample equivalent to 10 randomly selected years.

normal distribution, since the original distribution is so

$\mu_{\bar{x}} = \mu = 32.473$

$\sigma_{\bar{x}} = \sigma/\sqrt{n} = 5.601/\sqrt{10} = 1.771$

$P(\bar{x} < 29.00)$

$= P(z < -1.96)$

$= .5000 - .4750$

$= .0250$

c. When the original distribution is normal, the sampling distribution of the $\bar{x}$'s is normal regardless of the sample size.

13. a. normal distribution, by the Central Limit Theorem

$\mu_{\bar{x}} = \mu = 12.00$

$\sigma_{\bar{x}} = \sigma/\sqrt{n} = .09/\sqrt{36} = .015$

$P(\bar{x} \geq 12.29)$

$= P(z \geq 19.33)$

$= .5000 - .4999$

$= .0001$

b. It appears that the cans are NOT being filled with a mean of 12.00 oz and a standard deviation of .09 oz. Since the cans appear to contain more than 12.00 oz, and that poses no difficulties for consumers, consumers are not being cheated.

15. a. normal distribution

$\mu = 509$

$\sigma = 112$

$P(x \geq 590)$

$= P(z \geq .72)$

$= .5000 - .2642$

$= .2358$

b. normal distribution, since the original distribution is so

$\mu_{\bar{x}} = \mu = 509$

$\sigma_{\bar{x}} = \sigma/\sqrt{n} = 112/\sqrt{16} = 28$

$P(\bar{x} \geq 590)$

$= P(z \geq 2.89)$

$= .5000 - .4981$

$= .0019$

c. When the original distribution is normal, the sampling distribution of the $\bar{x}$'s is normal regardless of the sample size.

d. Yes; since .0019 < .05 by quite a bit, there is strong evidence to support the claim that the course is effective -- assuming, as stated, that the 16 participants were truly a random sample (and not self-selected or otherwise chosen in some biased manner).

17. normal distribution, since the original distribution is so
$\mu_{\bar{x}} = \mu = 27.44$
$\sigma_{\bar{x}} = \sigma/\sqrt{n} = 12.46/\sqrt{4872} = .179$
$P(\bar{x} > 27.88)$
$\quad = P(z > 2.46)$
$\quad = .5000 - .4931$
$\quad = .0069$

The system is currently acceptable and can expect to be overloaded only $(.0069)(52) = .36$ weeks a year -- or about once every three years.

19. $\mu = .9147$
$\sigma = .0369$
a. This cannot be answered without knowing the shape of the distribution.
NOTE: If the weights of the M&M's are normally distributed, then Table A-2 can be used and $P(x > .9085) = P(z > -.17) = .0675 + .5000 = .5675$.

b. normal distribution, by the Central Limit Theorem
$\mu_{\bar{x}} = \mu = .9147$
$\sigma_{\bar{x}} = \sigma/\sqrt{n} = .0369/\sqrt{1498} = .000953$
$P(\bar{x} > .9085)$
$\quad = P(z > -6.50)$
$\quad = .4999 + .5000$
$\quad = .9999$

c. Yes; the company is providing at least the amount claimed on the label with almost absolute certainty.

21. normal distribution, since the original distribution is so
NOTE: Since $16/300 = .0533 > .05$, use the finite population correction factor
$\mu_{\bar{x}} = \mu = 143$
$\sigma_{\bar{x}} = [\sigma/\sqrt{n}] \cdot \sqrt{(N-n)/(N-1)}$
$\quad = [29/\sqrt{16}] \cdot \sqrt{(300-16)/(300-1)}$
$\quad = [29/\sqrt{16}] \cdot \sqrt{(284)/(299)}$
$\quad = 7.066$
$P(\bar{x} > 157.5)$
$\quad = P(z > 2.05)$
$\quad = .5000 - .4798$
$\quad = .0202$

Since $(12) \cdot (365) \cdot (.0202) = 88.48$, the elevator is likely to be overloaded about 88 times a year -- i.e., about once every 4 days. If the weights of the members of the Atlanta Women's Club are distributed as the given population, the club should consider limiting the elevator's capacity to 15.

23. The finite population correction factor is $\sqrt{(N-n)/(N-1)}$.
If $n = .05 \cdot N$, this becomes $\sqrt{(.95N)/(N-1)} = \sqrt{.95/(1-1/N)}$.
For $N \geq 600$, this is essentially $\sqrt{.95} = .975$.

5-6 Normal Distribution as Approximation to Binomial Distribution

NOTE: As in the previous sections, P(E) represents the probability of an event E; this manual uses $P_c(E)$ to represent the probability of an event E with the continuity correction applied.

1. the area to the right of 57.5
   in symbols, $P(x > 57) = P_c(x > 57.5)$

3. the area to the left of 56.5
   in symbols, $P(x < 57) = P_c(x < 56.5)$

5. the area to the left of 24.5
   in symbols, $P(x \leq 24) = P_c(x < 24.5)$

7. the area from 15.5 to 22.5
   in symbols, $P(16 \leq x \leq 22) = P_c(15.5 < x < 22.5)$

IMPORTANT NOTE: As in the previous sections, store σ in the calculator so that it may be recalled with complete accuracy whenever it is needed in subsequent calculations.

9. binomial: n = 10 and p = .80
   a. from Table A-1, $P(x = 7) = .201$
   b. normal approximation <u>not</u> appropriate since
        $n(1-p) = 10(.20) = 2 < 5$

11. binomial: n = 15 and p = .60
    a. from Table A-1, $P(x \geq 9) = P(x = 9) + P(x = 10) + ... + P(x = 15)$
        $= .207 + .186 + .127 + .063 + .022 + .005 + 0^+$
        $= .610$
    b. normal approximation appropriate since
        $np = 15(.60) = 9 \geq 5$
        $n(1-p) = 15(.40) = 6 \geq 5$
        $\mu = np = 15(.60) = 9$
        $\sigma = \sqrt{np(1-p)} = \sqrt{15(.60)(.40)} = 1.897$
        $P(x \geq 9)$
            $= P_c(x > 8.5)$
            $= P(z > -.26)$
            $= .1026 + .5000$
            $= .6026$

13. let x = the number of girls born
    binomial: n = 100 and p = .50
    normal approximation appropriate since
        $np = 100(.50) = 50 \geq 5$
        $n(1-p) = 100(.50) = 50 \geq 5$
        $\mu = np = 100(.50) = 50$
        $\sigma = \sqrt{np(1-p)} = \sqrt{100(.50)(.50)} = 5.000$
        $P(x \geq 52)$
            $= P_c(x > 51.5)$
            $= P(z > .30)$
            $= .5000 - .1179$
            $= .3821$

15. let x = the number of correct responses
binomial: n = 50 and p = .50
normal approximation appropriate since
$\quad$ np = 50(.50) = 25 ≥ 5
$\quad$ n(1-p) = 50(.50) = 25 ≥ 5
μ = np = 50(.50) = 25
$\sigma = \sqrt{np(1-p)} = \sqrt{50(.50)(.50)}$ = 3.536
P(x ≥ 30)
$\quad$ = P$_c$(x > 29.5)
$\quad$ = P(z > 1.27)
$\quad$ = .5000 - .3980
$\quad$ = .1020

17. let x = the number of wrong numbers
binomial: n = 50 and p = .15
normal approximation appropriate since
$\quad$ np = 50(.15) = 7.5 ≥ 5
$\quad$ n(1-p) = 50(.85) = 42.5 ≥ 5
μ = np = 50(.15) = 7.5
$\sigma = \sqrt{np(1-p)} = \sqrt{50(.15)(.85)}$ = 2.525
P(x < 3) = P$_c$(x < 2.5)
$\qquad$ = P(z < -1.98)
$\qquad$ = .5000 - .4761
$\qquad$ = .0239

Since .0239 < .05, getting fewer than 3 error would be an unusual event if the error rate were really 15%. Conclude that the error rate for AT&T is less than 15%.

19. let x = the number of graduates
binomial: n = 500 and p = .94
normal approximation appropriate since
$\quad$ np = 500(.94) = 470 ≥ 5
$\quad$ n(1-p) = 500(.06) = 30 ≥ 5
μ = np = 500(.94) = 470
$\sigma = \sqrt{np(1-p)} = \sqrt{500(.94)(.06)}$ = 5.310
P(x < 456) = P$_c$(x < 455.5)
$\qquad\quad$ = P(z < -2.73)
$\qquad\quad$ = .5000 - .4968
$\qquad\quad$ = .0032

Yes; since .0032 < .05, getting fewer than 456 who graduate would be an unusual event if the graduation rate were really 94%.

21. let x = the number of booked passengers that arrive
binomial: n = 400 and p = .85
normal approximation appropriate since
$\quad$ np = 400(.85) = 340 ≥ 5
$\quad$ n(1-p) = 400(.15) = 60 ≥ 5
μ = np = 400(.85) = 340
$\sigma = \sqrt{np(1-p)} = \sqrt{400(.85)(.15)}$ = 7.141
P(x > 350) = P$_c$(x > 350.5)
$\qquad\quad$ = P(z > 1.47)
$\qquad\quad$ = .5000 - .4292
$\qquad\quad$ = .0708

Yes; the airline would probably be willing to allow overbooking to occur 7% of the time (i.e., about 1 in every 14 flights) in order to fly with fewer empty seats.

23. let x = the number of women hired
    binomial: n = 62 and p = .50
    normal approximation appropriate since
    $\quad$ np = 62(.50) = 31 ≥ 5
    $\quad$ n(1-p) = 62(.50) = 31 ≥ 5
    μ = np = 62(.50) = 31
    $\sigma = \sqrt{np(1-p)} = \sqrt{62(.50)(.50)} = 3.937$
    $P(x \leq 21) = P_c(x < 21.5)$
    $\quad\quad\quad\quad\quad = P(z < -2.41)$
    $\quad\quad\quad\quad\quad = .5000 - .4920$
    $\quad\quad\quad\quad\quad = .0080$

Yes; since the resulting probability is so small, it does support a charge of discrimination based on gender.

25. let x = the number that test positive
    binomial: n = 150 and p = .038
    normal approximation appropriate since
    $\quad$ np = 150(.038) = 5.7 ≥ 5
    $\quad$ n(1-p) = 150(.962) = 144.3 ≥ 5
    μ = np = 150(.038) = 5.7
    $\sigma = \sqrt{np(1-p)} = \sqrt{150(.038)(.962)} = 2.342$
    $P(x \geq 10) = P_c(x > 9.5)$
    $\quad\quad\quad\quad\quad = P(z > 1.62)$
    $\quad\quad\quad\quad\quad = .5000 - .4474$
    $\quad\quad\quad\quad\quad = .0526$

No; since .0526 > .05, the number with positive test results is not unusually high.

27. let x = the number with group O blood
    binomial: n = 400 and p = .45
    normal approximation appropriate since
    $\quad$ np = 400(.45) = 180 ≥ 5
    $\quad$ n(1-p) = 400(.55) = 220 ≥ 5
    μ = np = 400(.45) = 180
    $\sigma = \sqrt{np(1-p)} = \sqrt{400(.45)(.55)} = 9.950$
    $P(x \geq 177) = P_c(x > 176.5)$
    $\quad\quad\quad\quad\quad = P(z > -.35)$
    $\quad\quad\quad\quad\quad = .1368 + .5000$
    $\quad\quad\quad\quad\quad = .6368$

29. let x = the number of times Marc wins $35 because 7 occurs
    binomial: n = 200 and p = 1/38
    normal approximation appropriate since
    $\quad$ np = 200(1/38) = 5.26 ≥ 5
    $\quad$ n(1-p) = 200(37/38) = 194.76 ≥ 5
    μ = np = 200(1/38) = 5.26
    $\sigma = \sqrt{np(1-p)} = \sqrt{200(1/38)(37/38)} = 2.264$
    $\quad$ for a profit, Marc needs at least 6 wins @ $35 each
    $P(x \geq 6) = P_c(x > 5.5)$
    $\quad\quad\quad\quad = P(z > .10)$
    $\quad\quad\quad\quad = .5000 - .0398$
    $\quad\quad\quad\quad = .4602$

31. let t = time (in years) of TV life before replacement
normal: $\mu = 8.2$ and $\sigma = 1.1$
$P(t > 10.0) = P(z > 1.64)$
$= .5000 - .4495$
$= .0505$

let x = the number of TV's with $t > 10$
binomial: $n = 250$ and $p = .0505$
normal approximation appropriate since
$np = 250(.0505) = 12.625 \geq 5$
$n(1-p) = 250(.9495) = 237.375 \geq 5$
$\mu = np = 250(.0505) = 12.625$
$\sigma = \sqrt{np(1-p)} = \sqrt{250(.0505)(.9495)} = 3.462$
$P(x \geq 15) = P_c(x > 14.5)$
$= P(z > .54)$
$= .5000 - .2054$
$= .2946$

33. The normal approximation to the binomial applies in all parts of this exercise.
a. $\mu = np = 20(.5) = 10$
$\sigma = \sqrt{np(1-p)} = \sqrt{20(.5)(.5)} = 2.236$

without the correction
$P(x \geq 11) = P(z > .45)$
$= .5000 - .1736$
$= .3264$

with the correction
$P(x \geq 11) = P_c(x > 10.5)$
$= P(z > .22)$
$= .5000 - .0871$
$= .4129$

The difference is $.4129 - .3264 = .0865$. Assuming the value obtained using the correction is closer to the correct value, the percent error is $.0865/.4129 = 20.9\%$.

b. $\mu = np = 40(.5) = 20$
$\sigma = \sqrt{np(1-p)} = \sqrt{40(.5)(.5)} = 3.162$

without the correction
$P(x \geq 22) = P(z > .63)$
$= .5000 - .2357$
$= .2643$

with the correction
$P(x \geq 22) = P_c(x > 21.5)$
$= P(z > .47)$
$= .5000 - .1808$
$= .3192$

The difference is $.3192 - .2643 = .0549$. Assuming the value obtained using the correction is closer to the correct value, the percent error is $.0549/.3192 = 17.2\%$.

c. $\mu = np = 400(.5) = 200$
$\sigma = \sqrt{np(1-p)} = \sqrt{400(.5)(.5)} = 10.000$

without the correction
$P(x \geq 220) = P(z > 2.00)$
$= .5000 - .4772$
$= .0228$

with the correction
$P(x \geq 220) = P_c(x > 219.5)$
$= P(z > 1.95)$
$= .5000 - .4744$
$= .0256$

The difference is $.0256 - .0228 = .0028$. Assuming the value obtained using the correction is closer to the correct value, the percent error is $.0028/.0256 = 10.9\%$.

As the sample size increases, the difference between the continuity-corrected and uncorrected normal approximation to the binomial grows smaller -- both in absolute size and in relative size.

5-7 Determining Normality

1. not normal; the points exhibit a systematic "snaking" pattern around a straight line passing through the data

3. not normal; it would require a "broken" line of four segments with four different slopes to pass through the data

5. No. The frequency distribution and histogram for the Wednesday rain data are given below. The distribution is not normal because it is not symmetric -- being positively skewed and having no values below the modal class.

rainfall	frequency
-.05 - .05	41
.05 - .15	7
.15 - .25	1
.25 - .35	2
.35 - .45	0
.45 - .55	0
.55 - .65	2
	53

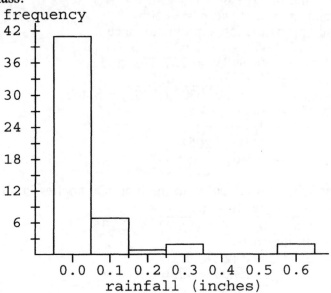

7. No. The frequency distribution and histogram for the duration data are given below. The distribution is not normal because it is not symmetric -- being negatively skewed, having a secondary mode around 125, and having a "low spot" (even dropping to zero) in the middle.

duration	frequency
87.5 - 112.5	3
112.5 - 137.5	8
137.5 - 162.5	1
162.5 - 187.5	0
187.5 - 212.5	1
212.5 - 237.5	11
237.5 - 262.5	13
262.5 - 287.5	13
	50

The information needed to construct normal quantile plots for exercises 9-12 is given on the following page. Exercise #11 most clearly shows the process: *Each of the 50 scores is 2% of the data set. The first score covers .00-.02, the midpoint of which is .01.* [See also exercise #13.]

#	9. rainfall x	cp	z	10. M&M weight x	cp	z	11. duration x	cp	z	12. sugar weight x	cp	z	x	cp	z
1	.00	.009	-2.35	.856	.015	-2.17	102	.01	-2.33	3.407	.007	-2.45	3.588	.507	0.02
2	.00	.028	-1.91	.858	.045	-1.69	103	.03	-1.88	3.450	.021	-2.03	3.590	.521	0.05
3	.00	.047	-1.67	.860	.076	-1.43	105	.05	-1.64	3.464	.026	-1.80	3.598	.536	0.09
4	.00	.066	-1.51	.866	.106	-1.25	113	.07	-1.48	3.468	.050	-1.64	3.600	.550	0.13
5	.00	.085	-1.37	.867	.136	-1.10	114	.09	-1.34	3.475	.064	-1.52	3.601	.564	0.16
6	.00	.104	-1.26	.871	.167	-0.97	120	.11	-1.22	3.482	.079	-1.41	3.604	.579	0.20
7	.00	.123	-1.16	.875	.197	-0.85	120	.13	-1.13	3.491	.093	-1.32	3.604	.593	0.23
8	.00	.142	-1.07	.876	.227	-0.75	122	.15	-1.04	3.492	.107	-1.24	3.611	.607	0.27
9	.00	.160	-0.99	.889	.257	-0.65	124	.17	-0.95	3.494	.121	-1.17	3.617	.621	0.31
10	.00	.179	-0.92	.897	.288	-0.56	127	.19	-0.88	3.494	.136	-1.10	3.621	.636	0.35
11	.00	.198	-0.85	.898	.318	-0.47	134	.21	-0.81	3.506	.150	-1.04	3.622	.650	0.39
12	.00	.217	-0.78	.900	.348	-0.39	140	.23	-0.74	3.507	.164	-0.98	3.625	.664	0.42
13	.00	.236	-0.72	.902	.379	-0.31	203	.25	-0.67	3.508	.179	-0.92	3.632	.679	0.46
14	.00	.255	-0.66	.902	.409	-0.23	214	.27	-0.61	3.511	.193	-0.87	3.635	.693	0.50
15	.00	.274	-0.60	.904	.439	-0.15	218	.29	-0.55	3.516	.207	-0.82	3.635	.707	0.55
16	.00	.292	-0.55	.909	.470	-0.08	226	.31	-0.50	3.521	.221	-0.77	3.638	.721	0.59
17	.00	.311	-0.49	.909	.500	0.00	227	.33	-0.44	3.522	.236	-0.72	3.639	.736	0.63
18	.00	.330	-0.44	.914	.530	0.08	228	.35	-0.39	3.522	.250	-0.67	3.643	.750	0.67
19	.00	.349	-0.39	.914	.561	0.15	232	.37	-0.33	3.526	.264	-0.63	3.643	.764	0.72
20	.00	.368	-0.34	.919	.591	0.23	233	.39	-0.28	3.531	.279	-0.59	3.645	.779	0.77
21	.00	.387	-0.29	.920	.621	0.31	235	.41	-0.23	3.532	.293	-0.55	3.647	.793	0.82
22	.00	.406	-0.24	.921	.652	0.39	237	.43	-0.18	3.535	.307	-0.50	3.654	.807	0.87
23	.00	.425	-0.19	.923	.682	0.47	237	.45	-0.13	3.542	.321	-0.46	3.660	.821	0.92
24	.00	.443	-0.14	.928	.712	0.56	237	.47	-0.08	3.545	.336	-0.42	3.665	.836	0.98
25	.00	.462	-0.09	.930	.742	0.65	238	.49	-0.03	3.548	.350	-0.39	3.666	.850	1.04
26	.00	.481	-0.05	.930	.773	0.75	238	.51	0.03	3.569	.364	-0.35	3.667	.864	1.10
27	.00	.500	0.00	.932	.803	0.85	239	.53	0.08	3.673	.379	-0.31	3.671	.879	1.17
28	.00	.519	0.05	.936	.833	0.97	240	.55	0.13	3.576	.393	-0.27	3.673	.893	1.24
29	.00	.538	0.09	.955	.864	1.10	241	.57	0.18	3.577	.407	-0.23	3.678	.907	1.32
30	.00	.557	0.14	.965	.894	1.25	241	.59	0.23	3.580	.421	-0.20	3.687	.921	1.41
31	.00	.575	0.19	.976	.924	1.43	241	.61	0.28	3.582	.436	-0.16	3.688	.936	1.52
32	.00	.594	0.24	.988	.955	1.69	243	.63	0.33	3.582	.450	-0.13	3.718	.950	1.64
33	.00	.613	0.29	1.033	.985	2.17	245	.65	0.39	3.583	.464	-0.09	3.723	.964	1.80
34	.00	.632	0.34				248	.67	0.44	3.585	.479	-0.05	3.725	.979	2.03
35	.00	.651	0.39				249	.69	0.50	3.588	.493	-0.02	3.726	.993	2.45
36	.00	.670	0.44				250	.71	0.55						
37	.00	.698	0.49				258	.73	0.61						
38	.01	.708	0.55				264	.75	0.67						
39	.01	.726	0.60				265	.77	0.74						
40	.01	.745	0.66				267	.79	0.81						
41	.02	.764	0.72				267	.81	0.88						
42	.06	.783	0.78				268	.83	0.95						
43	.06	.802	0.85				270	.85	1.04						
44	.06	.821	0.92				270	.87	1.13						
45	.08	.840	0.99				270	.89	1.23						
46	.08	.858	1.07				271	.91	1.34						
47	.12	.877	1.16				272	.93	1.48						
48	.14	.896	1.26				275	.95	1.64						
49	.18	.915	1.37				275	.97	1.88						
50	.27	.934	1.51				276	.99	2.33						
51	.31	.953	1.67												
52	.64	.972	1.91												
53	.64	.991	2.35												

Normal quantile plots for exercises 9-12 may be constructed by hand by graphing the column labeled "x" on the horizontal axis and "z" on the vertical axis, where

x = the original scores in order

cp = the cumulative probability values 1/2n, 3/2n, 5/2n, ..., (2n-1)/2n

z = the z scores having the given cp values to their left

9. Since the points of the normal quantile plot do not lie close to a straight line, conclude that the population distribution is not approximately normal.

11. Since the points of the normal quantile plot do not lie close to a straight line, conclude that the population distribution is not approximately normal.

13. Start by arranging the scores in order and placing them in the "x" column of a table like the one below.

   a. These values are given in the "cp" column below and represent cumulative probability.

   b. These values are given in the "z" column below, and are the z-scores corresponding to the cumulative probability values from part (a).

x	cp	z	[from Table A-2]
102	1/14 = .0714	-1.47	[A = .5000 - .0714 = .4286]
103	3/14 = .2143	-0.79	[A = .5000 - .2143 = .2857]
137	5/14 = .3571	-0.37	[A = .5000 - .3571 = .1429]
142	7/14 = .5000	0.00	[A = .0000]
154	9/14 = .6429	0.37	[A = .6429 - .5000 = .1429]
157	11/14 = .7857	0.79	[A = .7857 - .5000 = .2857]
188	13/14 = .9286	1.47	[A = .9286 - .5000 = .4286]

   When converting "cp" to "z" use the closest entry in Table A-2. The picture/technique is given below for the first case.

   c. The normal quantile plot is given below. It is constructed by plotting "x" on the horizontal axis and "z" on the vertical axis.

Review Exercises

1. normal: $\mu = 63.6$ and $\sigma = 2.5$
   a. P(58 < x < 73)
   $= P(-2.24 < z < 3.76)$
   $= .4875 + .4999$
   $= .9874$

   b. For the shortest 1%, A = .4900 [.4901] and z = -2.33.
   $x = \mu + z\sigma$
   $= 63.6 + (-2.33)(2.5)$
   $= 63.6 - 5.8$
   $= 57.8$

   For the tallest 1%, A = .4900 [.4901] and z = 2.33.
   $x = \mu + z\sigma$
   $= 63.6 + (2.33)(2.5)$
   $= 63.6 + 5.8$
   $= 69.4$

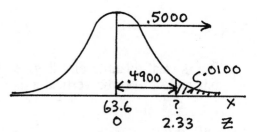

   c. normal, since the original distribution is so
   $\mu_{\bar{x}} = \mu = 63.6$
   $\sigma_{\bar{x}} = \sigma/\sqrt{n} = 2.5/\sqrt{9} = .833$
   $P(63.0 < \bar{x} < 65.0)$
   $= P(-.72 < z < 1.68)$
   $= .2642 + .4535$
   $= .7177$

2. normal: $\mu = 69.0$ and $\sigma = 2.8$
   a. P(x ≥ 74)
   $= P(z > 1.79)$
   $= .5000 - .4633$
   $= .0367$

b. P(62.0 < x < 76.0)
   = P(-2.50 < z < 2.50)
   = .4938 + .4938
   = .9876

c. For the shortest 2%, A = .4800 [.4798] and z = -2.05.
   x = μ + zσ
   = 69.0 + (-2.05)(2.8)
   = 69.0 - 5.7
   = 63.3

For the tallest 2%, A = .4800 [.4798] and z = 2.05.
   x = μ + zσ
   = 69.0 + (2.05)(2.8)
   = 69.0 + 5.7
   = 74.7

d. normal, since the original distribution is so
   $\mu_{\bar{x}} = \mu = 69.0$
   $\sigma_{\bar{x}} = \sigma/\sqrt{n} = 2.8/\sqrt{25} = .560$
   $P(\bar{x} > 68.0)$
   = P(z > -1.79)
   = .4633 + .5000
   = .9633

3. normal: μ = 132.6 and σ = 5.4
   a. P(x > 140)
   = P(z > 1.37)
   = .5000 - .4147
   = .0853

b. For $D_2 = P_{20}$, A = .3000 [.2995] and z = -.84.

$$x = \mu + z\sigma$$
$$= 132.6 + (-.84)(5.4)$$
$$= 132.6 - 4.5$$
$$= 128.1$$

c. normal, since the original distribution is so

$$\mu_{\bar{x}} = \mu = 132.6$$
$$\sigma_{\bar{x}} = \sigma/\sqrt{n} = 5.4/\sqrt{35} = .913$$
$$P(133.0 < \bar{x} < 134.0)$$
$$= P(.44 < z < 1.53)$$
$$= .4370 - .1700$$
$$= .2670$$

4. binomial: n = 500 and p = .26
normal approximation appropriate since
$$np = 500(.26) = 130 \geq 5$$
$$n(1-p) = 500(.74) = 370 \geq 5$$
$$\mu = np = 500(.26) = 130$$
$$\sigma = \sqrt{np(1-p)} = \sqrt{500(.26)(.74)} = 9.808$$
$$P(125 \leq x \leq 150) = P_c(124.5 < x < 150.5)$$
$$= P(-.56 < z < 2.09)$$
$$= .2123 + .4817$$
$$= .6940$$

5. normal distribution with $\mu = 0$ and $\sigma = 1$
[Because this is a standard normal, the variable is already expressed in terms of z scores.]

a. $P(0 < z < 1.00)$
$$= .3413$$

b. $P(z > 2.00)$
$$= .5000 - .4772$$
$$= .0228$$

c. P(z > -1.50)
   = .4332 + .5000
   = .9332

d. normal, since the original distribution is so
   $\mu_{\bar{x}} = \mu = 0$
   $\sigma_{\bar{x}} = \sigma/\sqrt{n} = 1/\sqrt{25} = .200$
   $P(.25 < \bar{x} < .50)$
      = P(1.25 < z < 2.50)
      = .4938 - .3944
      = .0994

e. For $D_4 = P_{40}$, A = .1000 [.0987] and z = -.25.
   [Because this is a standard normal, the variable is already expressed in terms of z scores.]

6. uniform distribution. Since the total area must be 1.00 and the width of the rectangle is .4, the height of the rectangle is 1/.4 = 2.5. Probability corresponds to area, and the area of a rectangle is (width)·(height).

a. P(x ≥ 12.0)
      = (width)·(height)
      = (12.2 - 12.0)·(2.5)
      = (.2)·(2.5)
      = .5

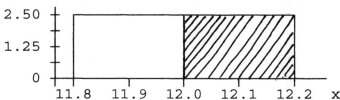

b. P(11.9 < x < 12.1)
      = (width)·(height)
      = (12.1 - 11.9)·(2.5)
      = (.2)·(2.5)
      = .5

c. P(x < 11.9)
      = (width)·(height)
      = (11.9 - 11.8)·(2.5)
      = (.1)·(2.5)
      = .25

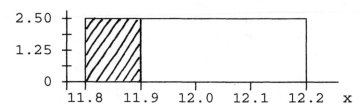

7. normal: $\mu = 9.1$ and $\sigma = 2.1$

   a. $P(x > 8.5)$
       $= P(z > -.29)$
       $= .1141 + .5000$
       $= .6141$

   b. $P(7.5 < x < 9.0)$
       $= P(-.76 < z < -.05)$
       $= .2764 - .0199$
       $= .2565$

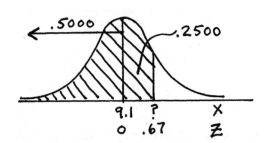

   c. For $Q_3 = P_{75}$, $A = .2500$ [.2486] and $z = .67$.
      $x = \mu + z\sigma$
        $= 9.1 + (.67)(2.1)$
        $= 9.1 + 1.4$
        $= 10.5$

   d. For $Q_1 = P_{25}$, $A = .2500$ [.2486] and $z = -.67$.
      $x = \mu + z\sigma$
        $= 9.1 + (-.67)(2.1)$
        $= 9.1 - 1.4$
        $= 7.7$

   e. normal, since the original distribution is so
     $\mu_{\bar{x}} = \mu = 9.1$
     $\sigma_{\bar{x}} = \sigma/\sqrt{n} = 2.1/\sqrt{35} = .355$
     $P(\bar{x} > 9.0)$
       $= P(z > -.28)$
       $= .1103 + .5000$
       $= .6103$

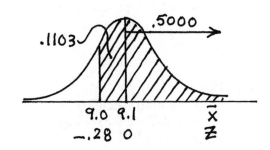

8. binomial: n = 800 and p = .18
   normal approximation appropriate since
   $np = 800(.18) = 144 \geq 5$
   $n(1-p) = 800(.82) = 656 \geq 5$
   $\mu = np = 800(.18) = 144$
   $\sigma = \sqrt{np(1-p)} = \sqrt{800(.18)(.82)} = 10.866$
   $P(x \geq 150) = P_c(x > 149.5)$
   $\qquad = P(z > .51)$
   $\qquad = .5000 - .1950$
   $\qquad = .3050$

9. normal: $\mu = 35,600$ and $\sigma = 4275$
   a. $P(x > 30,000)$
   $\qquad = P(z > -1.31)$
   $\qquad = .4049 + .5000$
   $\qquad = .9049$

   b. normal, since the original distribution is so
   $\mu_{\bar{x}} = \mu = 35,600$
   $\sigma_{\bar{x}} = \sigma/\sqrt{n} = 4275/\sqrt{40} = 675.9$
   $P(\bar{x} > 35,000)$
   $\qquad = P(z > -.89)$
   $\qquad = .3133 + .5000$
   $\qquad = .8133$

   c. For the lowest 3%, A = .4700 [.4699] and z = -1.88.
   $x = \mu + z\sigma$
   $\qquad = 35,600 + (-1.88)(4275)$
   $\qquad = 35,600 - 8037$
   $\qquad = 27,563$

   A guarantee of 27,000 or 27,500 miles should be offered.

10. binomial: n = 36 and p = 1/6
    normal approximation appropriate since
    $np = 36(1/6) = 6 \geq 5$
    $n(1-p) = 36(5/6) = 30 \geq 5$
    $\mu = np = 36(1/6) = 6$
    $\sigma = \sqrt{np(1-p)} = \sqrt{36(1/6)(5/6)} = 2.236$
    $P(x \geq 13) = P_c(x > 12.5)$
    $\qquad = P(z > 2.91)$
    $\qquad = .5000 - .4982$
    $\qquad = .0018$

Yes; since .0018 < .05, this is strong evidence that he has the ability to predict the outcomes.
NOTE: Statistically, the result of 13 or more correct predictions if the assumptions that he was
guessing at random is true is very remote -- and so we conclude he was not guessing at
random. Cynically, a person may wish to note there were other assumptions (e.g., the die was
balanced) -- and perhaps the conclusion should be that one of those is false.

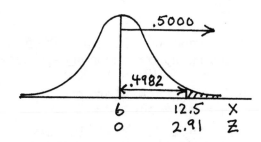

Cumulative Review Exercises

1. A. tally of the scores is given at the right.
   summary values: $n = 43$, $\Sigma x = 2624$, $\Sigma x^2 = 160,672$

   a. $\bar{x} = (\Sigma x)/n = (2624)/43 = 61.023$ rounded to 61.0 mm

   b. $\tilde{x} = 61$ mm

   c. $M = 60$ mm

   d. $s^2 = [n(\Sigma x^2) - (\Sigma x)^2]/[n(n-1)]$
   $= [43(160,672) - (2624)^2]/[43(42)]$
   $= 23520/1806$
   $= 13.023$
   $s = 3.609$ rounded to 3.6 mm

   e. While the tally of the original scores does not appear
   bell-shaped, the frequency distribution given below
   the tally does. Yes; based on the grouped data,
   it does seem that the values are from a population
   with a distribution that is approximately normal.

   f. $z = (x-\bar{x})/s$
   $z_{66} = (66 - 61.023)/3.609 = 1.38$

   g. r.f. of scores greater than 59 is $31/43 = .7209 = 72\%$

   h. normal: $\mu = 61.023$ and $\sigma = 3.609$
   $P(x > 59) = P(z > -.56)$
   $= .2123 + .5000$
   $= .7123$

   i. ratio, since distances between scores are meaningful and
   there is a natural zero

   j. continuous, since length can be any value on a continuum

| 52-1 |
| 53- |
| 54-1 |
| 55-1 |
| 56-111 |
| 57-1 |
| 58-1 |
| 59-1111 |
| 60-111111111 |
| 61-11 |
| 62-11111 |
| 63-11111 |
| 64-11 |
| 65-1 |
| 66-11111 |
| 67-11 |

width (mm)	frequency
51 - 53	1
54 - 56	5
57 - 59	6
60 - 62	16
63 - 65	8
66 - 68	7
	43

2. a. Let $L = $ a person is left-handed
   $P(L) = .10$, for each random selection
   $P(L_1 \text{ and } L_2 \text{ and } L_3) = P(L_1) \cdot P(L_2) \cdot P(L_3) = (.10) \cdot (.10) \cdot (.10) = .001$

   b. Let $N = $ a person is not left-handed
   $P(N) = .90$, for each random selection
   $P(\text{at least one left-hander}) = 1 - P(\text{no left-handers})$
   $= 1 - P(N_1 \text{ and } N_2 \text{ and } N_3)$
   $= 1 - P(N_1) \cdot P(N_2) \cdot P(N_3)$
   $= 1 - (.90) \cdot (.90) \cdot (.90)$
   $= 1 - .729$
   $= .271$

   c. binomial: $n = 3$ and $p = P(\text{left-hander}) = .10$
   normal approximation <u>not</u> appropriate since
   $np = 3(.10) = 0.3 < 5$

   d. binomial: $n = 50$ and $p = .10$
   $\mu = np = 50(.10) = 5$

   e. binomial: <u>$n = 50$ and $p = .10$</u>
   $\sigma = \sqrt{np(1-p)} = \sqrt{50(.10)(.90)} = 2.121$

   f. An unusual score is one that is more than 2 standard deviations from the mean.
   Since $z = (x-\mu)/\sigma$, $z_8 = (8-5)/2.212 = 1.41$.
   Since 8 is 1.41 standard deviations from the mean, it is <u>not</u> an unusual score.

Chapter 6

Estimates and Sample Sizes

6-2 Estimating a Population Mean: Large Samples

1. let L = the lower confidence limit
   U = the upper confidence limit
   $\bar{x} = (L + U)/2 = (72.6 + 82.6)/2 = 155.2/2 = 77.6$
   $E = (U - L)/2 = (82.6 - 72.6)/2 = 10.0/2 = 5.0$

3. let L = the lower confidence limit
   U = the upper confidence limit
   $\bar{x} = (L + U)/2 = (207 + 212)/2 = 419/2 = 209.5$
   $E = (U - L)/2 = (212 - 207)/2 = 5/2 = 2.5$

5. $\alpha = .01, \alpha/2 = .005$
   $A = .5000 - .005 = .4950$
   $z_{.005} = 2.575$

7. $\alpha = .03, \alpha/2 = .015$
   $A = .5000 - .015 = .4850$
   $z_{.015} = 2.17$

9. $n > 30$, use z (with s for σ)
   $\alpha = .05, \alpha/2 = .025, z_{.025} = 1.96$
   a. $E = z_{.025} \cdot \sigma/\sqrt{n}$
   $= 1.96 \cdot (12,345)/\sqrt{60}$
   $= 3,124$ dollars
   b. $\bar{x} \pm E$
   $85,678 \pm 3,124$
   $82,554 < \mu < 88,802$ (dollars)

11. $n > 30$, use z (with s for σ)
    $\alpha = .01, \alpha/2 = .005, z_{.005} = 2.33$
    a. $E = z_{.005} \cdot \sigma/\sqrt{n}$
    $= 2.575 \cdot (2.72)/\sqrt{150}$
    $= .57$ seconds
    b. $\bar{x} \pm E$
    $7.60 \pm .57$
    $7.03 < \mu < 8.17$ (seconds)

13. We are 95% confident that the interval from 133.35 mm to 135.65 mm contains the true value of μ, the mean maximum skull breadth of all Egyptian males who lived around 1850 B.C.

15. a. n > 30, use z (with s for σ)
   $\bar{x} \pm z_{.025} \cdot \sigma/\sqrt{n}$
   $205.5 \pm 1.96 \cdot 35.4/\sqrt{40}$
   205.5 ± 11.0
   $194.5 < \mu < 216.5$

   b. n > 30, use z (with s for σ)
   $\bar{x} \pm z_{.025} \cdot \sigma/\sqrt{n}$
   $185.0 \pm 1.96 \cdot 28.4/\sqrt{40}$
   185.0 ± 8.8
   $176.2 < \mu < 193.8$

   c. We are 95% certain that the interval from 194.5 to 216.5 contains the true mean score for oiled lanes, and we are 95% certain that the interval from 176.2 to 193.8 contains the true mean score for dry lanes. Since these intervals do not overlap, we can be fairly certain that the means are truly different and that the lane surface does have an effect on bowling scores -- with the oily surface producing higher scores.
   NOTE: From probability theory, the confidence we have in both statements simultaneously is $(.95) \cdot (.95) = .9025$ -- about 90% confidence.

17. preliminary values: n = 36, Σx = 5866.98, Σx^2 = 1832952.2176
   $\bar{x} = (\Sigma x)/n = (5866.98)/36 = 162.972$
   $s^2 = [n(\Sigma x^2) - (\Sigma x)^2]/[n(n-1)] = [36(1832952.2176) - (5866.98)^2]/[36(35)] = 25051.448$
   $s = 158.276$

   n > 30, use z (with s for σ)
   $\bar{x} \pm z_{.005} \cdot \sigma/\sqrt{n}$
   $162.972 \pm 2.575 \cdot 158.276/\sqrt{36}$
   162.972 ± 67.927
   $95.045 < \mu < 230.898$

   NOTE: Here and in similar problems, $\bar{x}$ and s are shown with appropriate accuracy, but the values were stored and recalled with full accuracy for use in subsequent calculations.

19. preliminary values: n = 40, Σx = 38.023, Σx^2 = 36.143980
   $\bar{x} = (\Sigma x)/n = (38.023)/40 = .9506$
   $s^2 = [n(\Sigma x^2) - (\Sigma x)^2]/[n(n-1)] = [40(36.143980) - (38.023)^2]/[40(39)] = .0000068$
   $s = .0026$

   n > 30, use z (with s for σ)
   $\bar{x} \pm z_{.05} \cdot \sigma/\sqrt{n}$
   $.9506 \pm 1.645 \cdot .0026/\sqrt{40}$
   $.9506 \pm .0006$
   $.9499 < \mu < .9513$

21. a. preliminary values: n = 36, Σx = 29.4056, Σx^2 = 24.02112012
   $\bar{x} = (\Sigma x)/n = (29.4056)/36 = .81682$
   $s^2 = [n(\Sigma x^2) - (\Sigma x)^2]/[n(n-1)] = [36(24.02112012) - (29.4056)^2]/[36(35)] = .0000564$
   $s = .00751$

   n > 30, use z (with s for σ)
   $\bar{x} \pm z_{.025} \cdot \sigma/\sqrt{n}$
   $.81682 \pm 1.96 \cdot .00751/\sqrt{36}$
   $.81682 \pm .00245$
   $.81437 < \mu < .81927$

b. preliminary values: $n = 36$, $\Sigma x = 28.2526$, $\Sigma x^2 = 22.17315828$

$\bar{x} = (\Sigma x)/n = (28.2526)/36 = .78479$

$s^2 = [n(\Sigma x^2) - (\Sigma x)^2]/[n(n-1)] = [36(22.17315828) - (28.2526)^2]/[36(35)] = .0000193$

$s = .00439$

$n > 30$, use $\underline{z}$ (with s for σ)

$\bar{x} \pm z_{.025} \cdot \sigma/\sqrt{n}$

$.78479 \pm 1.96 \cdot .00439/\sqrt{36}$

$.78479 \pm .00143$

$.78336 < \mu < .78623$

c. We are 95% certain that the interval from .81437 to .81927 contains the true mean weight for regular Coke, and we are 95% certain that the interval from .78336 to .78623 contains the true mean weight for diet Coke. Since these intervals do not overlap, we can be fairly certain that the means are truly different and that the diet Coke weighs less.

NOTE: From probability theory, the confidence we have in both statements simultaneously is $(.95) \cdot (.95) = .9025$ -- about 90% confidence.

23.

plastic	paper
$n = 62$	$n = 62$
$\Sigma x = 118.47$	$\Sigma x = 584.54$
$\Sigma x^2 = 295.6233$	$\Sigma x^2 = 6570.8217$
$\bar{x} = (\Sigma x)/n = (118.47)/62 = 1.911$	$\bar{x} = (\Sigma x)/n = (584.54)/62 = 9.428$
$s^2 = [n(\Sigma x^2) - (\Sigma x)^2]/[n(n-1)]$	$s^2 = [n(\Sigma x^2) - (\Sigma x)^2]/[n(n-1)]$
$= [62(295.6233) - (118.47)^2]/[62(61)]$	$= [62(6570.8217) - (584.54)^2]/[62(61)]$
$= 1.1352$	$= 17.3728$
$s = 1.065$	$s = 4.168$
$n > 30$, use $\underline{z}$ (with s for σ)	$n > 30$, use $\underline{z}$ (with s for σ)
$\bar{x} \pm z_{.04} \cdot \sigma/\sqrt{n}$	$\bar{x} \pm z_{.04} \cdot \sigma/\sqrt{n}$
$1.911 \pm 1.75 \cdot 1.065/\sqrt{62}$	$9.428 \pm 1.75 \cdot 4.168/\sqrt{62}$
$1.911 \pm .237$	$9.428 \pm .926$
$1.674 < \mu < 2.148$ (lbs)	$8.502 < \mu < 10.354$ (lbs)

By weight, discarded paper is a much larger ecological problem than discarded plastic. Since paper is less dense than plastic, it would also be a larger problem by volume. Since paper is more degradable and recyclable than plastic, however, it would be difficult to make an unqualified statement.

25. The usual formulas express the two values $\bar{x}$ and s^2 in terms of the two quantities Σx and Σx^2. Solving the usual formulas "in reverse" yields $\Sigma x = n\bar{x}$ and $\Sigma x^2 = (n-1)s^2 + n\bar{x}^2$.
In this problem, use the given values of $\bar{x}$ and s^2 to determine Σx and Σx^2, subtract out 98.6 and 98.6^2, and then add in 986 and 986^2.

original $\Sigma x = n\bar{x} = 106 \cdot (98.20) = 10409.2$

original $\Sigma x^2 = (n-1)s^2 + n\bar{x}^2 = 105 \cdot (.62)^2 + 106 \cdot (98.20)^2 = 1022223.802$

new $\Sigma x = 10409.2 - 98.6 + 986 = 11296.6$

new $\Sigma x^2 = 1022223.802 - 98.6^2 + 986^2 = 1984697.84$

a. $\bar{x} = (\Sigma x)/n = (11296.6)/106 = 106.57$

$s^2 = [n(\Sigma x^2) - (\Sigma x)^2]/[n(n-1)] = [106(1984697.84) - (11296.6)^2]/[106(105)] = 7436.19$

$s = 86.233$

$n > 30$, use $\underline{z}$ (with s for σ)

$\bar{x} \pm z_{.025} \cdot \sigma/\sqrt{n}$

$106.57 \pm 1.96 \cdot 86.233/\sqrt{106}$

106.57 ± 16.42

$90.16 < \mu < 122.99$

b. In this problem, the outlier changed the center of the confidence interval from 98.20 to 106.57 and the width of the confidence interval from 98.32-98.08 = .24 to 122.99-90.16 = 32.83. Both changes were significant, and they demonstrate that confidence intervals can be very sensitive to outliers.

c. Based on part (b), values determined to be outliers should probably not be used in further statistical analyses -- including the construction of confidence intervals.

27. let L = the lower confidence limit
U = the upper confidence limit

a. $\bar{x}$ = (L + U)/2 = (430 + 470)/2 = 900/2 = 450

b. E = (U - L)/2 = (470 - 430)/2 = 40/2 = 20
$E = z_{.025} \cdot \sigma/\sqrt{n}$
$20 = 1.96 \cdot s/\sqrt{100}$ [since n > 30, s is used for σ]
20 = .196s
s = 20/.196 = 102

c. $\bar{x} \pm z_{.005} \cdot \sigma/\sqrt{n}$
$450 \pm 2.575 \cdot 102/\sqrt{100}$
450 ± 26
$424 < \mu < 476$

d. E = (U - L)/2 = (468 - 432)/2 = 36/2 = 18
$E = z_{\alpha/2} \cdot \sigma/\sqrt{n}$
$18 = z_{\alpha/2} \cdot 102/\sqrt{100}$
$18 = z_{\alpha/2} \cdot 10.2$
$z_{\alpha/2} = 18/10.2 = 1.76$, A = .4608
$\alpha/2 = .5000 - .4608 = .0392$, $\alpha = 2 \cdot (.0392) = .0784$
confidence level = 1 - α = 1 - .0784 = .9216 = 92%

6-3 Estimating a Population Mean:
Small Samples

IMPORTANT NOTE: This manual uses the following conventions.
(1) The designation "df" stands for "degrees of freedom."
(2) Since the t value depends on both the degrees of freedom and the probability lying beyond it, double subscripts are used to identify points on t distributions. The t distribution with 15 degrees of freedom and .025 beyond it, for example, is designated $t_{15,.025}$ = 2.132.
(3) When df≥30 the difference between the t and z distributions is negligible and Table A-3 uses one final row of z values to cover all such cases. Consequently, the z scores for certain "popular" α and $\alpha/2$ values may be found by reading Table A-3 "frontwards" instead of reading Table A-2 "backwards." This is not only easier but also more accurate, since Table A-3 includes one more decimal place. The manual uses this technique from this point on.

1. population approximately normal, n ≤ 30, σ unknown: use t
n = 20; df = 19
α = .05; $\alpha/2$ = .025
$t_{19,.025}$ = 2.093

3. population approximately normal, n ≤ 30, σ unknown: use t
n = 25; df = 14
α = .01; $\alpha/2$ = .005
$t_{14,.005}$ = 2.977

5. population skewed, n ≤ 30: neither the t nor the z applies

7. population approximately normal, n ≤ 30, σ unknown: use t
   n = 9; df = 8
   α = .02; α/2 = .01
   $t_{8,.01}$ = 2.896

9. population approximately normal, n ≤ 30, σ unknown: use t
   a. E = $t_{14,.025} \cdot s/\sqrt{n}$
      = $2.145 \cdot 108/\sqrt{15}$
      = 60
   b. $\bar{x} \pm E$
      496 ± 60
      436 < μ < 556

11. 99 < μ < 109 [rounded to agree with the accuracy of $\bar{x}$]
    We are 95% certain that the interval from 99 to 109 contains the true mean IQ score for professional athletes.

13. n ≤ 30 and σ unknown, use t
    $\bar{x} \pm t_{9,.025} \cdot s/\sqrt{n}$
    124 ± $2.262 \cdot 18/\sqrt{10}$
    124 ± 13
    111 < μ < 137 (beats per minute)
    Since the corresponding interval for manual snow shoveling was 164 < μ < 186, it appears that the maximum heart rates are lower during automated snow shoveling.

15. n ≤ 30 and σ unknown, use t
    $\bar{x} \pm t_{11,.025} \cdot s/\sqrt{n}$
    26,227 ± $2.201 \cdot (15,873)/\sqrt{12}$
    26,227 ± 10,085
    16,142 < μ < 36,312 (dollars)
    We are 95% certain that the interval from $16,142 to $36,312 contains the true mean repair cost for repairing Dodge Vipers under the specified conditions.

17. preliminary values: n = 7, Σx = .85, Σx^2 = .1123
    $\bar{x}$ = $(\Sigma x)/n$ = (.85)/7 = .121
    s^2 = $[n(\Sigma x^2) - (\Sigma x)^2]/[n(n-1)]$ = $[7(.1123) - (.85)^2]/[7(6)]$ = .00151
    s = .039
    n ≤ 30 and σ unknown, use t
    $\bar{x} \pm t_{6,.01} \cdot s/\sqrt{n}$
    .121 ± $3.143 \cdot (.039)/\sqrt{7}$
    .121 ± .046
    .075 < μ < .168

19. n > 30, use z (with s for σ)
    $\bar{x} \pm z_{.05} \cdot \sigma/\sqrt{n}$
    8.2 ± $1.645 \cdot (1.1)/\sqrt{75}$
    8.2 ± .2
    8.0 < μ < 8.4 (years)
    No; these sets were made at least 8 years ago, and the technology has changed since then.

21. a. preliminary values: n = 26, Σx = 23.849, Σx^2 = 21.904543

$\bar{x}$ = $(\Sigma x)/n$ = (23.849)/26 = .9173

s^2 = $[n(\Sigma x^2) - (\Sigma x)^2]/[n(n-1)]$ = $[26(21.904543) - (23.849)^2]/[26(25)]$ = .0011436

s = .0338

n ≤ 30 and σ unknown, use t

$\bar{x} \pm t_{25,.05} \cdot s/\sqrt{n}$

.9173 ± 1.708 · .0338/$\sqrt{26}$

.9173 ± .0113

.9059 < μ < .9286 (grams)

   b. preliminary values: n = 33, Σx = 30.123, Σx^2 = 27.546793

$\bar{x}$ = $(\Sigma x)/n$ = (30.123)/33 = .9128

s^2 = $[n(\Sigma x^2) - (\Sigma x)^2]/[n(n-1)]$ = $[33(27.546793) - (30.123)^2]/[33(32)]$ = .0015616

s = .0395

n > 30, use z (with s for σ)

$\bar{x} \pm z_{.05} \cdot \sigma/\sqrt{n}$

.9128 ± 1.645 · .0395/$\sqrt{33}$

.9128 ± .0113

.9015 < μ < .9241 (grams)

   c. The larger sample size in part (b) allowed the use of the z distribution instead of the t. While both intervals have essentially the same width, the brown M&M's in part (b) may be slightly lighter -- but not significantly so, as the intervals overlap considerably.

23. a. preliminary values: n = 25, Σx = 1972.45, Σx^2 = 160239.3228

$\bar{x}$ = $(\Sigma x)/n$ = (1972.45)/25 = 78.898

s^2 = $[n(\Sigma x^2) - (\Sigma x)^2]/[n(n-1)]$ = $[25(160239.3228) - (1972.45)^2]/[25(24)]$ = 192.37

s = 13.87

n ≤ 30 and σ unknown, use t

$\bar{x} \pm t_{24,.01} \cdot s/\sqrt{n}$

78.898 ± 2.492 · 13.87/$\sqrt{25}$

78.898 ± 6.913

71.985 < μ < 85.811 (dollars)

   b. preliminary values: n = 15, Σx = 632.25, Σx^2 = 30139.2424

$\bar{x}$ = $(\Sigma x)/n$ = (632.25)/15 = 42.150

s^2 = $[n(\Sigma x^2) - (\Sigma x)^2]/[n(n-1)]$ = $[15(30139.2424) - (632.25)^2]/[15(14)]$ = 249.28

s = 15.79

n ≤ 30 and σ unknown, use t

$\bar{x} \pm t_{14,.01} \cdot s/\sqrt{n}$

42.150 ± 2.625 · 15.79/$\sqrt{15}$

42.150 ± 10.701

31.119 < μ < 52.851 (dollars)

   c. The paperbacks cost significantly less. Since there was more variability and a smaller sample size for the paperbacks, the confidence interval in part (b) is considerably wider than the one in part (a).

25. For any α, the z value is smaller than the corresponding t value (although the difference decreases as n increases). This creates a smaller E and a narrower confidence interval than one is entitled to -- i.e., it does not take into consideration the extra uncertainty created by using the sample s instead of the true population σ.

27. preliminary values: $n = 5$, $\Sigma x = 44.2$, $\Sigma x^2 = 397.52$
$\bar{x} = (\Sigma x)/n = (44.2)/5 = 8.84$
$s^2 = [n(\Sigma x^2) - (\Sigma x)^2]/[n(n-1)] = [5(397.52) - (44.2)^2]/[5(4)] = 1.698$
$s = 1.30$
$n \leq 30$ and σ unknown, use t
$E = t_{4,\alpha/2} \cdot s/\sqrt{n}$
$1.24 = t_{4,\alpha/2} \cdot (1.30)/\sqrt{5}$
$1.24 = t_{4,\alpha/2} \cdot (5828)$
$t_{4,\alpha/2} = 1.24/.5828 = 2.128$ [closest entry = 2.132], $\alpha/2 = .05$, $\alpha = .10$
The level of confidence is $1 - \alpha = 1 - .10 = .90 = 90\%$.

6-4 Determining Sample Size Required to Estimate μ

1. $E = 1.5$ and $\alpha = .02$
$n = [z_{.01} \cdot \sigma/E]^2$
$= [2.327 \cdot 15/1.5]^2$
$= 541.49$ rounded up to 542

3. $E = 2$ and $\alpha = .04$
$n = [z_{.02} \cdot \sigma/E]^2$
$= [2.05 \cdot 12.46/2]^2$
$= 163.11$ rounded up to 164

5. $E = 500$ and $\alpha = .05$
$n = [z_{.025} \cdot \sigma/E]^2$
$= [1.960 \cdot 6250/500]^2$
$= 600.25$ rounded up to 601

7. $E = .25$ and $\alpha = .04$
$n = [z_{.02} \cdot \sigma/E]^2$
$= [2.05 \cdot 1.87/.25]^2$
$= 235.13$ rounded up to 236

9. estimated maximum age = 10 years; estimated minimum age = 0 years; $R = 10 - 0 = 10$
$\sigma \approx R/4 = 10/4 = 2.5$
$E = .25$ and $\alpha = .04$
$n = [z_{.02} \cdot \sigma/E]^2$
$= [2.05 \cdot 2.5/.25]^2$
$= 420.25$ rounded up to 421

11. a. for the $n = 53$ Wednesdays, $s = .1346$
$E = .02$ and $\alpha = .02$
$n = [z_{.01} \cdot \sigma/E]^2$
$= [2.327 \cdot .1346/.02]^2$
$= 245.26$ rounded up to 246
b. for the $n = 52$ Sundays, $s = .2000$
$E = .02$ and $\alpha = .02$
$n = [z_{.01} \cdot \sigma/E]^2$
$= [2.327 \cdot .2000/.02]^2$
$= 541.49$ rounded up to 542

c. The results of parts (a) and (b) indicate that the required sample size can vary significantly according to the estimate used for σ. This illustrates the importance and necessity of planning for the worst possible scenario by using a large estimate for σ. It happens that Wednesday (s = .1346) was the least variable day. One might be tempted to use the most variable day, Saturday (s = .2900), to produce a conservative estimate of n = 1139. The best estimate, however, would probably be obtained by using all 365 days (s = .2125).

13. E = 1.5
    α = .02
    N = 200
    n = $[N\sigma^2(z_{.01})^2]/[(N-1)E^2 + \sigma^2(z_{.01})^2]$
      = $[(200)(15)^2(2.327)^2]/[(199)(1.5)^2 + (15)^2(2.327)^2]$
      = [243671.8]/[1666.109]
      = 146.25 rounded up to 147

15. for the n = 70 sugar weights, s = .0740
    the required sample size formula uses z (with s for σ)
    E = $z_{.025} \cdot \sigma/\sqrt{n}$
      = $1.96 \cdot (.0740)/\sqrt{843}$
      = .005 grams

6-5 Estimating a Population Proportion

IMPORTANT NOTE: When calculating confidence intervals using the formula
    $\hat{p} \pm E$
    $\hat{p} \pm z_{\alpha/2}\sqrt{\hat{p}\hat{q}/n}$
do not round off in the middle of the problem. This may be accomplished conveniently on most calculators having a memory as follows.
(1) Calculate $\hat{p}$ = x/n and STORE the value
(2) Calculate E as 1 - RECALL = * RECALL = $\div$ n = $\sqrt{}$ * $z_{\alpha/2}$ =
(3) With the value of E showing on the display, the upper confidence limit is calculated by + RECALL.
(4) With the value of the upper confidence limit showing on the display, the lower confidence limit is calculated by - RECALL $\pm$ + RECALL
You must become familiar with your own calculator. [Do your homework using the same type of calculator you will be using for the exams.] The above procedure works on most calculators; make certain you understand why it works and verify whether it works on your calculator. If it does not seem to work on your calculator, or if your calculator has more than one memory so that you can STORE both $\hat{p}$ and E at the same time, ask your instructor for assistance.

NOTE: It should be true that $0 \le \hat{p} \le 1$ and that E $\le$ .5 [usually, much less than .5]. If such is not the case, an error has been made.

1. let L = the lower confidence limit
        U = the upper confidence limit
    $\hat{p}$ = (L + U)/2 = (.800 + .840)/2 = 1.640/2 = .820
    E = (U - L)/2 = (.840 - .800)/2 = .040/2 = .020

3. let L = the lower confidence limit
        U = the upper confidence limit
    $\hat{p}$ = (L + U)/2 = (.432 + .455)/2 = .4435
    E = (U - L)/2 = (.455 - .432)/2 = .023/2 = .0115

5. $\hat{p} = x/n = 400/500 = .80$
 $\alpha = .05$
 $E = z_{.025}\sqrt{\hat{p}\hat{q}/n}$
 $= 1.960\sqrt{(.80)(.20)/500}$
 $= .0351$

7. $\hat{p} = x/n = 237/1068 = .222$
 $\alpha = .02$
 $E = z_{.01}\sqrt{\hat{p}\hat{q}/n}$
 $= 2.327\sqrt{(.222)(.778)/1068}$
 $= .0296$

9. $\hat{p} = x/n = .35$
 $\alpha = .05$
 $E = z_{.025}\sqrt{\hat{p}\hat{q}/n}$
 $= 1.960\sqrt{(.35)(.65)/200}$
 $= .0661$

11. $\hat{p} = x/n = 150/750 = .200$
 $\alpha = .01$
 $\hat{p} \pm z_{.001}\sqrt{\hat{p}\hat{q}/n}$
 $.200 \pm 2.575\sqrt{(.200)(.800)/750}$
 $.200 \pm .038$
 $.162 < p < .238$

13. $\hat{p} = x/n = 222/1357 = .164$
 $\alpha = .05$
 $\hat{p} \pm z_{.025}\sqrt{\hat{p}\hat{q}/n}$
 $.164 \pm 1.960\sqrt{(.164)(.836)/1357}$
 $.164 \pm .020$
 $.144 < p < .183$

15. $\hat{p}$ unknown, use $\hat{p} = .5$
 $n = [(z_{.005})^2\hat{p}\hat{q}]/E^2$
 $= [(2.575)^2(.5)(.5)]/(.025)^2$
 $= 2652.25$, rounded up to 2653

17. $\hat{p} = .15$
 $n = [(z_{.025})^2\hat{p}\hat{q}]/E^2$
 $= [(1.960)^2(.15)(.82)]/(.03)^2$
 $= 544.23$, rounded up to 545

19. a. We are 99% certain that the interval from 93.053% to 94.926% contains the true percentage of U.S. households having telephones.

 b. Based on the preceding result, pollsters should not be too concerned about results from surveys conducted by telephone if (1) proper techniques are used and (2) issues that would elicit differing responses from phone and non-phone households are not at stake. Proper techniques involve follow up of non-responses (due to work schedules, etc.), making certain the appropriate person in the household (and not just whoever answers the phone, etc.) is responding, etc. Issues involving agricultural policies, for example, would miss the opinions of the Amish (who are primarily involved in farming and do not have phones for religious reasons). In general, however, one can probably assume that those households with phones would be the households most likely to buy, use or be familiar with the subject of the survey.

21. a. $\hat{p} = x/n = \underline{111}/1233 = .090 = 9.0\%$
    b. $\hat{p} \pm z_{.025}\sqrt{\hat{p}\hat{q}/n}$
       $.090 \pm 1.960\sqrt{(.090)(.910)/1233}$
       $.090 \pm .016$
       $.074 < p < .106$
       $7.4\% < p < 10.6\%$

23. a. $\hat{p} = x/n = 5.2\% = .052$
       $\alpha = .05$
       $\hat{p} \pm z_{.025}\sqrt{\hat{p}\hat{q}/n}$
       $.052 \pm 1.960\sqrt{(.052)(.948)/8411}$
       $.052 \pm .005$
       $.047 < p < .057$
       $4.7\% < p < 5.7\%$
       NOTE: Since $\hat{p} = x/n = 5.2\%$ was given to only the nearest tenth of a percent, the usual 3 significant digit confidence interval is not possible. Additional accuracy cannot by obtained without knowing the true value of x -- and any x from 434 to 441 inclusive would give an x/8411 that rounds to .052.
    b. Yes; since the intervals are far from overlapping, the results are substantially different from those in part (a). In the presence of an explosion or fire on the ground, the likelihood of a pilot fatality is significantly increased.

25. a. $\hat{p} = x/n = 7/221 = .032$
       NOTE: Since $\hat{p} = x/n = 3.2\%$ was given, we may assume $x = \hat{p} \cdot n = .032 \cdot 221 = 7$. No other integer value x/221 rounds to .032.
       $\alpha = .01$
       $\hat{p} \pm z_{.005}\sqrt{\hat{p}\hat{q}/n}$
       $.032 \pm 2.575\sqrt{(.032)(.968)/221}$
       $.032 \pm .030$
       $.00134 < p < .0620$
       $0.134\% < p < 6.20\%$
    b. Since 1.8% is within the interval in part (a), it cannot be ruled out as the correct value for the true proportion of Ziac users that experience dizziness. The test does not provide evidence that Ziac users experience any more dizziness than non-users -- i.e., the test does not provide evidence that dizziness is an adverse reaction to Ziac.

27. a. $\hat{p} = .82$
       $n = [(z_{.01})^2\hat{p}\hat{q}]/E^2$
       $= [(2.327)^2(.82)(.18)]/(.05)^2$
       $= 319.70$, rounded up to 320
    b. $\hat{p}$ unknown, use $\hat{p} = .5$
       $n = [(z_{.01})^2\hat{p}\hat{q}]/E^2$
       $= [(2.327)^2(.5)(.5)]/(.05)^2$
       $= 541.49$, rounded up to 542

29. a. $\hat{p} = .86$
       $n = [(z_{.03})^2\hat{p}\hat{q}]/E^2$
       $= [(1.88)^2(.86)(.14)]/(.03)^2$
       $= 472.82$, rounded up to 473
    b. $\hat{p}$ unknown, use $\hat{p} = .5$
       $n = [(z_{.03})^2\hat{p}\hat{q}]/E^2$
       $= [(1.88)^2(.5)(.5)]/(.03)^2$
       $= 981.78$, rounded up to 982

31. a. $\hat{p} = x/n = 7/80 = .0875$
    $\alpha = .10$
    $\hat{p} \pm z_{.01}\sqrt{\hat{p}\hat{q}/n}$
    $.0875 \pm 1.645\sqrt{(.0875)(.9125)/80}$
    $.0875 \pm .0520$
    $.036 < p < .139$
 b. $\hat{p} = .0875$
    $n = [(z_{.02})^2\hat{p}\hat{q}]/E^2$
    $= [(2.05)^2(.0875)(.9125)]/(.03)^2$
    $= 372.83$, rounded up to 373
 c. Yes; since $0.25\% = .0025$ is not in within the confidence interval in part (a), we may safely conclude that women have a lower rate of red/green color blindness than men do.

32. a. $\hat{p} = x/n = 22\% = .22$
    $\alpha = .03$
    $\hat{p} \pm z_{.015}\sqrt{\hat{p}\hat{q}/n}$
    $.22 \pm 2.17\sqrt{(.22)(.78)/4000}$
    $.22 \pm .01$
    $.21 < p < .23$
    NOTE: Since $\hat{p} = x/n = 22\%$ was given to only the nearest whole percent, the usual 3 significant digit confidence interval is not possible. Additional accuracy cannot by obtained without knowing the true value of x -- and any x between 860 and 900 would give an x/4000 that rounds to .22.
 b. $\hat{p}$ unknown, use $\hat{p} = .5$
    $n = [(z_{.005})^2\hat{p}\hat{q}]/E^2$
    $= [(2.575)^2(.5)(.5)]/(.005)^2$
    $= 66306.25$, rounded up to 66307
 c. Yes; since $11\% = .11$ is not in within the confidence interval in part (a), we may safely conclude that 60 Minutes had a greater share of the audience than the wrestling special did. No; exposing professional wrestling was probably not necessary.

33. $\hat{p} = x/n = 21/100 = .210$
    $\alpha = .05$
    $\hat{p} \pm z_{.025}\sqrt{\hat{p}\hat{q}/n}$
    $.210 \pm 1.960\sqrt{(.210)(.790)/100}$
    $.210 \pm .080$
    $.130 < p < .290$
    $13.0\% < p < 29.0\%$
    Yes; this result is consistent with the 20% rate reported by the candy maker.

35. $\hat{p}$ unknown, use $\hat{p} = .5$
    $n = [N\hat{p}\hat{q}(z_{.005})^2]/[\hat{p}\hat{q}(z_{.005})^2 + (N-1)E^2]$
    $= [(5000)(.5)(.5)(2.575)^2]/[(.5)(.5)(2.575)^2 + (4999)(.005)^2]$
    $= [8288.281]/[1.782631]$
    $= 4649.46$, rounded up to 4650

37. $\hat{p} = x/n = 80\% = .80$
    $\alpha = .05$
    $\hat{p} \pm z_{.025}\sqrt{\hat{p}\hat{q}/n}$
    $.80 \pm 1.960\sqrt{(.80)(.20)/2250}$
    $.80 \pm .02$
    $.78 < p < .82$
    $78\% < p < 82\%$
    NOTE: Since $\hat{p} = x/n = 80\%$ was given to only the nearest whole percent, the usual 3

significant digit confidence interval is not possible. Additional accuracy cannot by obtained without knowing the true value of x -- and any x between 1789 and 1811 inclusive would give an x/4000 that rounds to .80.

The result is not valid because the sample was self-selected and is not necessarily representative of the population -- whether the intended population was all "people" (households?, phone owners?) or all people with Internet access (the population from which he actually sampled).

39. $\hat{p} = x/n = 630/750 = .840$
$\alpha = .05$
$\hat{p} - z_{.05}\sqrt{\hat{p}\hat{q}/n}$
$.840 - 1.645\sqrt{(.840)(.160)/750}$
$.840 - .022$
$.818 < p$
expressed as $p > 81.8\%$; the desired figure is 81.8%

41. $\hat{p} = x/n = 95/100 = .950$
$\alpha = .01$
$\hat{p} \pm z_{.005}\sqrt{\hat{p}\hat{q}/n}$
$.950 \pm 2.575\sqrt{(.950)(.050)/100}$
$.950 \pm .056$
$.894 < p < 1.006$, with the upper limit truncated to $.894 < p < 1.000$
NOTE: This occurs because the normal approximation to the binomial is only marginally appropriate here since $np = 100(.95) = 5 \geq 5$ (just barely!). The upper inequality is still "$<$" and not "$\leq$" since the 5 times that tails came up indicate that the probability of heads is definitely less than 100%.

6-6 Estimating a Population Variance

1. $\chi_L^2 = \chi_{29,.975}^2 = 16.047$
$\chi_R^2 = \chi_{29,.025}^2 = 45.722$

3. $\chi_L^2 = \chi_{49,.995}^2 = 27.991$
$\chi_R^2 = \chi_{49,.005}^2 = 79.490$
NOTE: Use df $= 50$, the closest entry.

5. $(n-1)s^2/\chi_{14,.025}^2 < \sigma^2 < (n-1)s^2/\chi_{14,.975}^2$
$(14)(108)^2/26.119 < \sigma^2 < (14)(108)^2/5.629$
$6252 < \sigma^2 < 29009$
$79 < \sigma < 170$

7. $(n-1)s^2/\chi_{24,.05}^2 < \sigma^2 < (n-1)s^2/\chi_{24,.95}^2$
$(24)(12)^2/36.415 < \sigma^2 < (24)(12)^2/13.848$
$94.9 < \sigma^2 < 249.6$
$10 < \sigma < 16$

9. From the upper right section of Table 6-3, n $= 767$.

11. From the lower left section of Table 6-3, n $= 1,401$.

NOTE: When raw scores are available, $\bar{x}$ and s should be calculated as the primary descriptive statistics -- but use the unrounded value of s^2 in the confidence interval formula. In addition, always make certain that the confidence interval for σ includes the calculated value of s.

13. summary information

$n = 18$	$\bar{x} = 3787.0$
$\Sigma x = 68,166$	$s^2 = 3066.82$
$\Sigma x^2 = 258,196,778$	$s = 55.4$

$$(n-1)s^2/\chi^2_{17,.005} < \sigma^2 < (n-1)s^2/\chi^2_{17,.995}$$
$$(17)(3066.82)/35.718 < \sigma^2 < (17)(3066.82)/5.697$$
$$1459.65 < \sigma^2 < 9151.47$$
$$38.2 < \sigma < 95.7 \quad (mL)$$

No; the interval indicates 99% confidence that $\sigma > 30$.

15. a. $s \approx (x_n - x_1)/4 = (1.033 - .856)/4 = .177/4 = .04425$

b. summary information

$n = 33$	$\bar{x} = .9128$
$\Sigma x = 30.123$	$s^2 = .0015616$
$\Sigma x^2 = 27.546793$	$s = .0395$

$$(n-1)s^2/\chi^2_{32,.01} < \sigma^2 < (n-1)s^2/\chi^2_{32,.99}$$
$$(32)(.0015616)/50.892 < \sigma^2 < (32)(.0015616)/14.954$$
$$.0009819 < \sigma^2 < .0033416$$
$$.0313 < \sigma < .0578 \quad (grams)$$

NOTE: The χ^2 values used are those for 30 df, which is the closest entry from Table A-4. Since df = n-1 is a part of the confidence interval formula, and since the expected value of the χ^2 statistic is n-1, some instructors and text books recommend using in the formula a value for n-1 that agrees with the df of the tabled χ^2

c. Yes; .04425 is within the confidence interval.

17. a. summary information

$n = 10$	$\bar{x} = 7.15$
$\Sigma x = 71.5$	$s^2 = .2272$
$\Sigma x^2 = 513.27$	$s = .48$

$$(n-1)s^2/\chi^2_{9,.025} < \sigma^2 < (n-1)s^2/\chi^2_{9,.975}$$
$$(9)(.2272)/19.023 < \sigma^2 < (9)(.2272)/2.700$$
$$.1075 < \sigma^2 < .7573$$
$$.33 < \sigma < .87 \quad (minutes)$$

b. summary information

$n = 10$	$\bar{x} = 7.15$
$\Sigma x = 71.5$	$s^2 = 3.3183$
$\Sigma x^2 = 541.09$	$s = 1.82$

$$(n-1)s^2/\chi^2_{9,.025} < \sigma^2 < (n-1)s^2/\chi^2_{9,.975}$$
$$(9)(3.3183)/19.023 < \sigma^2 < (9)(3.3183)/2.700$$
$$1.5699 < \sigma^2 < 11.0610$$
$$1.25 < \sigma < 3.33 \quad (minutes)$$

c. Yes, there is a difference. The multiple line system exhibits more variability among the waiting times. The greater consistency among the single line waiting times seems fairer to the customers and more professional.

19. a. summary information

$n = 25$	$\bar{x} = 78.898$
$\Sigma x = 1972.45$	$s^2 = 192.373$
$\Sigma x^2 = 160239.3228$	$s = 13.87$

$$(n-1)s^2/\chi^2_{24,.01} < \sigma^2 < (n-1)s^2/\chi^2_{24,.99}$$
$$(24)(192.373)/42.980 < \sigma^2 < (24)(192.373)/10.856$$
$$107.421 < \sigma^2 < 425.291$$
$$10.364 < \sigma < 20.623 \quad \text{(dollars)}$$

b. summary information

$n = 15$	$\bar{x} = 42.150$
$\Sigma x = 632.25$	$s^2 = 249.279$
$\Sigma x^2 = 30139.2424$	$s = 15.79$

$$(n-1)s^2/\chi^2_{14,.01} < \sigma^2 < (n-1)s^2/\chi^2_{14,.99}$$
$$(14)(249.279)/29.141 < \sigma^2 < (14)(249.279)/4.660$$
$$119.759 < \sigma^2 < 748.907$$
$$10.943 < \sigma < 27.366 \quad \text{(dollars)}$$

c. There does not appear to be a significant difference in the variability of the two groups.

21. a. The given interval $2.8 < \sigma < 6.0$
$$7.84 < \sigma^2 < 36.00$$
and the usual calculations $(n-1)s^2/\chi^2_{19,\alpha/2} < \sigma^2 < (n-1)s^2/\chi^2_{19,1-\alpha/2}$
$$(19)(3.8)^2/\chi^2_{19,\alpha/2} < \sigma^2 < (19)(3.8)^2/\chi^2_{19,1-\alpha/2}$$
$$274.36/\chi^2_{19,\alpha/2} < \sigma^2 < 274.36/\chi^2_{19,1-\alpha/2}$$

imply that $7.84 = 274.37/\chi^2_{19,\alpha/2}$ and $36.00 = 274.36/\chi^2_{19,1-\alpha/2}$
$$\chi^2_{19,\alpha/2} = 274.36/7.84 \qquad\qquad \chi^2_{19,1-\alpha/2} = 274.36/36.00$$
$$= 34.99 \qquad\qquad\qquad\qquad = 7.62$$

The closest entries in Table A-4 are $\chi^2_{19,\alpha/2} = 34.805$ and $\chi^2_{19,1-\alpha/2} = 7.633$
which imply

$\alpha/2 = .01$	$1 - \alpha/2 = .99$
$\alpha = .02$	$\alpha/2 = .01$
	$\alpha = .02$

The level of confidence is therefore is $1-\alpha = 98\%$.

b. $(n-1)s^2/\chi^2_{11,.025} < \sigma^2 < (n-1)s^2/\chi^2_{11,.975}$

using the lower endpoint   OR   using the upper endpoint
$$(11)s^2/21.920 = (19.1)^2 \qquad\qquad (11)s^2/3.816 = (45.8)^2$$
$$s^2 = 726.97 \qquad\qquad\qquad\qquad s^2 = 727.69$$
$$s = 27.0 \qquad\qquad\qquad\qquad\quad s = 27.0$$

Review Exercises

1. a. $n > 30$, use $\underline{z}$ (with s for σ)
$$\bar{x} \pm z_{.01} \cdot \sigma/\sqrt{n}$$
$$4.4038 \pm 2.327 \cdot 2.8440/\sqrt{52}$$
$$4.4038 \pm .9177$$
$$3.5 < \mu < 5.3$$

b. $n > 30$, use $\underline{z}$ (with s for σ)
$$\bar{x} \pm z_{.01} \cdot \sigma/\sqrt{n}$$
$$1.7308 \pm 2.327 \cdot 1.7502/\sqrt{52}$$
$$1.7308 \pm .5648$$
$$1.2 < \mu < 2.3$$

c. Since the intervals do not overlap we can be confident that the population values are truly different. The Dvorak ratings are significantly lower than the QWERTY ratings.

d. The methods of Section 6-6 require that the populations have normal distributions. These population distributions are not normal -- they are truncated on the left at 0 and skewed to the right.

2. $\hat{p}$ unknown, use $\hat{p} = .5$
$n = [(z_{.015})^2\hat{p}\hat{q}]/E^2$
$\quad = [(2.17)^2(.5)(.5)]/(.02)^2$
$\quad = 2943.06$, rounded up to 2944

3. a. $E = 2$ and $\alpha = .03$
$n = [z_{.015} \cdot \sigma/E]^2$
$\quad = [2.17 \cdot 15.7/2]^2$
$\quad = 290.17$, rounded up to 291

   b. The closest colleges are not necessarily representative of all colleges, and they may differ in ways that would affect the scores -- size of the classes, admission standard, types of majors offered, etc. A convenience sample is not an appropriate substitute for a random sample.

4. a. $n \le 30$ and σ unknown, use t
$\bar{x} \pm t_{24,.025} \cdot s/\sqrt{n}$
$7.01 \pm 2.064 \cdot 3.741/\sqrt{25}$
7.01 ± 1.54
$5.47 < \mu < 8.55$ (years)

   b. $\quad (n-1)s^2/\chi^2_{24,.025} < \sigma^2 < (n-1)s^2/\chi^2_{24,.975}$
$(24)(3.74)^2/39.364 < \sigma^2 < (24)(3.74)^2/12.401$
$\quad\quad 8.528 < \sigma^2 < 27.071$
$\quad\quad 2.92 < \sigma < 5.20$ (years)

5. $\hat{p} = x/n = 61/94 = .649$
$\alpha = .05$
$\hat{p} \pm z_{.025}\sqrt{\hat{p}\hat{q}/n}$
$.649 \pm 1.960\sqrt{(.649)(.351)/94}$
$.649 \pm .096$
$.552 < p < .745$
$55.2\% < p < 74.5\%$
We are 95% certain that the interval from 55.2% to 74.5% contains the true proportion of automobile air bag fatalities that involved improper belting.

6. $\hat{p} = .93$
$n = [(z_{.01})^2\hat{p}\hat{q}]/E^2$
$\quad = [(2.327)^2(.93)(.07)]/(.04)^2$
$\quad = 220.32$, rounded up to 221

7. preliminary information: $n = 16$, $\Sigma x = 1162.6$, $\Sigma x^2 = 84577.34$
$\bar{x} = (\Sigma x)/n = 1162.6/16 = 72.66$ [do not round, store all the digits in the calculator]
$s^2 = [n(\Sigma x^2) - (\Sigma x)^2]/[n(n-1)]$
$\quad = [16(84577.34) - (1162.6)^2]/[16(15)]$
$\quad = 6.661$
$s = 2.581$ [do not round, store all the digits in the calculator]
$n \le 30$ and σ unknown, use t
$\bar{x} \pm t_{15,.025} \cdot s/\sqrt{n}$
$72.66 \pm 2.132 \cdot 2.581/\sqrt{16}$
72.66 ± 1.38
$71.29 < \mu < 74.04$ (cm)

8. $n > 30$, use z (with s for σ)
   $\bar{x} \pm z_{.005} \cdot \sigma/\sqrt{n}$
   $558 \pm 2.575 \cdot 139/\sqrt{67}$
   558 ± 44
   $514 < \mu < 602$

9. $E = 4$ and $\alpha = .05$
   $n = [z_{.025} \cdot \sigma/E]^2$
   $\quad = [1.960 \cdot 41.0/4]^2$
   $\quad = 403.61$, rounded up to 404

10. $\hat{p} = .24$
    $\alpha = .01$
    $\hat{p} \pm z_{.005}\sqrt{\hat{p}\hat{q}/n}$
    $.24 \pm 2.575\sqrt{(.24)(.76)/1998}$
    $.24 \pm .02$
    $.22 < p < .26$
    $22\% < p < 26\%$
    NOTE: Since $\hat{p}$ was given with only two decimal accuracy and the actual value of x was not given, the final answer is limited to two decimal accuracy. Any x from 470 to 489 inclusive rounds to 24% with varying digits for the next decimal point.

Cumulative Review Exercises

1. Begin by making a stem-and-leaf plot and calculating summary statistics.

   ```
   10 | 5
   11 |
   11 | 599                n = 9
   12 | 3                  Σx = 1089
   12 | 5788               Σx² = 132223
   ```

   $n = 9$
   $\Sigma x = 1089$
   $\Sigma x^2 = 132223$

   a. $\bar{x} = (\Sigma x)/n = (1089)/9 = 121.0$ lbs
   b. $\tilde{x} = 123.0$ lbs
   c. M = 119, 128 (bi-modal)
   d. m.r. = $(105 + 128)/2 = 116.5$ lbs
   e. R = 128 - 105 = 23 lbs
   f. $s^2 = [n(\Sigma x^2) - (\Sigma x)^2]/[n(n-1)]$
      $\quad = [9(132223) - (1089)^2]/[9(8)]$
      $\quad = 56.75$ lbs^2
   g. s = 7.5 lbs
   h. for $Q_1 = P_{25}$, $L = (25/100)(9) = 2.25$, round up to 3
      $Q_1 = x_3 = 119$ lbs
   i. for $Q_2 = P_{50}$, $L = (50/100)(9) = 4.50$, round up to 5
      $Q_2 = x_5 = 123$ lbs
   j. for $Q_3 = P_{75}$, $L = (75/100)(9) = 6.75$, round up to 7
      $Q_3 = x_7 = 127$ lbs
   k. ratio, since differences are consistent and there is a meaningful zero
   l.

```
   105                        119    123    127 128
```

m. $n \le 30$ and σ unknown, use t

$\bar{x} \pm t_{8,.005} \cdot s/\sqrt{n}$

$121.0 \pm 3.355 \cdot 7.5/\sqrt{9}$

121.0 ± 8.4

$112.6 < \mu < 129.4$ (lbs)

n. Since both the stem-and-leaf plot and the boxplot indicate that the weights do not appear to come from a normal population, the methods of Section 6-6 are not appropriate. It appears the distribution of the weights of supermodels is truncated at some upper limit and skewed to the left. Even though a confidence interval for σ should not be constructed, the following interval is given as a review of the technique.

$(n-1)s^2/\chi^2_{8,.005} < \sigma^2 < (n-1)s^2/\chi^2_{8,.995}$

$(8)(56.75)/21.955 < \sigma^2 < (8)(56.75)/1.344$

$20.68 < \sigma^2 < 337.80$

$4.5 < \sigma < 18.4$ (lbs)

o. $E = 2$ and $\alpha = .01$

$n = [z_{.005} \cdot \sigma/E]^2$

$= [2.575 \cdot 7.5/2]^2$

$= 94.07$, rounded up to 95

p. For the general female population, an unusually low weight would be one below 85 lbs (i.e., more than 2σ below μ). Individually, none of the supermodels has an unusually low weight. As a group, however, they are each well below the general population mean weight and appear to weigh substantially less than the general population -- as evidenced by their mean weight as given by the point estimate in part (a) and the interval estimate in part (m).

2. a. binomial: $n = 200$ and $p = .25$

the normal approximation is appropriate since

$np = 200(.25) = 50 \ge 5$

$n(1-p) = 200(.75) = 150 \ge 5$

use $\mu = np = 200(.25) = 50$

$\sigma = \sqrt{n(p)(1-p)} = \sqrt{200(.25)(.75)} = 6.214$

$P(x \ge 65)$

$= P_c(x > 64.5)$

$= P(z > 2.37)$

$= .5000 - .4911$

$= .0089$

b. $\hat{p} = 65/200 = .325$

$\alpha = .05$

$\hat{p} \pm z_{.025}\sqrt{\hat{p}\hat{q}/n}$

$.325 \pm 1.960\sqrt{(.325)(.675)/200}$

$.325 \pm .065$

$.260 < p < .390$

c. No; the expert's value of .25 does not seem correct for two reasons.

From part (a): if the expert is correct, then the probability of getting the sample obtained is very small -- less than 1%.

From part (b): according to the sample obtained, there is 95% confidence that the interval from .26 to .39 includes the true value -- that is, there is 95% confidence that the value is not .25.

Chapter 7

Hypothesis Testing

7-2 Fundamentals of Hypothesis testing

1. Conclude that the coin is not fair. Since the occurrence of 27 heads in 30 tosses would be a rare event if the coin were fair, we reject the claim that it is fair.

3. Do not conclude that women who eat blue M&M's have a better chance of having a baby boy. Since the occurrence of 27 boys in 50 births is not a rare event if the candy has no effect, it is not significant evidence to support the stated claim.

5. claim: Lisa's cars are stolen at the "normal" rate.
   conclusion: Lisa's cars are stolen at a higher rate than normal, which may be for any number of reasons -- the models she buys or places she drives, her carefulness (or lack thereof), etc.

7. claim: The experimental vaccine is effective against Lyme disease.
   conclusion: The evidence is not convincing. The difference between the 2.2% rate of infection for those receiving the placebo and the 1.8% rate for those receiving the vaccine is not significant -- such a difference in the sample rates could easily occur by chance when the population rates are identical and the vaccine has no effect.

9. original statement: $\mu > 30$
   competing idea: $\mu \leq 30$ (contains the equality; must be H_o)
   H_o: $\mu \leq 30$
   H_1: $\mu > 30$

11. original statement: $p < .5$
    competing idea: $p \geq .5$ (contains the equality; must be H_o)
    H_o: $p \geq .5$
    H_1: $p < .5$

13. original statement: $\sigma < 2.8$
    competing idea: $\sigma \geq 2.8$ (contains the equality; must be H_o)
    H_o: $\sigma \geq 2.8$
    H_1: $\sigma < 2.8$

15. original statement: $\mu \geq 12$ (contains the equality; must be H_o)
    competing idea: $\mu < 12$
    H_o: $\mu \geq 12$
    H_1: $\mu < 12$

17. two-tailed test; place $\alpha/2$ in each tail
    use A = .5000 - $\alpha/2$ = .5000 -.0250 = .4750
    critical values are $\pm z_{\alpha/2}$ = $\pm z_{.025}$ = ± 1.960

19. right-tailed test; place α in the upper tail
    use A = .5000 - α = .5000 - .0200 = .4800 [closest entry = .4798]
    critical value is $+z_\alpha$ = $+z_{.02}$ = $+2.05$

21. two-tailed test; place $\alpha/2$ in each tail
   use A = .5000 - $\alpha/2$ = .5000 -.0200 = .4800 [closest entry = .4798]
   critical values are $\pm z_{\alpha/2} = \pm z_{.02} = \pm 2.05$

23. right-tailed test; place α in the upper tail
   use A = .5000 - α = .5000 -.0600 = .4400 [half-way between .4394 and .4406]
   critical value is $+z_{\alpha} = +z_{.06} = +1.555$

25. $z_{\bar{x}} = (\bar{x} - \mu)/(\sigma/\sqrt{n})$ [use s to estimate σ]
   $= (.83 - .21)/(.24/\sqrt{32})$
   $= (.62)/(.0424)$
   $= 14.614$

27. $z_{\bar{x}} = (\bar{x} - \mu)/(\sigma/\sqrt{n})$ [use s to estimate σ]
   $= (.83 - 1.39)/(.16/\sqrt{123})$
   $= (-.56)/(.0144)$
   $= -38.817$

29. original claim: $\mu = 60$ (contains the equality; must be H_o)
   competing idea: $\mu \neq 60$
      H_o: $\mu = 60$
      H_1: $\mu \neq 60$
   initial conclusion: Reject H_o.
   final conclusion: There is sufficient evidence to reject the claim that women have a mean height of 60.0 inches.

31. original claim: $\mu < 12$
   competing idea: $\mu \geq 12$ (contains the equality; must be H_o)
      H_o: $\mu \geq 12$
      H_1: $\mu < 12$
   initial conclusion: Fail to reject H_o.
   final conclusion: There is not sufficient evidence to support the claim that the mean amount of Coke in 12 ounce cans is actually less than 12 ounces.

33. original claim: $\mu = 12$ (contains the equality; must be H_o)
   competing idea: $\mu \neq 12$
      H_o: $\mu = 12$
      H_1: $\mu \neq 12$
   type I error: Reject the original claim that $\mu = 12$ when $\mu = 12$ is actually true.
   type II error: Fail to reject the original claim that $\mu = 12$ when $\mu \neq 12$ is actually true.

35. original claim: $\mu < 100$
   competing idea: $\mu \geq 100$ (contains the equality; must be H_o)
      H_o: $\mu \geq 100$
      H_1: $\mu < 100$
   type I error: Reject the null hypothesis, and conclude the original claim that $\mu < 100$ is true when $\mu \geq 100$ is actually true.
   type II error: Fail to reject the null hypothesis, and fail to conclude the original claim that $\mu < 100$ is true when $\mu < 100$ is actually true.

37. The statistical test is based on rejecting results that represent an extreme departure from H_o. When testing the null hypothesis that $\mu \geq 100$, it is necessary to conduct the test based on the assumption that $\mu = 100$ instead of $\mu \geq 100$ for two reasons:
   (1) The test assumption must be specific so that we can decide exactly what we expect and measure how much the results depart from that expectation. While $\mu = 100$ is specific and

we know to expect $\bar{x}$'s normally distributed around 100, $\mu \geq 100$ is not specific.
(2) Testing $\mu = 100$ is sufficient. If there is enough evidence to reject $\mu = 100$ to conclude that $\mu < 100$, there is certainly enough evidence to reject any other specific value greater than 100.

39. Mathematically, in order for α to equal 0 the magnitude of the critical value would have to be infinite. Practically, the only way never to make a type I error is to always fail to reject H_o. From either perspective, the only way to achieve $\alpha = 0$ is to never reject H_o no matter how extreme the sample data might be.

7-3 Testing a Claim about a Mean: Large Samples

1. H_o: $\mu = 100$
   H_1: $\mu \neq 100$
   the test is two-tailed
   P-value = P(getting a sample result at least as extreme as the one obtained)
   $= 2 \cdot P(z < -1.03)$
   $= 2 \cdot (.5000 - .3485)$
   $= 2 \cdot (.1515)$
   $= .3030$

3. H_o: $\mu \geq 500$
   H_1: $\mu < 500$
   the test is left-tailed
   P-value = P(getting a sample result at least as extreme as the one obtained)
   $= P(z < -12.67)$
   $= .5000 - .4999$
   $= .0001$

5. original claim: $\mu = 75$ [n > 30, use z (with s for σ)]
   H_o: $\mu = 75$
   H_1: $\mu \neq 75$
   $\alpha = .05$
   a. test statistic: $z_{\bar{x}} = (\bar{x} - \mu)/\sigma_{\bar{x}} = (77 - 75)/(15/\sqrt{500}) = 2/.671 = 2.981$
   b. critical values: $z = \pm z_{.025} = \pm 1.960$
   c. P-value: $2 \cdot P(z > 2.98) = 2 \cdot (.5000 - .4986) = 2 \cdot (.0014) = .0028$
   d. final conclusion: Reject H_o; there is sufficient evidence to reject the claim that $\mu = 75$ and conclude that $\mu \neq 75$ (in fact, that $\mu > 75$).

7. original claim: $\mu > 2.25$ [n > 30, use z (with s for σ)]
   H_o: $\mu \leq 2.50$
   H_1: $\mu > 2.50$
   $\alpha = .02$
   a. test statistic: $z_{\bar{x}} = (\bar{x} - \mu)/\sigma_{\bar{x}} = (2.35 - 2.25)/(.80/\sqrt{150}) = .10/.0653 = 1.53$
   b. critical value: $z = +z_{.02} = +2.05$
   c. P-value: $P(z > 1.53) = .5000 - .4370 = .0630$
   d. final conclusion: Do not reject H_o; there is not sufficient evidence to support the claim that $\mu > 2.25$.

9. original claim: $\mu < 98.6$ [$n > 30$, use z (with s for σ)]
 H_o: $\mu \geq 98.6$
 H_1: $\mu < 98.6$
 $\alpha = .01$
 C.R. $z < -z_{.01} = -2.327$
 calculations:
   $z_{\bar{x}} = (\bar{x} - \mu)/\sigma_{\bar{x}}$
   $= (98.20 - 98.6)/(.62/\sqrt{106})$
   $= -.40/.0602$
   $= -6.642$
 conclusion:
   Reject H_o; there is sufficient evidence to conclude that $\mu < 98.6$.

11. original claim: $\mu \neq 92.84$ [$n > 30$, use z (with s for σ)]
 H_o: $\mu = 92.84$
 H_1: $\mu \neq 92.84$
 $\alpha = .05$
 C.R. $z < -z_{.025} = -1.960$
    $z > z_{.025} = 1.960$
 calculations:
   $z_{\bar{x}} = (\bar{x} - \mu)/\sigma_{\bar{x}}$
   $= (92.67 - 92.84)/(1.79/\sqrt{40})$
   $= -.17/.283$
   $= -.601$
 conclusion:
   Do not reject H_o; there is not sufficient evidence to conclude that $\mu \neq 92.84$.
   No; the new balls do not appear to be significantly different.

13. original claim: $\mu = 600$ [$n > 30$, use z (with s for σ)]
 H_o: $\mu = 600$
 H_1: $\mu \neq 600$
 $\alpha = .01$
 C.R. $z < -z_{.005} = -2.575$
    $z > z_{.005} = 2.575$
 calculations:
   $z_{\bar{x}} = (\bar{x} - \mu)/\sigma_{\bar{x}}$
   $= (593 - 600)/(21/\sqrt{40})$
   $= -7/3.320$
   $= -2.108$
 conclusion:
   Do not reject H_o; there is not sufficient evidence to reject the claim that $\mu = 600$.

While there is not enough evidence to reject the claim that $\mu = 600$ at the .01 level of significance, the results are certainly not evidence that $\mu = 600$ is true. As the error of 7 mg is only about 1% of the advertised amount, however, the product is probably not dangerous or unacceptable to use and may well be within acceptable limits of quality control. Based on this information alone, there is no reason not to buy this medicine.

15. original claim: $\mu > 0$ [n > 30, use z (with s for σ)]
   $H_o: \mu \le 0$
   $H_1: \mu > 0$
   $\alpha = .05$
   calculations:

   $z_{\bar{x}} = (\bar{x} - \mu)/\sigma_{\bar{x}}$
   $\qquad = (.6 - 0)/(3.8/\sqrt{75}\,)$
   $\qquad = .6/.439$
   $\qquad = 1.367$
   P-value $= P(z > 1.37) = .5000 - .4147 = .0853$
   conclusion:
   Do not reject H_o; there is not sufficient evidence to conclude that $\mu > 0$.
   No; people should not take the course. Even if the mean increase of .6 had been statistically significant, it would not be of practical significance.

17. original claim: $\mu \neq 69.5$ [n > 30, use z (with s for σ)]
   $H_o: \mu = 69.5$
   $H_1: \mu \neq 69.5$
   $\alpha = .05$
   calculations:

   $z_{\bar{x}} = (\bar{x} - \mu)/\sigma_{\bar{x}}$
   $\qquad = (73.4 - 69.5)/(8.7/\sqrt{35}\,)$
   $\qquad = 3.9/1.471$
   $\qquad = 2.652$
   P-value $= 2 \cdot P(z > 2.651) = 2 \cdot (.5000 - .4960) = 2 \cdot (.0040) = .0080$
   conclusion:
   Reject H_o; there is sufficient evidence to conclude that $\mu \neq 69.5$ (in fact, that $\mu > 69.5$).

19. original claim: $\mu = 5.670$ [n > 30, use z (with s for σ)]
   $H_o: \mu = 5.670$
   $H_1: \mu \neq 5.670$
   $\alpha = .01$
   calculations:

   $z_{\bar{x}} = (\bar{x} - \mu)/\sigma_{\bar{x}}$
   $\qquad = (5.622 - 5.670)/(.068/\sqrt{50}\,)$
   $\qquad = -.048/.00962$
   $\qquad = -4.991$
   P-value $= 2 \cdot P(z < -4.99) = 2 \cdot (.5000 - .4999) = 2 \cdot (.0001) = .0002$
   conclusion:
   Reject H_o; there is sufficient evidence to conclude that $\mu \neq 5.670$ (in fact, $\mu < 5.670$).
   That the mean weight of worn quarters found in circulation is less than 5.670 is not evidence that the mean weight of uncirculated quarters straight from the mint is different from 5.670.

21. original claim: $\mu = 3.5$ [n > 30, use z (with s for σ)]
   summary statistics: n = 70, $\Sigma x = 251.023$, $\Sigma x^2 = 900.557369$, $\bar{x} = 3.5860$, s = .07403
   $H_o: \mu = 3.5$
   $H_1: \mu \neq 3.5$
   $\alpha = .05$ [assumed]
   C.R. $z < -z_{.025} = -1.960$
   $\qquad z > z_{.025} = 1.960$
   calculations:

   $z_{\bar{x}} = (\bar{x} - \mu)/\sigma_{\bar{x}}$
   $\qquad = (3.5860 - 3.5)/(.07403/\sqrt{70}\,)$
   $\qquad = .0860/.008848$
   $\qquad = 9.725$

conclusion:
  Reject H_o; there is sufficient evidence to reject the claim that $\mu = 3.5$ and conclude that $\mu \neq 3.5$ (in fact, that $\mu > 3.5$).
P-value $= 2 \cdot P(z > 9.72) = 2 \cdot (.5000 - .4999) = 2 \cdot (.0001) = .0002$

23. original claim: $\mu > 12$ [$n > 30$, use z (with s for σ)]
summary statistics: $n = 36$, $\Sigma x = 442.5$, $\Sigma x^2 = 5439.35$, $\bar{x} = 12.29$, $s = .0906$
H_o: $\mu \leq 12$
H_1: $\mu > 12$
$\alpha = .05$ [assumed]
C.R. $z > z_{.05} = 1.645$
calculations:
  $z_{\bar{x}} = (\bar{x} - \mu)/\sigma_{\bar{x}}$
    $= (12.29 - 12)/(.0906/\sqrt{36}\,)$
    $= .2917/.0151$
    $= 19.309$

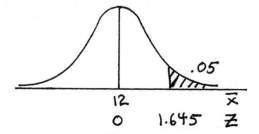

conclusion:
  Reject H_o; there is sufficient evidence to conclude that $\mu > 12$.
P-value $= P(z > 19.31) = .5000 - .4999 = .0001$
Yes; If the company were to cut back slightly on the volume per can and keep the price the same, Pepsi could probably be supplied in a way that is more profitable to the company. But it is probably true that (1) the company is willing to continue overfilling to minimize the occurrence of underfilling and (2) the company bases its cost on what it actually supplies and not what the label indicates.

25. NOTE: Minitab calls the calculated statistic T instead of Z because it uses the notation and techniques of the next section, 7-4.
original claim: $\mu < .9085$ [$n > 30$, use z (with s for σ)]
H_o: $\mu \geq .9085$
H_1: $\mu < .9085$
$\alpha = .05$ [assumed]
C.R. $z < -z_{.05} = -1.645$
calculations:
  $z_{\bar{x}} = (\bar{x} - \mu)/\sigma_{\bar{x}}$
    $= (.91282 - .9085)/(.03952/\sqrt{33}\,)$
    $= .00432/.00688$
    $= .63$ [from Minitab]

conclusion:
  Do not reject H_o; there is not sufficient evidence to conclude that $\mu < .9085$.
P-value $= P(z < .63) = .73$ [from Minitab]
No; consumers are not being cheated.

27. original claim: $\mu = 98.6$ [$n > 30$, use z (with s for σ)]
H_o: $\mu = 98.6$
H_1: $\mu \neq 98.6$
$\alpha = .05$ [assumed]
C.R. $z < z_{.025} = -1.960$
     $z > z_{.025} = 1.960$
calculations:
  $z_{\bar{x}} = (\bar{x} - \mu)/\sigma_{\bar{x}}$
    $= (98.1263 - 98.6)/(.7554/\sqrt{38}\,)$
    $= -3.865$ [from TI-83 Plus]

conclusion:
    Reject H_o; there is sufficient evidence the reject the claim that $\mu = 98.6$ and to conclude that $\mu \neq 98.6$ (in fact, that $\mu < 98.6$).

P-value $= 2 \cdot P(z < -3.865) = 1.109E-4 = .0001109$ [from TI-83 Plus]

Yes; these results are consistent with those found in the example. Yes; the large number of missing values could be a problem -- not because of their quantity (the remaining sample size is large enough), but because they represent those unable to give an early morning reading. It is possible that those who were unable to be present at 8 am were those who were still in bed, and the desire to stay in bed later may conceivably be related to body temperature -- meaning that those who showed up were a biased sample.

29. original claim: $\mu = 100$ [n > 30, use z (with s for σ)]
    H_o: $\mu = 100$
    H_1: $\mu \neq 100$
    $\alpha = .01$
    C.R. $z < -z_{.005} = -2.575$
         $z > z_{.005} = 2.575$
    calculations:

      $z_{\bar{x}} = (\bar{x} - \mu)/\sigma_{\bar{x}}$
      $2.575 < (103.6 - 100)/(s/\sqrt{62})$
         $s < (103.6 - 100)/(2.575/\sqrt{62})$
         $s < 3.6/.327$
         $s < 11.008$
The largest possible standard deviation is 11.0.

NOTE: There is no conclusion because this exercise is not asking for the completion of a test of hypotheses.

31. original claim: $\mu = 98.6$ [n > 30, use z (with s for σ)]
    H_o: $\mu = 98.6$
    H_1: $\mu \neq 98.6$
    $\alpha = .05$
    C.R. $z < -z_{.025} = -1.960$
         $z > z_{.025} = 1.960$
    calculations:

    $z_{\bar{x}} = (\bar{x} - \mu)/\sigma_{\bar{x}}$ $z_{\bar{x}} = (\bar{x} - \mu)/\sigma_{\bar{x}}$
$-1.960 = (\bar{x} - 98.6)/(.62/\sqrt{106})$ $1.960 = (\bar{x} - 98.6)/(.62/\sqrt{106})$
    $\bar{x} = 98.482$ $\bar{x} = 98.718$

NOTE: There is no conclusion because this exercise is not asking for the completion of a test of hypotheses.

a. $\beta = P(98.482 < \bar{x} < 98.718 \mid \mu = 98.7)$
    $= P(-3.62 < z < .30)$
    $= .4999 + .1179$
    $= .6178$

b. $\beta = P(98.482 < \bar{x} < 98.718 \mid \mu = 98.4)$
    $= P(1.36 < z < 5.28)$
    $= .4999 - .4131$
    $= .0868$

7-4 Testing a Claim about a Mean: Small Samples

1. original claim: $\mu = 75$ [$n \leq 30$ and σ unknown, use t]
   H_o: $\mu = 75$
   H_1: $\mu \neq 75$
   $\alpha = .05$
   C.R. $t < -t_{15,.025} = -2.132$
   $\qquad t > t_{15,.025} = 2.132$
   calculations:
   $\qquad t_{\bar{x}} = (\bar{x} - \mu)/s_{\bar{x}}$
   $\qquad\quad = (77 - 75)/(15/\sqrt{16})$
   $\qquad\quad = 2/3.75$
   $\qquad\quad = .533$
   conclusion:
   $\qquad$ Do not reject H_o; there is not sufficient evidence to reject that claim that $\mu = 75$.

3. original claim: $\mu > 2.00$ [$n \leq 30$ and σ unknown, use t]
   H_o: $\mu \leq 2.00$
   H_1: $\mu > 2.00$
   $\alpha = .01$
   C.R. $t > t_{23,.01} = 2.500$
   calculations:
   $\qquad t_{\bar{x}} = (\bar{x} - \mu)/s_{\bar{x}}$
   $\qquad\quad = (2.35 - 2.00)/(.70/\sqrt{24})$
   $\qquad\quad = .35/.1428$
   $\qquad\quad = 2.449$
   conclusion:
   $\qquad$ Do not reject H_o; there is not sufficient evidence to conclude that $\mu > 2.00$.

5. since the calculated $t = .533$ is to the left of $t_{15,.25} = .691$,
   the P-value for this two-tailed test is greater than $2 \cdot (.25)$
   -- i.e., P-value $> .50$

7. since the calculated $t = 2.449$ is between $t_{23,.01} = 2.500$ and $t_{23,.025} = 2.069$,
   the P-value for this one-tailed test is between .01 and .025
   -- i.e., $.01 <$ P-value $< .025$

9. original claim: $\mu = 98.6$ [$n \leq 30$ and σ unknown, use t]
   H_o: $\mu = 98.6$
   H_1: $\mu \neq 98.6$
   $\alpha = .05$
   C.R. $t < -t_{24,.025} = -2.064$
   $\qquad t > t_{24,.025} = 2.064$
   calculations:
   $\qquad t_{\bar{x}} = (\bar{x} - \mu)/s_{\bar{x}}$
   $\qquad\quad = (98.24 - 98.6)/(.56/\sqrt{25})$
   $\qquad\quad = -.36/.112$
   $\qquad\quad = -3.214$
   conclusion:
   $\qquad$ Reject H_o; there is sufficient evidence to reject the claim that $\mu = 98.6$ and to conclude that $\mu \neq 98.6$ (in fact, that $\mu < 98.6$).

11. original claim: $\mu < 1000$ [$n \le 30$ and σ unknown, use t]
summary statistics: $n = 5$, $\Sigma x = 3837$, $\Sigma x^2 = 3268799$, $\bar{x} = 767.4$, $s = 284.7$
H_o: $\mu \ge 1000$
H_1: $\mu < 1000$
$\alpha = .05$
C.R. $t < -t_{4,.05} = -2.132$
calculations:

$$t_{\bar{x}} = (\bar{x} - \mu)/s_{\bar{x}}$$
$$= (767.4 - 1000)/(284.7/\sqrt{5})$$
$$= -232.6/127.3$$
$$= -1.827$$
conclusion:
Do not reject H_o; there is not sufficient evidence to conclude that $\mu < 1000$.
No; since results like those obtained are not considered unusual if $\mu \ge 1000$, the company can not advertise with confidence that $\mu < 1000$.

13. original claim: $\mu \ne 243.5$ [$n \le 30$ and σ unknown, use t]
H_o: $\mu = 243.5$
H_1: $\mu \ne 243.5$
$\alpha = .05$
C.R. $t < -t_{21,.025} = -2.080$
$\quad\quad t > t_{21,.025} = 2.080$
calculations:

$$t_{\bar{x}} = (\bar{x} - \mu)/s_{\bar{x}}$$
$$= (170.2 - 243.5)/(35.3/\sqrt{22})$$
$$= -73.3/7.526$$
$$= -9.740$$
conclusion:
Reject H_o; there is sufficient evidence to conclude that $\mu \ne 243.5$ (in fact, that $\mu < 243.5$).
Yes; based on this result the VDT test scores are significantly lower than those for the printed version -- the VDT test is not fair to those choosing that format and should be revised.

15. original claim: $\mu \ne 41.9$ [$n \le 30$ and σ unknown, use t]
H_o: $\mu = 41.9$
H_1: $\mu \ne 41.9$
$\alpha = .01$
C.R. $t < -t_{14,.005} = -2.977$
$\quad\quad t > t_{14,.005} = 2.977$
calculations:

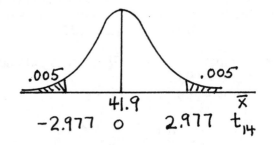

$$t_{\bar{x}} = (\bar{x} - \mu)/s_{\bar{x}}$$
$$= (31.0 - 41.9)/(10.5/\sqrt{15})$$
$$= -10.9/2.711$$
$$= -4.021$$
conclusion:
Reject H_o; there is sufficient evidence to conclude that $\mu \ne 41.9$ (in fact, that $\mu < 41.9$).

17. original claim: $\mu > .9085$ [$n > 30$, use z (with s for σ)]
H_o: $\mu \le .9085$
H_1: $\mu > .9085$
$\alpha = .05$
C.R. $z > z_{.05} = 1.645$
calculations:

$$z_{\bar{x}} = (\bar{x} - \mu)/\sigma_{\bar{x}}$$
$$= (.9128 - .9085)/(.0395/\sqrt{33})$$
$$= .0043/.00688$$
$$= .625$$

conclusion:
    Do not reject H_o; there is not sufficient evidence to conclude that $\mu > .9085$.
    No; based on this sample we cannot conclude that the packages contain more than the claimed weight printed on the label. Even if the conclusion had been that $\mu > .9085$, that would not be evidence that the packages contain more than the printed weight because (1) the test involved only brown M&M's and (2) the assumption that packages contain 1498 M&M's is probably not correct -- it is more likely that the bags are filled by weight and not by count.

19. original claim: $\mu > 62.2$ [$n \le 30$ and σ unknown, use t]
    summary statistics: n = 9, $\Sigma x = 579.4$, $\Sigma x^2 = 37334.60$, $\bar{x} = 64.38$, s = 2.065
    H_o: $\mu \le 62.2$
    H_1: $\mu > 62.2$
    $\alpha = .01$
    C.R. $t > t_{8,.01} = 2.896$
    calculations:

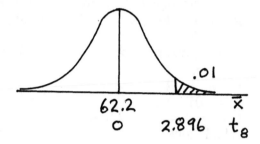

$$t_{\bar{x}} = (\bar{x} - \mu)/s_{\bar{x}}$$
$$= (64.38 - 62.2)/(2.065/\sqrt{9})$$
$$= 2.178/.6884$$
$$= 3.164$$
    conclusion:
    Reject H_o; there is sufficient evidence to conclude that $\mu > 62.2$.

21. original claim: $\mu > 12$ [$n > 30$, use z (with s for σ)]
    summary statistics: n = 36, $\Sigma x = 439.2$, $\Sigma x^2 = 5358.36$, $\bar{x} = 12.20$, s = .05855
    H_o: $\mu \le 12$
    H_1: $\mu > 12$
    $\alpha = .05$ [assumed]
    C.R. $z > z_{.05} = 1.645$
    calculations:

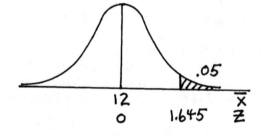

$$z_{\bar{x}} = (\bar{x} - \mu)/\sigma_{\bar{x}}$$
$$= (12.20 - 12)/(.05855/\sqrt{36})$$
$$= .200/.009759$$
$$= 20.494$$
    conclusion:
    Reject H_o; there is sufficient evidence to conclude that $\mu > 12$.

23. original claim: $\mu = 3.39$ [$n \le 30$ and σ unknown, use t]
    summary statistics: n = 16, $\Sigma x = 58.80$, $\Sigma x^2 = 222.5710$, $\bar{x} = 3.675$, s = .657
    H_o: $\mu = 3.39$
    H_1: $\mu \ne 3.39$
    $\alpha = .05$
    C.R. $t < -t_{15,.025} = -2.132$
        $t > t_{15,.025} = 2.132$
    calculations:

$$t_{\bar{x}} = (\bar{x} - \mu)/s_{\bar{x}}$$
$$= (3.675 - 3.39)/(.657/\sqrt{16})$$
$$= .285/.1643$$
$$= 1.734$$
    conclusion:
    Do not reject H_o; there is not sufficient evidence to conclude that $\mu \ne 3.39$ -- i.e., there is not sufficient evidence to conclude that the vitamin supplement has an effect on birth weight.

25. original claim: $\mu < 1000$ [$n \leq 30$ and σ unknown, use t]
    H_o: $\mu \geq 1000$
    H_1: $\mu < 1000$
    $\alpha = .05$ [assumed]
    C.R. $t < -t_{20,.05} = -1.725$
    calculations:

   $t_{\bar{x}} = (\bar{x} - \mu)/s_{\bar{x}}$
       $= (985.67 - 1000)/(46.52/\sqrt{21})$
       $= -1.412$ [from TI-83 Plus]
    conclusion:
       Do not reject H_o; there is not sufficient evidence to conclude that $\mu < 1000$.
    P-value $= P(t_{20} < -1.412) = .08666$ [from TI-83 Plus]

27. original claim: $\mu > 12$ [$n \leq 30$ and σ unknown, use t]
    H_o: $\mu \leq 12$
    H_1: $\mu > 12$
    $\alpha = .01$
    C.R. $t > t_{10,.01} = 2.764$
    calculations:

   $t_{\bar{x}} = (\bar{x} - \mu)/s_{\bar{x}}$
       $= (13.286 - 12)/(1.910/\sqrt{11})$
       $= 1.286/.576$
       $= 2.23$ [from Minitab]
    conclusion:
       Do not reject H_o; there is not sufficient evidence to conclude that $\mu > 12$.
    P-value $= .025$ [from Minitab]

29. Only the P-value is changed. The summary values n, $\bar{x}$ and s do not change. The standard error of the mean, which is $s_{\bar{x}} = s/\sqrt{n}$, does not change. The calculated t, which is $t_{\bar{x}} = (\bar{x} - \mu)/s_{\bar{x}}$, does not change. Since the test is now two-tailed, the P-value doubles and is now $2 \cdot P(t_{14} > 3.17) = 2 \cdot (.0034) = .0068$.

31. $A = z_{.05} \cdot [(8 \cdot df + 3)/(8 \cdot df + 1)]$
       $= 1.645 \cdot [(8 \cdot 9 + 3)/(8 \cdot 9 + 1)]$
       $= 1.645 \cdot [75/73]$
       $= 1.690$
    $t = \sqrt{df \cdot (e^{A \cdot A/df} - 1)}$
       $= \sqrt{9 \cdot (e^{(1.690)(1.690)/9} - 1)}$
       $= \sqrt{9 \cdot (e^{.317} - 1)}$
       $= \sqrt{9 \cdot (.3735)}$
       $= 1.833$
    This agrees exactly with $t_{9,.05} = 1.833$ given in Table A-3.

7-5 Testing a Claim about a Proportion

NOTE: To be consistent with the notation of the previous sections, and thereby reinforcing the patterns and concepts presented in those sections, the manual uses the "usual" z formula written to apply to $\hat{p}$'s

$$z_{\hat{p}} = (\hat{p} - \mu_{\hat{p}})/\sigma_{\hat{p}}$$

When the normal approximation to the binomial applies, the $\hat{p}$'s are normally distributed

with $\mu_{\hat{p}} = p$ and $\sigma_{\hat{p}} = \sqrt{pq/n}$

And so the formula for the z statistic may also be written as

$$z_{\hat{p}} = (\hat{p} - p)/\sqrt{pq/n}$$

1. original claim: p < .20 [normal approximation to the binomial, use z]

$\hat{p} = x/n = 149/880 = .169$

H_o: p ≥ .20

H_1: p < .20

$\alpha = .05$

C.R. z < $-z_{.05}$ = -1.645

calculations:

$$z_{\hat{p}} = (\hat{p} - \mu_{\hat{p}})/\sigma_{\hat{p}}$$
$$= (.169 - .20)/\sqrt{(.20)(.80)/880}$$
$$= -.0307/.0135$$
$$= -2.275$$

conclusion:

Reject H_o; there is sufficient evidence to conclude that p < .20.

P-value = P(z < -2.285) = .5000 - .4887 = .0113.

No; since technology and the use of technology is changing so rapidly, any figures from 1997 would no longer be valid today. [By the time you are reading this, e-mail could be so common that essentially 100% of the households use it -- or it could have been replaced by something newer and be so out-dated that essentially 0% of the households still use it!]

3. original claim: p > .50 [normal approximation to the binomial, use z]

$\hat{p} = x/n = 544/850 = .640$

H_o: p ≤ .50

H_1: p > .50

$\alpha = .02$

C.R. z > $z_{.02}$ = 2.05

calculations:

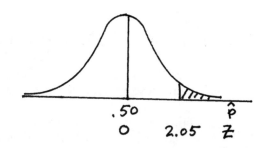

$$z_{\hat{p}} = (\hat{p} - \mu_{\hat{p}})/\sigma_{\hat{p}}$$
$$= (.640 - .50)/\sqrt{(.50)(.50)/850}$$
$$= .140/.0171$$
$$= 8.163$$

conclusion:

Reject H_o; there is sufficient evidence to conclude that p > .50.

P-value = P(z > 8.16) = .5000 - .4999 = .0001.

This test provides evidence that most drivers change tapes or CD's while driving, but it does not address the question as to whether that act is a driving hazard. The actual percent of drivers engaging in some practice is not necessarily what determines whether it is a hazard. If 2% of the drivers are intoxicated, that would be a hazard; if 98% of the drivers are chewing gum, that would not be a hazard.

5. original claim: p = .43 [normal approximation to the binomial, use z]
   $\hat{p}$ = x/n = 308/611 = .504
   H_o: p = .43
   H_1: p ≠ .43
   α = .04
   C.R. z < -$z_{.02}$ = -2.05
      z > $z_{.02}$ = 2.05
   calculations:

   z_p = ($\hat{p}$ - μ_p)/σ_p
      = (.504 - .43)/$\sqrt{(.43)(.57)/611}$
      = .0741/.0200
      = 3.699
   conclusion:
      Reject H_o; there is sufficient evidence reject the claim that p = .43 and to conclude that
      p ≠ .43 (in fact, that p > .43).
   P-value = 2·P(z > 3.70) = 2·(.5000 - .4999) = 2·(.0001) = .0002
   The results suggest that people are not telling the truth. Most likely (1) some people who voted
   for the loser don't want to reveal their true vote and be identified with a loser and (2) some
   people who didn't vote at all won't admit that fact and claim to have voted for the winner.

7. original claim: p = .01 [normal approximation to the binomial, use z]
   $\hat{p}$ = x/n = 20/1234 = .0162
   H_o: p = .01
   H_1: p ≠ .01
   α = .05
   C.R. z < -$z_{.025}$ = -1.960
      z > $z_{.025}$ = 1.960
   calculations:

   z_p = ($\hat{p}$ - μ_p)/σ_p
      = (.0162 - .01)/$\sqrt{(.01)(.99)/1234}$
      = .0062/.0028
      = 2.192
   conclusion:
      Reject H_o; there is sufficient evidence reject the claim that p = .01 and to conclude that
      p ≠ .01 (in fact, that p > .01).
   P-value = 2·P(z > 2.19) = 2·(.5000 - .4857) = 2·(.0143) = .0286
   No; based on these results, scanners do not appear to help consumers avoid overcharges.

9. original claim: p > .50 [normal approximation to the binomial, use z]
   NOTE: A "success" is <u>not</u> being knowledgeable about the Holocaust.
   $\hat{p}$ = x/n = 268/506 = .530
   H_o: p ≤ .50
   H_1: p > .50
   α = .05
   C.R. z > $z_{.05}$ = 1.645
   calculations

   z_p = ($\hat{p}$ - μ_p)/σ_p
      = (.530 - .50)/$\sqrt{(.50)(.50)/506}$
      = .0296/.0222
      = 1.334
   conclusion:
      Do not reject H_o; there is not sufficient evidence to conclude that p > .50.
   P-value = P(z > 1.33) = .5000 - .4082 = .0918
   No; we cannot be 95% certain that p > .50 and that the curriculum needs to be revised.

11. original claim: p > .50 [normal approximation to the binomial, use z]
   NOTE: A "success" is <u>not</u> having quit smoking one year later.
   $\hat{p} = x/n = 39/71 = .5493$
   H_o: p ≤ .50
   H_1: p > .50
   α = .10
   C.R. z > $z_{.10}$ = 1.282
   calculations
   $z_{\hat{p}} = (\hat{p} - \mu_{\hat{p}})/\sigma_{\hat{p}}$
   $= (.5493 - .50)/\sqrt{(.50)(.50)/71}$
   $= .0493/.0593$
   $= .831$

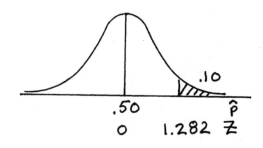

   conclusion:
   Do not reject H_o; there is not sufficient evidence to conclude that p > .50.
   P-value = P(z > .83) = .5000 - .2967 = .2033
   No; while the sample success rate is less than ½, that value must be seen in context -- would you call a major league baseball player with a .475 batting average a failure because his success rate is less than ½? The success rate must be compared to success rates of other programs (or to a control group) and to the fact that helping even a few is of some effect.

13. original claim: p = .75 [normal approximation to the binomial, use z]
   $\hat{p} = x/n = 410/500 = .82$ [NOTE: x = (.82)(500) = 410]
   H_o: p = .75
   H_1: p ≠ .75
   α = .05 [assumed]
   C.R. z < $-z_{.025}$ = -1.960
      z > $z_{.025}$ = 1.960
   calculations:
   $z_{\hat{p}} = (\hat{p} - \mu_{\hat{p}})/\sigma_{\hat{p}}$
   $= (.82 - .75)/\sqrt{(.75)(.25)/500}$
   $= .07/.0194$
   $= 3.615$

   conclusion:
   Reject H_o; there is sufficient evidence reject the claim that p = .75 and to conclude that p ≠ .75 (in fact, that p > .75).
   P-value = 2·P(z > 3.61) = 2·(.5000 - .4999) = 2·(.0001) = .0002

15. original claim: p < .10 [normal approximation to the binomial, use z]
   $\hat{p} = x/n = 28/400 = .07$ [NOTE: x = (.07)(400) = 28]
   H_o: p ≥ .10
   H_1: p < .10
   α = .01
   C.R. z < $-z_{.01}$ = -2.327
   calculations:
   $z_{\hat{p}} = (\hat{p} - \mu_{\hat{p}})/\sigma_{\hat{p}}$
   $= (.07 - .10)/\sqrt{(.10)(.90)/400}$
   $= -.03/.0150$
   $= -2.000$

   conclusion:
   Do not reject H_o; there is not sufficient evidence to conclude that p < .10.
   P-value = P(z < -2.00) = .5000 - .4772 = .0228
   The conclusion cannot be applied to all U.S. college students. The students surveyed were from one particular college and were not a random sample of all U.S. college students.

17. original claim: $p > .75$ [normal approximation to the binomial, use z]
$\hat{p} = x/n = 455/500 = .91$ [NOTE: $x = (.91)(500) = 455$]
$H_o: p \le .75$
$H_1: p > .75$
$\alpha = .01$
C.R. $z > z_{.01} = 2.327$
calculations:

$$z_{\hat{p}} = (\hat{p} - \mu_{\hat{p}})/\sigma_{\hat{p}}$$
$$= (.91 - .75)/\sqrt{(.75)(.25)/500}$$
$$= .16/.0194$$
$$= 8.262$$
conclusion:
    Reject H_o; there is sufficient evidence to conclude that $p > .75$.
P-value $= P(z > 8.26) = .5000 - .4999 = .0001$
Yes; based on the sample results, the funding will be approved.

19. original claim: $p = .10$ [normal approximation to the binomial, use z]
$\hat{p} = x/n = 5/100 = .05$
$H_o: p = .10$
$H_1: p \ne .10$
$\alpha = .05$ [assumed]
C.R. $z < -z_{.025} = -1.960$
      $z > z_{.025} = 1.960$
calculations:

$$z_{\hat{p}} = (\hat{p} - \mu_{\hat{p}})/\sigma_{\hat{p}}$$
$$= (.05 - .10)/\sqrt{(.10)(.90)/100}$$
$$= -.05/.030$$
$$= -1.667$$
conclusion:
    Do no reject H_o; there is not sufficient evidence reject the claim that $p = .10$
P-value $= 2 \cdot P(z < -1.67) = 2 \cdot (.5000 - .4525) = 2 \cdot (.0475) = .0950$

21. a. For the claim $p > .0025$, we have $H_o: p \le .0025$.
    Since $np = (80)(.0025) = 0.2 < 5$, the normal approximation to the binomial is not
    appropriate and the methods of this section cannot be used.
    b. binomial: $n = 80$, $p = .0025$
    $P(x) = [n!/x!(n-x)!]p^x(1-p)^{n-x}$
    $P(x = 0) = [80!/0!80!](.0025)^0(.9975)^{80} = .818525754$
    $P(x = 1) = [80!/1!79!](.0025)^1(.9975)^{79} = .164115439$
    $P(x = 2) = [80!/2!78!](.0025)^2(.9975)^{78} = .016247017$
    $P(x = 3) = [80!/3!77!](.0025)^3(.9975)^{77} = .001058703$
    $P(x = 4) = [80!/4!76!](.0025)^4(.9975)^{76} = .000051078$
    $P(x = 5) = [80!/5!75!](.0025)^5(.9975)^{75} = .000001946$
    $P(x = 6) = [80!/6!74!](.0025)^6(.9975)^{74} = \underline{.000000061}$
    $\phantom{P(x = 6) = [80!/6!74!](.0025)^6(.9975)^{74} = }.999999998$

    $P(x \ge 7) = 1 - P(x \le 6)$
    $ = 1 - .999999998$
    $ = .000000002$
    c. Based on the result from part (b), getting 7 or more males with this color blindness if
    $p = .0025$ would be an extremely rare event. This leads to reject the claim that $p = .0025$.
    NOTE: In a one-tailed test of $H_o: p = .0025$ vs. $H_1: p > .0025$, the observed value $x = 7$
    produces: P-value $= P$(getting a result as extreme as or more extreme than $x = 7$)
    $$= P(x \ge 7)$$
    $$= .000000002$$

ADDITIONAL NOTE: This problem may be worked in a format similar to that used in this section as follows.

original claim: $p > .0025$ [$np = (80)(.0025) = 0.2 < 5$; use binomial distribution]

$x = 7$

H_o: $p \le .0025$

H_1: $p > .0025$

$\alpha = .01$

C.R. $x > 2$

calculations

  $x = 7$

conclusion:

  Reject H_o; there is sufficient evidence to conclude $p > .0025$.

NOTE: Refer to this probability distribution for n=80 and p=.0025

x	P(x)
0	.818526
1	.164115
2	.016247
3	.001059
4	.000051
5	.000002
6	0^+
7	0^+
.	.
.	.
.	.
.	.
	1.000000

RATIONALE: We want to place .01 (or as close to it as possible without going over) in the upper tail of this one-tailed test.

$P(x > 1) = P(x = 2) + ...$

$\qquad\ = .016247 + ...$

$\qquad\ > .01$

$P(x > 2) = 1 - [P(x = 0) + P(x = 1) + P(x = 2)]$

$\qquad\ = 1 - [.818526 + .164115 + .016247]$

$\qquad\ = 1 - .998888$

$\qquad\ = .001112 \le .01$; hence this is the desired critical region

23. original claim: $p \le d$, a rate to be determined   [normal approximation to the binomial, use z]

$\hat{p} = x/n = .03$

H_o: $p \le d$

H_1: $p > d$

$\alpha = .05$

C.R. $z > z_{.05} = 1.645$

calculations:

$z_{\hat{p}} = (\hat{p} - \mu_{\hat{p}})/\sigma_{\hat{p}}$

$1.645 = (.03 - d)/\sqrt{(d)(1-d)/500}$

$1.645\sqrt{d(1-d)} = \sqrt{500}\ (.03 - d)$

$2.706(d-d^2) = 500(.0009 - .06d + d^2)$

$d - d^2 = 184.7729(.0009 - .06d + d^2)$

$d - d^2 = .1663 - 11.0864d + 184.7729d^2$

$185.7729d^2 - 12.0864d + .1163 = 0$

$d = [12.0864 \pm \sqrt{(-12.0864)^2 - 4(185.7729)(.1163)}]/2(185.7729)$

$\quad = [12.0864 \pm 4.7442]/371.5457$

$\quad = .01976$ or $.04530$ [with .04530 being the d for which .03 has a z score of -1.645]

The lowest defective rate he can claim and still have $\hat{p} = x/500 = .03$ <u>not</u> cause the claim to be rejected is .01977 (rounded up).

25. Assuming that each mouse is either a success or not a success, the number of successes must be a whole number between 0 and 20 inclusive. The possible success rates are all multiples of 5%: 0/20 = 0%, 1/20 = 5%, 2/20 = 10%, 3/20 = 15%, etc. A success rate of 47% is not possible.

7-6 Testing a Claim about a Standard Deviation or Variance

NOTE: Following the pattern used with the z and t distributions, this manual uses the closest entry from Table A-4 for χ^2 as if it were the precise value necessary and does not use interpolation. This procedure sacrifices very little accuracy -- and even interpolation does not yield precise values. When extreme accuracy is needed in practice, statisticians refer either to more accurate tables or to computer-produced values.

ADDITIONAL NOTE: The χ^2 distribution depends upon n, and it "bunches up" around df = n-1. In addition, the formula $\chi^2 = (n-1)s^2/\sigma^2$ used in the calculations contains df = n-1. When the exact df needed in the problem does not appear in the table and the closest χ^2 value is used to determine the critical region, some instructors recommend using the same df in the calculations that were used to determine the C.R. This manual typically uses the closest entry to determine the C.R. and the n from the problem in the calculations, even though this introduces a slight discrepancy.

1. a. $\chi^2_{9,.995} = 1.735$
    $\chi^2_{9,.005} = 23.589$
   b. $\chi^2_{26,.01} = 45.642$
   c. $\chi^2_{20,.95} = 10.851$

3. original claim: $\sigma = 15$
   H_o: $\sigma = 15$
   H_1: $\sigma \neq 15$
   $\alpha = .05$
   C.R. $\chi^2 < \chi^2_{9,.975} = 2.700$
     $\chi^2 > \chi^2_{9,.025} = 19.023$
   calculations:
     $\chi^2 = (n-1)s^2/\sigma^2$
       $= (9)(18.0)^2/(15)^2$
       $= 12.960$
   conclusion:

   Do not reject H_o; there is not sufficient evidence to reject the claim that $\sigma = 15$.

5. original claim: $\sigma > .04$
   H_o: $\sigma \leq .04$
   H_1: $\sigma > .04$
   $\alpha = .01$
   C.R. $\chi^2 > \chi^2_{39,.01} = 63.691$
   calculations:
     $\chi^2 = (n-1)s^2/\sigma^2$
       $= (39)(.31)^2/(.04)^2$
       $= 2342.4375$
   conclusion:

   Reject H_o; there is sufficient evidence to conclude that $\sigma > .04$.

We expect more variation for the weights of peanut M&M's. Plain M&M's are produced according to specifications and should show little variation from piece to piece. Peanut M&M's start with a peanut -- and peanuts are not produced to specifications, but occur with varying sizes. Even though the company rejects very large and very small peanuts, there will still be more variation from piece to piece than there would be from a controlled production process.

7. original claim: $\sigma \neq 43.7$
H_o: $\sigma = 43.7$
H_1: $\sigma \neq 43.7$
$\alpha = .05$
C.R. $\chi^2 < \chi^2_{80,.975} = 57.153$
$\quad \chi^2 > \chi^2_{80,.025} = 106.629$
calculations:
$\quad \chi^2 = (n-1)s^2/\sigma^2$
$\quad\quad = (80)(52.3)^2/(43.7)^2$
$\quad\quad = 114.586$

conclusion:
Reject H_o; there is sufficient evidence to conclude that $\sigma \neq 43.7$ (in fact, that $\sigma > 43.7$).
Since a larger σ indicates more variability, it appears that the new production method is worse.

9. original claim: $\sigma < 14.1$
H_o: $\sigma \geq 14.1$
H_1: $\sigma < 14.1$
$\alpha = .01$
C.R. $\chi^2 < \chi^2_{26,.99} = 12.198$
calculations:
$\quad \chi^2 = (n-1)s^2/\sigma^2$
$\quad\quad = (26)(9.3)^2/(14.1)^2$
$\quad\quad = 11.311$

conclusion:
Reject H_o; there is sufficient evidence to conclude that $\sigma < 14.1$.
No; the lower standard deviation does not suggest that the current class is doing better, but
that they are more similar to each other and show less variation. The standard deviation speaks
only about the spread of the scores and not their location.

11. original claim: $\sigma > 19.7$
H_o: $\sigma \leq 19.7$
H_1: $\sigma > 19.7$
$\alpha = .05$
C.R. $\chi^2 > \chi^2_{49,.05} = 67.505$
calculations:
$\quad \chi^2 = (n-1)s^2/\sigma^2$
$\quad\quad = (49)(23.4)^2/(19.7)^2$
$\quad\quad = 69.135$

conclusion:
Reject H_o; there is sufficient evidence to conclude that $\sigma > 19.7$.

13. summary statistics: n = 25, $\Sigma x = 1721.07$, $\Sigma x^2 = 118614.7057$, $\bar{x} = 68.843$, $s^2 = 5.476$
original claim: $\sigma \neq 2.9$
H_o: $\sigma = 2.9$
H_1: $\sigma \neq 2.9$
$\alpha = .05$ [assumed]
C.R. $\chi^2 < \chi^2_{24,.975} = 12.401$
$\quad \chi^2 > \chi^2_{24,.025} = 39.364$
calculations:
$\quad \chi^2 = (n-1)s^2/\sigma^2$
$\quad\quad = (24)(5.476)/(2.9)^2$
$\quad\quad = 15.627$

conclusion:
Do not reject H_o; there is not sufficient evidence to conclude that $\sigma \neq 2.9$.

15. summary statistics: $n = 9$, $\Sigma x = 1089$, $\Sigma x^2 = 132223$, $\bar{x} = 121.0$, $s^2 = 56.75$
    original claim: $\sigma < 29$
    H_o: $\sigma \geq 29$
    H_1: $\sigma < 29$
    $\alpha = .01$
    C.R. $\chi^2 < \chi^2_{8,.99} = 1.646$
    calculations:

$$\chi^2 = (n-1)s^2/\sigma^2$$
$$= (8)(56.75)/(29)^2$$
$$= .540$$
conclusion:
    Reject H_o; there is sufficient evidence to conclude that $\sigma < 29$.

17. a. To find the P-value for $\chi^2_9 = 12.960$ in a two-tailed test,
      note that $\chi^2_{9,.10} = 14.684$ is the smallest value above the "center" at $\chi^2_9 = 9.000$
      Therefore, ½(P-value) > .10 and P-value > .20.

    b. To find the P-value for $\chi^2_{39} = 2342.4375$ in a one-tailed greater than test,
      note that $\chi^2_{40,.005} = 66.766$ is the largest χ^2 in the row.
      Therefore, P-value < .005.

    c. To find the P-value for $\chi^2_8 = .540$ in a one-tailed less than test,
      note that $\chi^2_{8,.995} = 1.344$ is the smallest χ^2 in the row.
      Therefore, P-value < .005

19. a. upper $\chi^2 = df \cdot [1 - 2/(9 \cdot df) + z_{.025}\sqrt{2/(9 \cdot df)}]^3$
                $= 100 \cdot [1 - 2/(9 \cdot 100) + 1.960\sqrt{2/(9 \cdot 100)}]^3$
                $= 100 \cdot (1.0902)^3$
                $= 129.565$ [compare to $\chi^2_{100,.025} = 129.561$ from Table A-4]
        lower $\chi^2 = df \cdot [1 - 2/(9 \cdot df) - z_{.025}\sqrt{2/(9 \cdot df)}]^3$
                $= 100 \cdot [1 - 2/(9 \cdot 100) - 1.960\sqrt{2/(9 \cdot 100)}]^3$
                $= 100 \cdot (.9054)^3$
                $= 74.216$ [compare to $\chi^2_{100,.975} = 74.222$ from Table A-4]
    b. upper $\chi^2 = df \cdot [1 - 2/(9 \cdot df) + z_{.025}\sqrt{2/(9 \cdot df)}]^3$
                $= 149 \cdot [1 - 2/(9 \cdot 149) + 1.960\sqrt{2/(9 \cdot 149)}]^3$
                $= 149 \cdot (1.0742)^3$
                $= 184.690$
        lower $\chi^2 = df \cdot [1 - 2/(9 \cdot df) - z_{.025}\sqrt{2/(9 \cdot df)}]^3$
                $= 149 \cdot [1 - 2/(9 \cdot 149) - 1.960\sqrt{2/(9 \cdot 149)}]^3$
                $= 149 \cdot (.9228)^3$
                $= 117.093$

21. a. If there are disproportionately more 0's and 5's, there would be an overabundance of digits
       less than 6 (i.e., at the lower end of the scale). This would reduce the mean to something
       less than 4.5 and (since most of the digits are bunched within these narrower limits) reduce
       the standard deviation to less than 3.
    b. If H_o is true and $\sigma = 3$, we expect a uniform distribution of digits; if H_1 is true and $\sigma < 3$,
       we expect a distribution of digits that peaks at 0 and/or 5 and is positively skewed. In either
       instance, the data would not be coming from an normal distribution -- which is one of the
       assumptions of the test.

Review Exercises

1. a. Prefer a P-value of .001. This indicates only a .1% chance of achieving the observed effectiveness by chance alone if the new cure really has no effect.
   b. The original claim should be $\mu < 114.8$. This would lead to
   H_o: $\mu \geq 114.8$
   H_1: $\mu < 114.8$
   c. The original claim that $\mu < 90$ leads to
   H_o: $\mu \geq 90$
   H_1: $\mu < 90$
   d. Fail to reject H_o: $\mu \leq 12$; there is not enough evidence to conclude the original claim that $\mu > 12$.
   e. A type I error is the mistake of rejecting the null hypothesis when the null hypothesis is really true.

2. a. concerns μ:
   $n \leq 30$ and σ unknown, use t
   $-t_{9,.05} = -1.833$
   b. concerns p:
   normal approx. to the binomial, use z
   $\pm z_{.005} = \pm 2.575$
   c. concerns σ:
   use χ^2
   $\chi^2_{19,.975} = 8.907$ and $\chi^2_{19,.025} = 32.852$
   d. concerns μ:
   $n > 30$, use z (with s for σ)
   $\pm z_{.05} = \pm 1.645$
   e. concerns p:
   normal approx. to the binomial, use z
   $-z_{.01} = -2.327$

3. original claim: p > .50 [concerns p: normal approximation to the binomial, use z]
   $\hat{p} = x/n = x/504 = .57$ [NOTE: $x=(.57)(504)=287.28$; any $285 \leq x \leq 289$ rounds to 57%]
   H_o: p $\leq$ .50
   H_1: p > .50
   $\alpha = .01$
   C.R. z > $z_{.01} = 2.327$
   calculations:

   $z_{\hat{p}} = (\hat{p} - \mu_{\hat{p}})/\sigma_{\hat{p}}$
   $= (.57 - .50)/\sqrt{(.50)(.50)/504}$
   $= .07/.0223$
   $= 3.143$
   conclusion:
   Reject H_o; there is sufficient evidence to conclude that p > .50.
   P-value = $P(z > 3.14) = .5000 - .4999 = .0001$

4. original claim: $\mu < 12$ [concerns μ: n ≤ 30 and σ unknown, use t]
   H_o: $\mu \geq 12$
   H_1: $\mu < 12$
   $\alpha = .05$ [assumed]
   C.R. $t < -t_{23,.05} = -1.714$
   calculations:
   $$t_{\bar{x}} = (\bar{x} - \mu)/s_{\bar{x}}$$
   $$= (11.4 - 12)/(.62/\sqrt{24})$$
   $$= -.6/.1266$$
   $$= -4.741$$

conclusion:
   Reject H_o; there is sufficient evidence to conclude that $\mu < 12$.

No; assuming that proper random sampling techniques were used, his claim that the sample is too small to be meaningful is not valid. The t distribution adjusts for the various sample sizes to produce a valid test for any sample size n.

5. original claim: $\sigma < .75$ [concerns σ: use χ^2]
   H_o: $\sigma \geq .75$
   H_1: $\sigma < .75$
   $\alpha = .05$
   C.R. $\chi^2 < \chi^2_{60,.95} = 43.188$
   calculations:
   $$\chi^2 = (n-1)s^2/\sigma^2$$
   $$= (60)(.48)^2/(.75)^2$$
   $$= 24.576$$

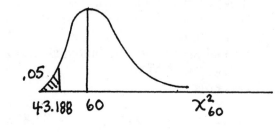

conclusion:
   Reject H_o; there is sufficient evidence to conclude that $\sigma < 29$.

The weights appear to be more consistent (i.e., they have a smaller standard deviation and less variation from bag to bag) now that the worn parts have been replaced.

6. original claim: $p < .10$ [concerns p: normal approximation to the binomial, use z]
   $\hat{p} = x/n = x/1248 = .08$ [NOTE: $x=(.08)(1248)=99.84$; any $94 \leq x \leq 106$ rounds to 8%]
   H_o: $p \geq .10$
   H_1: $p < .10$
   $\alpha = .01$
   C.R. $z < -z_{.01} = -2.327$
   calculations:
   $$z_{\hat{p}} = (\hat{p} - \mu_{\hat{p}})/\sigma_{\hat{p}}$$
   $$= (.08 - .10)/\sqrt{(.10)(.90)/1248}$$
   $$= -.02/.00849$$
   $$= -2.355$$

conclusion:
   Reject H_o; there is sufficient evidence to conclude that $p < .10$.

Yes; based on this result, the phrase "almost 1 out of 10" is justified. While "almost" is not well-defined, 10% is the closest round number to 8%. The test indicates we are confident that $p < .10$, but does not specify how much less.

ADDITIONAL NOTE: Using the upper possibility $\hat{p} = x/n = 106/1248 = .0849$ from the previous NOTE produces $z_{\hat{p}} = -1.774$, which is not enough evidence to reject H_o: $p \geq .10$. This not only supports the opinion that "almost 1 out of 10" is justified, but it also shows how sensitive tests may be to rounding procedures -- and that in this case the results are actually not reported with sufficient accuracy to conduct the test.

7. original claim: $\mu < 7124$ [concerns μ: $n > 30$, use z (with s for σ)]
   H_o: $\mu \geq 7124$
   H_1: $\mu < 7124$
   $\alpha = .01$
   C.R. $z < -z_{.01} = -2.327$
   calculations:
   $\quad z_{\bar{x}} = (\bar{x} - \mu)/\sigma_{\bar{x}}$
   $\qquad = (6047 - 7124)/(2944/\sqrt{750})$
   $\qquad = -1077/107.50$
   $\qquad = -10.019$

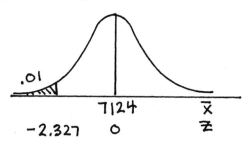

   conclusion:
      Reject H_o; there is sufficient evidence to conclude that $\mu < 7124$.
   Not necessarily; distance driven is only factor to take into consideration when determining insurance rates. Even though the 16-24 year olds drive less than those in the higher age bracket, their lack of experience and maturity may still make them higher insurance risks.

8. original claim: $p \geq .50$ [concerns p: normal approximation to the binomial, use z]
   $\hat{p} = x/n = x/1012 = .28$ [NOTE: $x=(.28)(1012)=283.36$; any $279 \leq x \leq 288$ rounds to 28%]
   H_o: $p \geq .50$
   H_1: $p < .50$
   $\alpha = .01$
   C.R. $z < -z_{.01} = -2.327$
   calculations:
   $\quad z_{\hat{p}} = (\hat{p} - \mu_{\hat{p}})/\sigma_{\hat{p}}$
   $\qquad = (.28 - .50)/\sqrt{(.50)(.50)/1012}$
   $\qquad = -.22/.0157$
   $\qquad = -13.997$

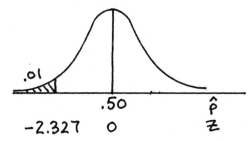

   conclusion:
      Reject H_o; there is sufficient evidence to reject the claim that $p \geq .50$ and to conclude that $p < .50$.
   Yes; based on this result, it appears that fruitcake producers should be concerned. If a product is not perceived as useful or appropriate for its intended purpose, its sales will probably fall if the economy weakens and/or new alternatives become available.

9. summary statistics: $n = 20$, $\Sigma x = 465.7$, $\Sigma x^2 = 11234.37$, $\bar{x} = 23.285$, $s = 4.534$
   original claim: $\mu = 20.0$ [concerns μ: $n \leq 30$ and σ unknown, use t]
   H_o: $\mu = 20.0$
   H_1: $\mu \neq 20.0$
   $\alpha = .01$
   C.R. $t < -t_{19,.005} = -2.861$
   $\qquad t > t_{19,.005} = 2.861$
   calculations:
   $\quad t_{\bar{x}} = (\bar{x} - \mu)/s_{\bar{x}}$
   $\qquad = (23.285 - 20.0)/(4.534/\sqrt{20})$
   $\qquad = 3.285/1.014$
   $\qquad = 3.240$

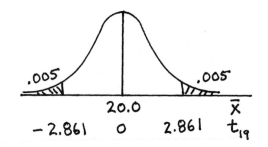

   conclusion:
      Reject H_o; there is sufficient evidence to reject the claim that $\mu = 20.0$ and to conclude that $\mu \neq 20.0$ (in fact, that $\mu > 20.0$).
   No; these pills are not acceptable at the .01 level because their mean amount appears to be higher than the stated 20.0 mg. Even if the hypothesis that $\mu = 20.0$ had not been rejected, the pills may be unacceptable because of their variability. While no specific criteria for variability were stated, note that the pills range from 165.0% (33 mg.) to 79.9% (15.9 mg.) of their stated amount.

10. original claim: $\mu > 30{,}000$ [concerns μ: $n > 30$, use z (with s for σ)]
H_o: $\mu \le 30{,}000$
H_1: $\mu > 30{,}000$
$\alpha = .01$
C.R. $z > z_{.01} = 2.327$
calculations:

$z_{\bar{x}} = (\bar{x} - \mu)/\sigma_{\bar{x}}$
$= (30122 - 30000)/(14276/\sqrt{150})$
$= 122/1165.6$
$= .105$

conclusion:
 Do not reject H_o; there is not sufficient evidence to conclude that $\mu > 30{,}000$.
No; her claim is not justified. Even though the sample results appear to support her claim, such results would not be unusual even if her claim were false -- and so the results are not enough to conclude with appropriate confidence that her claim is correct.

Cumulative Review Exercises

1. scores in order: 410 440 460 490 570 630 720 780
 summary statistics: $n = 8$, $\Sigma x = 4500$, $\Sigma x^2 = 2662000$
 a. $\bar{x} = (\Sigma x)/n = 4500/8 = 562.5$
 b. $\tilde{x} = (490 + 570)/2 = 530.0$
 c. $s^2 = [n(\Sigma x^2) - (\Sigma x)^2]/[n(n-1)] = [8(2662000) - (4500)^2]/[8(7)] = 18678.6$
  $s = 136.7$
 d. $s^2 = 18678.6$
 e. $R = 780 - 410 = 370$
 f. $\bar{x} = \pm t_{7,.025}s_{\bar{x}}$
  $562.5 \pm 2.365 \cdot (136.7/\sqrt{8})$
  562.5 ± 114.3
  $448.2 < \mu < 676.8$
 g. original claim: $\mu = 496$ [concerns μ: $n \le 30$ and σ unknown, use t]
  H_o: $\mu = 496$
  H_1: $\mu \ne 496$
  $\alpha = .05$
  C.R. $t < -t_{7,.025} = -2.365$
   $t > t_{7,.025} = 2.365$
  calculations:

  $t_{\bar{x}} = (\bar{x} - \mu)/s_{\bar{x}}$
   $= (562.5 - 496)/(136.7/\sqrt{8})$
   $= 66.5/48.3$
   $= 1.376$

  conclusion:
   Do not reject H_o; there is not sufficient evidence to reject the claim that $\mu = 496$.
 h. Based on the preceding results, it appears that the SAT math scores of female applicants to the College of Newport are about the same as those of the general female population. Both the confidence interval in (f) and the test in (g) fail to reject $\mu = 496$ as a possibility. In addition, ½ of the sample falls below 496 and ½ of the sample falls above 496.

2. normal distribution: $\mu = 496$, $\sigma = 108$
   a. $P(x > 500)$
      $= P(z > .04)$
      $= .5000 - .0160$
      $= .4840$

   b. let A = a selected score is above 500
      $P(A) = .4840$, for each selection
      $P(A_1 \text{ and } A_2 \text{ and } A_3 \text{ and } A_4 \text{ and } A_5)$
         $= P(A_1) \cdot P(A_2) \cdot P(A_3) \cdot P(A_4) \cdot P(A_4)$
         $= (.4840)^5$
         $= .0266$

   c. for samples of size $n = 5$,
      $\mu_{\bar{x}} = \mu = 496$
      $\sigma_{\bar{x}} = \sigma/\sqrt{n} = 108/\sqrt{5} = 48.30$
      normal distribution, since the original distribution is so
      $P(\bar{x} > 500)$
         $= P(z > .08)$
         $= .5000 - .0319$
         $= .4681$

   d. for P_{90}, $A = .4000$ and $z = 1.282$
      $x = \mu + z \cdot \sigma$
         $= 496 + (1.282)(108)$
         $= 496 + 138$
         $= 634$

3. summary statistics: $n = 16$, $\Sigma x = 1987.7$, $\Sigma x^2 = 254543.43$
   a. $\bar{x} = 124.23$, $s = 22.52$
   b. original claim: $\mu = 114.8$ [concerns μ: $n \leq 30$ and σ unknown, use t]
      H_o: $\mu = 114.8$
      H_1: $\mu \neq 114.8$
      $\alpha = .05$
      C.R. $t < -t_{15,.025} = -2.132$
           $t > t_{15,.025} = 2.132$
      calculations:
         $t_{\bar{x}} = (\bar{x} - \mu)/s_{\bar{x}}$
            $= (124.23 - 114.8)/(22.52/\sqrt{16})$
            $= 9.431/5.631$
            $= 1.675$

      conclusion:
         Do not reject H_o; there is not sufficient evidence to reject the claim that $\mu = 114.8$.

   c. $\bar{x} = \pm t_{15,.025} s_{\bar{x}}$
      $124.23 \pm 2.132 \cdot (22.52/\sqrt{16})$
      124.23 ± 12.00
      $112.23 < \mu < 136.24$
      Yes; the interval contains 114.8

d. original claim: $\sigma = 13.1$
   H_o: $\sigma = 13.1$
   H_1: $\sigma \neq 13.1$
   $\alpha = .05$
   C.R. $\chi^2 < \chi^2_{15,.975} = 6.262$
        $\chi^2 > \chi^2_{15,.025} = 27.488$
   calculations:
     $\chi^2 = (n-1)s^2/\sigma^2$
        $= (15)(22.52)^2/(13.1)^2$
        $= 44.339$

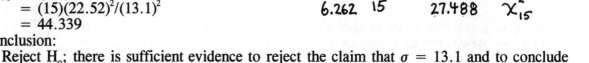

   conclusion:
     Reject H_o; there is sufficient evidence to reject the claim that $\sigma = 13.1$ and to conclude that $\sigma \neq 13.1$ (in fact, that $\sigma > 13.1$).

e. Yes; the new strain of viral infection seems to be associated with an increase in the variability among systolic blood pressures.

4. binomial distribution: $n = 25$ and $p = .25$

  a. $\mu = np = (25)(.25) = 6.25$

  b. $\sigma = \sqrt{np(1-p)} = \sqrt{25(.25)(.75)} = 2.165$

  c. Use the normal approximation (with $\mu = 6.25$ and $\sigma = 2.165$) to the binomial, since
     $np = 25(.25) = 6.25 \geq 5$ and $n(1-p) = 25(.75) = 18.75 \geq 5$.
    $P(x > 12) = P_c(x > 12.5)$
              $= P(z > 2.89)$
              $= .5000 - .4981$
              $= .0019$

  d. original claim: $p = .25$ [concerns p: normal approximation to the binomial, use z]
    $\hat{p} = $ (more than 12)/n $= 13/25 = .52$
    H_o: $p = .25$
    H_1: $p \neq .25$
    $\alpha = .05$
    C.R. $z < -z_{.025} = -1.960$
         $z > z_{.025} = 1.960$
    calculations:
     $z_{\hat{p}} = (\hat{p} - \mu_{\hat{p}})/\sigma_{\hat{p}}$
       $= (.52 - .25)/\sqrt{(.25)(.75)/25}$
       $= .27/.0866$
       $= 3.118$

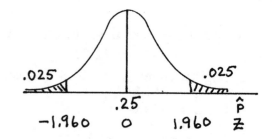

    conclusion:
     Reject H_o; there is sufficient evidence to reject the claim that $p = .25$ and to conclude that $p \neq .25$ (in fact, $p > .25$).

  NOTE: The above test of hypothesis follows the form and notation used in chapter 7. One could also consider the .0019 of part (c) above, calculate a P-value of $2 \cdot (.0019) = .0038$, and reject H_o because $.0038 < .05$. Part (c) uses the correction for continuity while the section on testing hypotheses concerning p ignores the correction.

  e. $\hat{p}$ unknown, use $\hat{p} = .5$
    $n = [(z_{.05})^2\hat{p}(1-\hat{p})]/E^2$
      $= [(1.645)^2(.5)(.5)]/(.04)^2$
      $= 422.8$, rounded up to 423

Chapter 8

Inferences from Two Samples

8-2 Inferences about Two Means: Independent and Large Samples

NOTE: To be consistent with the notation of the previous sections, and thereby reinforcing the patterns and concepts presented in those sections, the manual uses the "usual" z formula written to apply to $\bar{x}_1 - \bar{x}_2$'s

$$z_{\bar{x}_1 - \bar{x}_2} = (\bar{x}_1 - \bar{x}_2 - \mu_{\bar{x}_1 - \bar{x}_2}) / \sigma_{\bar{x}_1 - \bar{x}_2}$$

with $\mu_{\bar{x}_1 - \bar{x}_2} = \mu_1 - \mu_2$

and $\sigma_{\bar{x}_1 - \bar{x}_2} = \sqrt{\sigma_1^2/n_1 + \sigma_2^2/n_2}$

And so the formula for the z statistic may also be written as

$$z_{\bar{x}_1 - \bar{x}_2} = ((\bar{x}_1 - \bar{x}_2) - (\mu_1 - \mu_2)) / \sqrt{\sigma_1^2/n_1 + \sigma_2^2/n_2}$$

1. original claim: $\mu_1 - \mu_2 = 0$ [$n_1 > 30$ and $n_2 > 30$, use z (with s's for σ's)]
   $\bar{x}_1 - \bar{x}_2 = 7.00 - 6.00 = 1.00$
   $H_0: \mu_1 - \mu_2 = 0$
   $H_1: \mu_1 - \mu_2 \neq 0$
   $\alpha = .05$
   C.R. $z < -z_{.025} = -1.960$
   $\quad\quad z > z_{.025} = 1.960$
   calculations:

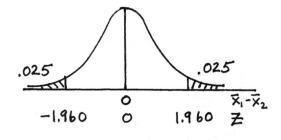

.025   .025
$-1.960 \quad 0 \quad 1.960$
0 ... $\bar{x}_1 - \bar{x}_2$
... z

$$z_{\bar{x}_1 - \bar{x}_2} = (\bar{x}_1 - \bar{x}_2 - \mu_{\bar{x}_1 - \bar{x}_2}) / \sigma_{\bar{x}_1 - \bar{x}_2}$$
$$= (1.00 - 0) / \sqrt{(1.00)^2/50 + (2.00)^2/100}$$
$$= 1.00 / .2449$$
$$= 4.082$$

conclusion:
   Reject H_0; there is sufficient evidence to reject the claim that $\mu_1 - \mu_2 = 0$ and to conclude that $\mu_1 - \mu_2 \neq 0$ (in fact, that $\mu_1 - \mu_2 > 0$).
   P-value $= 2 \cdot P(z > 4.08) = 2 \cdot (.5000 - .4999) = 2 \cdot (.0001) = .0002$

3. $(\bar{x}_1 - \bar{x}_2) \pm z_{.025} \sqrt{\sigma_1^2/n_1 + \sigma_2^2/n_2}$
   $1.00 \pm 1.960 \cdot \sqrt{(1.00)^2/50 + (2.00)^2/100}$
   $1.00 \pm .48$
   $.52 < \mu_1 - \mu_2 < 1.48$
   No; the confidence interval does not include 0. Since the confidence interval limits do not include 0, conclude that there is a real difference between the treatment and placebo groups.

5. a. original claim: $\mu_1 - \mu_2 = 0$ [$n_1 > 30$ and $n_2 > 30$, use z (with s's for σ's)]

$\bar{x}_1 - \bar{x}_2 = .81682 - .78479 = .03203$

$H_o: \mu_1 - \mu_2 = 0$

$H_1: \mu_1 - \mu_2 \neq 0$

$\alpha = .01$

C.R. $z < -z_{.005} = -2.575$

$\quad\quad z > z_{.005} = 2.575$

calculations:

$z_{\bar{x}_1 - \bar{x}_2} = (\bar{x}_1 - \bar{x}_2 - \mu_{\bar{x}_1 - \bar{x}_2})/\sigma_{\bar{x}_1 - \bar{x}_2}$

$\quad = (.03203 - 0)/\sqrt{(.007507)^2/36 + (.004391)^2/36}$

$\quad = .03203/.0014495$

$\quad = 22.098$

conclusion:

Reject H_o; there is sufficient evidence to reject the claim that $\mu_1 - \mu_2 = 0$ and to conclude that $\mu_1 - \mu_2 \neq 0$ (in fact, that $\mu_1 - \mu_2 > 0$).

P-value $= 2 \cdot P(z > 22.10) = 2 \cdot (.5000 - .4999) = 2 \cdot (.0001) = .0002$

b. $(\bar{x}_1 - \bar{x}_2) \pm z_{.005}\sqrt{\sigma_1^2/n_1 + \sigma_2^2/n_2}$

$.03203 \pm 2.575 \cdot \sqrt{(.007507)^2/36 + (.004391)^2/36}$

$.03203 \pm .00373$

$.02830 < \mu_1 - \mu_2 < .03576$

7. original claim: $\mu_1 - \mu_2 > 0$ [$n_1 > 30$ and $n_2 > 30$, use z (with s's for σ's)]

$\bar{x}_1 - \bar{x}_2 = 53.3 - 45.3 = 8.0$

$H_o: \mu_1 - \mu_2 \leq 0$

$H_1: \mu_1 - \mu_2 > 0$

$\alpha = .01$

C.R. $z > z_{.01} = 2.327$

calculations:

$z_{\bar{x}_1 - \bar{x}_2} = (\bar{x}_1 - \bar{x}_2 - \mu_{\bar{x}_1 - \bar{x}_2})/\sigma_{\bar{x}_1 - \bar{x}_2}$

$\quad = (8.0 - 0)/\sqrt{(11.6)^2/40 + (13.2)^2/40}$

$\quad = 8.0/2.778$

$\quad = 2.879$

conclusion:

Reject H_o; there is sufficient evidence to conclude that $\mu_1 - \mu_2 > 0$.

P-value $= P(z > 2.88) = .5000 - .4980 = .0020$

9. Let the younger men be group 1.

a. original claim: $\mu_1 - \mu_2 > 0$ [$n_1 > 30$ and $n_2 > 30$, use z (with s's for σ's)]

$\bar{x}_1 - \bar{x}_2 = 176 - 164 = 12$

$H_o: \mu_1 - \mu_2 \leq 0$

$H_1: \mu_1 - \mu_2 > 0$

$\alpha = .01$

C.R. $z > z_{.01} = 2.327$

calculations:

$z_{\bar{x}_1 - \bar{x}_2} = (\bar{x}_1 - \bar{x}_2 - \mu_{\bar{x}_1 - \bar{x}_2})/\sigma_{\bar{x}_1 - \bar{x}_2}$

$\quad = (12 - 0)/\sqrt{(35.0)^2/804 + (27.0)^2/1657}$

$\quad = 12/1.401$

$\quad = 8.563$

conclusion:

Reject H_o; there is sufficient evidence to conclude that $\mu_1 - \mu_2 > 0$.

P-value $= P(z > 8.56) = .5000 - .4999 = .0001$

b. $(\bar{x}_1 - \bar{x}_2) \pm z_{.005}\sqrt{\sigma_1^2/n_1 + \sigma_2^2/n_2}$

$12 \pm 2.575 \cdot \sqrt{(35.0)^2/804 + (27.0)^2/1657}$

12 ± 3.6

$8 < \mu_1 - \mu_2 < 16$

The confidence interval does not include the value 0. This indicates that there is a significant difference between the two means and agrees with the conclusion in part (a).

11. original claim: $\mu_1 - \mu_2 > 0$ [$n_1 > 30$ and $n_2 > 30$, use z (with s's for σ's)]

$\bar{x}_1 - \bar{x}_2 = 3214 - 3088 = 126$

H_o: $\mu_1 - \mu_2 \leq 0$
H_1: $\mu_1 - \mu_2 > 0$
$\alpha = .05$
C.R. $z > z_{.05} = 1.645$
calculations:

$\begin{aligned} z_{\bar{x}_1 - \bar{x}_2} &= (\bar{x}_1 - \bar{x}_2 - \mu_{\bar{x}_1 - \bar{x}_2})/\sigma_{\bar{x}_1 - \bar{x}_2} \\ &= (126 - 0)/\sqrt{(669)^2/294 + (728)^2/286} \\ &= 126/58.1 \\ &= 2.169 \end{aligned}$

conclusion:

Reject H_o; there is sufficient evidence to conclude that $\mu_1 - \mu_2 > 0$.

P-value = $P(z > 2.17) = .5000 - .4850 = .0150$

13. let the University of Massachusetts new texts be group 1

group 1: UMASS (n = 40)   group 2: DCC (n = 35)
$\Sigma x = 2604.70$ $\Sigma x = 2016.85$
$\Sigma x^2 = 190378.5648$ $\Sigma x^2 = 123549.5728$
$\bar{x} = 65.12$ $\bar{x} = 57.62$
$s^2 = 532.487$ $s^2 = 215.589$

original claim: $\mu_1 - \mu_2 = 0$ [$n_1 > 30$ and $n_2 > 30$, use z (with s's for σ's)]

$\bar{x}_1 - \bar{x}_2 = 65.12 - 57.62 = 7.493$

H_o: $\mu_1 - \mu_2 = 0$
H_1: $\mu_1 - \mu_2 \neq 0$
$\alpha = .05$ [assumed]
C.R. $z < -z_{.025} = -1.960$
     $z > z_{.025} = 1.960$
calculations:

$\begin{aligned} z_{\bar{x}_1 - \bar{x}_2} &= (\bar{x}_1 - \bar{x}_2 - \mu_{\bar{x}_1 - \bar{x}_2})/\sigma_{\bar{x}_1 - \bar{x}_2} \\ &= (7.493 - 0)/\sqrt{532.487/40 + 215.589/35} \\ &= 7.493/4.413 \\ &= 1.698 \end{aligned}$

conclusion:

Do not reject H_o; there is not sufficient evidence to reject the claim that $\mu_1 - \mu_2 = 0$.

P-value = $2 \cdot P(z > 1.70) = 2 \cdot (.5000 - .4554) = 2 \cdot (.0446) = .0892$

15. let the non-smokers not exposed to second-hand smoke (NOETS) be group 1

group 1: NOETS (n = 50)   group 2: ETS (n = 50)
$\Sigma x = 20.25$ $\Sigma x = 204.88$
$\Sigma x^2 = 80.2869$ $\Sigma x^2 = 5949.2360$
$\bar{x} = .4050$ $\bar{x} = 4.0976$
$s^2 = 1.471$ $s^2 = 104.280$

original claim: $\mu_1-\mu_2 = 0$ [$n_1 > 30$ and $n_2 > 30$, use z (with s's for σ's)]

$\overline{x}_1-\overline{x}_2 = .4050 - 4.0976 = -3.6926$

H_o: $\mu_1-\mu_2 = 0$

H_1: $\mu_1-\mu_2 \neq 0$

$\alpha = .05$ [assumed]

C.R. $z < -z_{.025} = -1.960$

$\qquad z > z_{.025} = 1.960$

calculations:

$z_{\overline{x}_1-\overline{x}_2} = (\overline{x}_1-\overline{x}_2 - \mu_{\overline{x}_1-\overline{x}_2})/\sigma_{\overline{x}_1-\overline{x}_2}$

$\qquad = (-3.6926 - 0)/\sqrt{1.471/50 + 104.280/50}$

$\qquad = -3.6926/1.454$

$\qquad = -2.539$

conclusion:

Reject H_o; there is sufficient evidence to reject the claim that $\mu_1-\mu_2 = 0$ and to conclude that $\mu_1-\mu_2 \neq 0$ (in fact, that $\mu_1-\mu_2 < 0$ -- i.e., that non-smokers not exposed to second-hand smoke have significantly less cotinine in their system than non-smokers so exposed).

P-value $= 2 \cdot P(z < -2.54) = 2 \cdot (.5000 - .4945) = 2 \cdot (.0055) = .0110$

17. original claim: $\mu_1-\mu_2 < 0$ [$n_1 > 30$ and $n_2 > 30$, use z (with s's for σ's)]

$\overline{x}_1-\overline{x}_2 = 152.0739 - 154.9669 = -2.893$

H_o: $\mu_1-\mu_2 \geq 0$

H_1: $\mu_1-\mu_2 < 0$

$\alpha = .05$

C.R. $z < -z_{.05} = -1.645$

calculations:

$z_{\overline{x}_1-\overline{x}_2} = (\overline{x}_1-\overline{x}_2 - \mu_{\overline{x}_1-\overline{x}_2})/\sigma_{\overline{x}_1-\overline{x}_2}$

$\qquad = (-2.893 - 0)/\sqrt{438.5388/50 + 239.1461/50}$

$\qquad = -.7858$ [from Excel]

conclusion:

Do not reject H_o; there is no sufficient evidence to conclude that $\mu_1-\mu_2 < 0$. While the sample mean was less for treatment group, it is not unlike that this could have occurred by chance and there was not enough evidence to conclude that this pattern applies in the populations.

P-value $= P(z < -.7858) = .215991$ [from Excel]

19. original claim: $\mu_1-\mu_2 > 0$ [$n_1 > 30$ and $n_2 > 30$, use z (with s's for σ's)]

$\overline{x}_1-\overline{x}_2 = 68.898 - 63.862 = 5.036$

H_o: $\mu_1-\mu_2 \leq 0$

H_1: $\mu_1-\mu_2 > 0$

$\alpha = .05$ [assumed]

C.R. $z > z_{.05} = 1.645$

calculations:

$z_{\overline{x}_1-\overline{x}_2} = (\overline{x}_1-\overline{x}_2 - \mu_{\overline{x}_1-\overline{x}_2})/\sigma_{\overline{x}_1-\overline{x}_2}$

$\qquad = 9.964$ [from TI-83 Plus]

conclusion:

Reject H_o; there is sufficient evidence to conclude that $\mu_1-\mu_2 > 0$. The probability of obtaining our sample result if the mean height for men was not greater than the mean height for women is so small, that we conclude as indicated.

P-value $= p(z > 9.964) = 1.11E-23 = .00000000000000000000000111$ [from TI-83 Plus]

21. FIRST: use the data sets as given
let the .0111 in. cans be group 1

group 1: .0111 in. (n = 175)   group 2: .0109 in (n = 175)
$\Sigma x = 49316$ $\Sigma x = 46745$
$\Sigma x^2 = 14031760$ $\Sigma x^2 = 12571335$
$\bar{x} = 281.8$ $\bar{x} = 267.1$
$s^2 = 771.43$ $s^2 = 488.95$

original claim: $\mu_1 - \mu_2 > 0$ [$n_1 > 30$ and $n_2 > 30$, use z (with s's for σ's)]
$\bar{x}_1 - \bar{x}_2 = 281.8 - 267.1 = 14.7$

H_o: $\mu_1 - \mu_2 \leq 0$
H_1: $\mu_1 - \mu_2 > 0$
$\alpha = .05$ [assumed]
C.R. $z > z_{.05} = 1.645$
calculations:

$z_{\bar{x}_1 - \bar{x}_2} = (\bar{x}_1 - \bar{x}_2 - \mu_{\bar{x}_1 - \bar{x}_2})/\sigma_{\bar{x}_1 - \bar{x}_2}$
$= (14.7 - 0)/\sqrt{771.43/175 + 488.95/175}$
$= 14.7/2.684$
$= 5.474$

conclusion:
Reject H_o; there is sufficient evidence to conclude that $\mu_1 - \mu_2 > 0$.
P-value = $P(z > 5.47) = .5000 - .4999 = .0001$

NEXT: consider the 504 to be an outlier and remove it from group 1

group 1: .0111 in. (n = 174)   group 2: .0109 in (n = 175)
$\Sigma x = 48812$ $\Sigma x = 46745$
$\Sigma x^2 = 13777744$ $\Sigma x^2 = 12571335$
$\bar{x} = 280.5$ $\bar{x} = 267.1$
$s^2 = 488.87$ $s^2 = 488.95$

NOTE: The new mean for group 1 is slightly smaller than the old one, and the new standard deviation is noticeably smaller -- just the types of changes one should expect when removing a large outlier.

original claim: $\mu_1 - \mu_2 > 0$ [$n_1 > 30$ and $n_2 > 30$, use z (with s's for σ's)]
$\bar{x}_1 - \bar{x}_2 = 280.5 - 267.1 = 13.4$
H_o: $\mu_1 - \mu_2 \leq 0$
H_1: $\mu_1 - \mu_2 > 0$
$\alpha = .05$ [assumed]
C.R. $z > z_{.05} = 1.645$
calculations:

$z_{\bar{x}_1 - \bar{x}_2} = (\bar{x}_1 - \bar{x}_2 - \mu_{\bar{x}_1 - \bar{x}_2})/\sigma_{\bar{x}_1 - \bar{x}_2}$
$= (13.4 - 0)/\sqrt{488.87/174 + 488.95/175}$
$= 13.4/2.367$
$= 5.661$

conclusion:
Reject H_o; there is sufficient evidence to conclude that $\mu_1 - \mu_2 > 0$.
P-value = $P(z > 5.66) = .5000 - .4999 = .0001$
NOTE: Even though the difference between the means was smaller, the decreased variability actually resulted in a slightly higher calculated z and a more significant difference.

ALSO: answer the questions about boxplots and confidence intervals
boxplots:

old 5-number boxplot summary (group 1)
$L = x_1 = 205$
$.25(175) = 43.74; P_{25} = x_{44} = 275$
$.50(175) = 87.50; P_{50} = x_{88} = 285$
$.75(175) = 131.25; P_{75} = x_{132} = 295$
$H = x_{175} = 504$

new 5-number boxplot summary (group 1)
$L = x_1 = 205$
$.25(174) = 43.5; P_{25} = x_{44} = 275$
$.50(174) = 87; P_{50} = (x_{87}+x_{88})/2 = (285+285)/2 = 285$
$.75(174) = 130.5; P_{75} = x_{131} = 294$
$H = x_{174} = 317$

The new boxplot is given below. The only significant difference is the smaller length of the extension from the third quartile to the maximum value, which would extend all the way to 504 for the original data.

205   275 285 294   317

confidence intervals:

original 95% confidence interval $(\mu_1-\mu_2)$
$(\bar{x}_1-\bar{x}_2) \pm z_{.025}\sqrt{\sigma_1^2/n_1 + \sigma_2^2/n_2}$
$(281.8 - 267.1) \pm 1.960 \cdot 2.684$
14.7 ± 5.3
$9.4 < \mu_1-\mu_2 < 20.0$

new 95% confidence interval $(\mu_1-\mu_2)$
$(\bar{x}_1-\bar{x}_2) \pm z_{.025}\sqrt{\sigma_1^2/n_1 + \sigma_2^2/n_2}$
$(280.5 - 267.1) \pm 1.960 \cdot 2.367$
13.4 ± 4.6
$8.8 < \mu_1-\mu_2 < 18.0$

The decrease in variability realized by dropping the outlier created a slightly narrower confidence interval.

CONCLUSION: In general, this sample is so large that the single outlier does not have a major effect on the test of hypothesis or the confidence interval for the difference between the means. The only major effect the outlier has is on the spread (as seen in the standard deviation and boxplot) of group 1.

23. a. x = 5,10,15
$\mu = \Sigma x/n = 30/3 = 10$
$\sigma^2 = \Sigma(x-\mu)^2/n = [(-5)^2 + (0)^2 + (5)^2]/3 = 50/3$

b. y = 1,2,3
$\mu = \Sigma y/n = 6/3 = 2$
$\sigma^2 = \Sigma(y-\mu)^2/n = [(-1)^2 + (0)^2 + (1)^2]/3 = 2/3$

c. z = x-y = 4,3,2,9,8,7,14,13,12
$\mu = \Sigma z/n = 72/9 = 8$
$\sigma^2 = \Sigma(z-\mu)^2/n$
$= [(-4)^2 + (-5)^2 + (-6)^2 + (1)^2 + (0)^2 + (-1)^2 + (6)^2 + (5)^2 + (4)^2]/9 = 156/9 = 52/3$

d. $\sigma_{x-y}^2 = \sigma_x^2 + \sigma_y^2$
$52/3 = 50/3 + 2/3$
$52/3 = 52/3$

e. Let R stand for range.
R_{x-y} = highest$_{x-y}$ - lowest$_{x-y}$
= (highest x - lowest y) - (lowest x - highest y)
= highest x - lowest y - lowest x + highest y
= (highest x - lowest x) + (highest y - lowest y)
= $R_x + R_y$

That is, the range of all possible x-y values is the sum of the individual ranges of x and y.

NOTE: The problem refers to all possible x-y differences (where n_x and n_y might even be different) and not to x-y differences for paired data.

8-3 Inferences about Two Means: Matched Pairs

NOTE: To be consistent with the notation of the previous sections, and thereby reinforcing the patterns and concepts presented in those sections, the manual uses the "usual" t formula written to apply to $\bar{d}$'s

$$t_{\bar{d}} = (\bar{d} - \mu_{\bar{d}})/s_{\bar{d}}$$

with $\mu_{\bar{d}} = \mu_d$
and $s_{\bar{d}} = s_d/\sqrt{n}$
And so the formula for the t statistic may also be written as

$$t_{\bar{d}} = (\bar{d} - \mu_d)/(s_d/\sqrt{n})$$

1. $d = x - y$: 5 6 0 5 -2
   summary statistics: $n = 5$, $\Sigma d = 14$, $\Sigma d^2 = 90$
   a. $\bar{d} = (\Sigma d)/n = 14/5 = 2.8$
   b. $s_d^2 = [n \cdot \Sigma d^2 - (\Sigma d)^2]/[n(n-1)]$
      $= [5 \cdot 90 - (14)^2]/[5(4)] = 254/20 = 12.7$
      $s_d = 3.6$
   c. $t_{\bar{d}} = (\bar{d} - \mu_{\bar{d}})/s_{\bar{d}}$
      $= (2.8 - 0)/(3.564/\sqrt{5})$
      $= 2.8/1.594$
      $= 1.757$
   d. $\pm t_{4,.025} = \pm 2.776$

3. $\bar{d} \pm t_{4,.025} \cdot s_d/\sqrt{n}$
   $2.8 \pm 2.776 \cdot 3.564/\sqrt{5}$
   2.8 ± 4.4
   $-1.6 < \mu_d < 7.2$

5. $d = x - y$: 0.5 -0.1 0.2 1.6 1.9 -0.4 0.3 -0.3 0.6 1.8
   $n = 10$
   $\Sigma d = 6.1$ $\bar{d} = .61$
   $\Sigma d^2 = 10.41$ $s_d = .862$
   a. original claim: $\mu_d > 0$ [$n \le 30$ and σ_d unknown, use t]
      $H_o: \mu_d \le 0$
      $H_1: \mu_d > 0$
      $\alpha = .05$
      C.R. $t > t_{9,.05} = 1.833$
      calculations:
      $t_{\bar{d}} = (\bar{d} - \mu_{\bar{d}})/s_{\bar{d}}$
      $= (.61 - 0)/(.862/\sqrt{10})$
      $= .61/.273$
      $= 2.238$

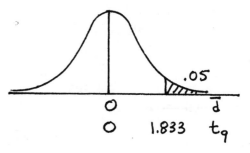

      conclusion:
         Reject H_o; there is sufficient evidence to conclude that $\mu_d > 0$.
   b. $\bar{d} \pm t_{9,.025} \cdot s_d/\sqrt{n}$
      $.61 \pm 2.262 \cdot .862/\sqrt{10}$
      $.61 \pm .62$
      $-.01 < \mu_d < 1.23$
      Since the confidence interval contains 0, there is no significant difference between the reported and measured heights. This does not contradict the result in part (a). Since the test of hypothesis was one-tailed test and the confidence interval is two-tailed, they are not equivalent. Together they show the value of one-tailed tests: a one-tailed test (remember: all claims must be made before seeing the data) can result in the rejection of the null hypothesis when a two-tailed test at the same level would not.

7. $d = x - y$: -20 0 10 -40 -30 -10 30 -20 -20 -10

    $n = 10$

    $\Sigma d = -110$ $\bar{d} = -11.0$

    $\Sigma d^2 = 4900$ $s_d = 20.248$

  a. original claim: $\mu_d < 0$ [$n \leq 30$ and σ_d unknown, use t]

    $H_o: \mu_d \geq 0$

    $H_1: \mu_d < 0$

    $\alpha = .05$

    C.R. $t < -t_{9,.05} = -1.833$

    calculations:

        $t_{\bar{d}} = (\bar{d} - \mu_{\bar{d}})/s_{\bar{d}}$

            $= (-11.0 - 0)/(20.248/\sqrt{10})$

            $= -11.0/6.40$

            $= -1.718$

    conclusion:

        Do not reject H_o; there is not sufficient evidence to conclude that $\mu_d < 0$.

  b. $\bar{d} \pm t_{9,.025} \cdot s_d/\sqrt{n}$

    $-11.0 \pm 2.262 \cdot 20.248/\sqrt{10}$

    -11.0 ± 14.5

    $-25.5 < \mu_d < 3.5$

  We have 95% confidence that the interval from -25.5 to 3.5 contains the true mean population difference. Since this interval includes 0, the mean before and after scores are not significantly different, and there is not enough evidence to say that the course has any effect.

9. $d = x_B - x_A$: -.2 4.1 1.6 1.8 3.2 2.0 2.9 9.6

    $n = 8$

    $\Sigma d = 25.0$ $\bar{d} = 3.125$

    $\Sigma d^2 = 137.46$ $s_d = 2.9114$

  a. asks for confidence interval for μ_d [$n \leq 30$ and σ_d unknown, use t]

    $\bar{d} \pm t_{7,.025} \cdot s_d/\sqrt{n}$

    $3.125 \pm 2.365 \cdot 2.9114/\sqrt{8}$

    3.125 ± 2.434

    $.69 < \mu_d < 5.56$

  b. original claim: $\mu_d > 0$

    $H_o: \mu_d \leq 0$

    $H_1: \mu_d > 0$

    $\alpha = .05$

    C.R. $t > t_{7,.05} = 1.895$

    calculations:

        $t_{\bar{d}} = (\bar{d} - \mu_{\bar{d}})/s_{\bar{d}}$

            $= (3.125 - 0)/(2.9114/\sqrt{8})$

            $= 3.125/1.029$

            $= 3.036$

    conclusion:

        Reject H_o; there is sufficient evidence to conclude that $\mu_d > 0$.

  c. Yes; hypnotism appears to be effective in reducing pain.

11. $d = x_8 - x_{12}$: given at the right; asks about μ_d [$n \le 30$ and σ_d unknown, use t]

$n = 11$

$\Sigma d = -8.6$ $\qquad\qquad$ $\bar{d} = -.782$

$\Sigma d^2 = 15.06$ $\qquad\qquad$ $s_d = .913$

sbj#	8am	12am	diff
19	97.0	97.7	-0.7
20	98.0	98.8	-0.8
22	96.4	98.0	-1.6
26	98.2	98.7	-0.5
71	98.8	98.0	0.8
78	98.6	98.5	0.1
80	97.8	98.3	-0.5
81	98.7	98.7	0.0
83	97.8	99.1	-1.3
98	96.4	98.2	-1.8
99	96.9	99.2	-2.3

a. $\bar{d} \pm t_{10,.025} \cdot s_d/\sqrt{n}$

$-.782 \pm 2.228 \cdot .913/\sqrt{11}$

$-.782 \pm .613$

$-1.40 < \mu_d < -.17$

b. original claim: $\mu_d = 0$

$H_o: \mu_d = 0$

$H_1: \mu_d \ne 0$

$\alpha = .05$

C.R. $t < -t_{10,.025} = -2.228$

$\quad\ t > t_{10,.025} = 2.228$

calculations:

$t_{\bar{d}} = (\bar{d} - \mu_{\bar{d}})/s_{\bar{d}}$

$\quad = (-.782 - 0)/(.913/\sqrt{11})$

$\quad = -.782/.275$

$\quad = -2.840$

conclusion:

Reject H_o; there is sufficient evidence to reject the claim that $\mu_d = 0$ and conclude that $\mu_d \ne 0$ (in fact, that $\mu_d < 0$.)

No; based on this result. morning and night body temperatures do not appear to be about the same. The 8 am (morning) temperatures are significantly lower.

13. $d = x_B - x_A$; asks about μ_d [$n \le 30$ and σ_d unknown, use t]

a. original claim: $\mu_d \ne 0$

$H_o: \mu_d = 0$

$H_1: \mu_d \ne 0$

$\alpha = .05$

C.R. $t < -t_{9,.025} = -2.262$

$\quad\ t > t_{9,.025} = 2.262$

calculations:

$t_{\bar{d}} = (\bar{d} - \mu_{\bar{d}})/s_{\bar{d}}$

$\quad = (-7.5 - 0)/(57.7/\sqrt{10})$

$\quad = -7.5/18.2$

$\quad = -.41$ [from Minitab]

conclusion:

Do not reject H_o; there is not sufficient evidence to conclude that $\mu_d \ne 0$.

No; based on this result, do not spend the money for the drug. NOTE: The conclusion could also have been reached by the P-value method. Since the Minitab's P-value $= 2 \cdot P(t_9 < -.41) = .69$ is greater than $\alpha = .05$, there is not sufficient evidence to reject H_o.

b. original claim: $\mu_d < 0$

$H_o: \mu_d \ge 0$

$H_1: \mu_d < 0$

$\alpha = .05$

C.R. $t < -t_{9,.05} = -1.833$

calculations:

$t_{\bar{d}} = (\bar{d} - \mu_{\bar{d}})/s_{\bar{d}}$

$\quad = (-7.5 - 0)/(57.7/\sqrt{10})$

$\quad = -7.5/18.2$

$\quad = -.41$ [from Minitab]

conclusion:

Do not reject H_o; there is not sufficient evidence to conclude that $\mu_d < 0$.

The calculations and the conclusion are the same as in part (a), but now the P-value = $P(t_9 < -.41) = \frac{1}{2}(.69) = .345$, which is still greater than $\alpha = .05$.

c. $\bar{d} \pm t_{9,.025} \cdot s_d/\sqrt{n}$

$-7.5 \pm 2.262 \cdot 57.7/\sqrt{10}$

-7.5 ± 41.3

$-48.8 < \mu_d < 33.8$

Yes; the confidence interval includes the value 0, indicating that the drug does not have a significant effect on vulnerability to motion sickness as measured.

15. summary statistics:

$n = 12$

$\Sigma d = 15.35 \qquad\qquad \bar{d} = 1.279$

$\Sigma d^2 = 75.0025 \qquad\qquad s_d = 2.244$

hypothesis test for the original claim: $\mu_d > 0$ [$n \leq 30$ and σ_d unknown, use t]

$H_o: \mu_d \leq 0$

$H_1: \mu_d > 0$

$\alpha = .05$

C.R. $t > t_{11,.05} = 1.796$

calculations:

$t_{\bar{d}} = (\bar{d} - \mu_{\bar{d}})/s_{\bar{d}}$

$= (1.279 - 0)/(2.244/\sqrt{12})$

$= 1.279/.6476$

$= 1.975$

conclusion:

Reject H_o; there is sufficient evidence to conclude that $\mu_d > 0$.

While the overall conclusion is the same, including the outlier changed the test considerably. The point estimate of μ_d almost doubled from .673 to 1.279, and the point estimate of σ_d almost tripled from .826 to 2.244. The increase in the estimate of the difference μ_d (which increases the significance of the sample result as measured by the calculated t) was offset by the increase in the estimate of the variability σ_d (which decreases the significance of the sample result as measured by the calculated t) and so the test statistic decreased from 2.701 to 1.975.

confidence interval:

$\bar{d} \pm t_{11,.025} \cdot s_d/\sqrt{n}$

$1.279 \pm 2.201 \cdot 2.244/\sqrt{12}$

1.279 ± 1.425

$-.146 < \mu_d < 2.705$

Even though the point estimate without the outlier was closer to 0, the confidence interval did not include 0 and suggested that there was a significant positive bias in the reported heights. Even though the point estimate including the outlier was farther from 0, the increased variability made the interval wide enough to include the value 0.

Yes; in general, an outlier can have a dramatic effect on the hypothesis test and the confidence interval.

NOTE: In order to obtain the best analysis possible and to avoid future outliers, most practicing statisticians will try to determine why a particular outlier may have occurred. In this instance it is plausible that the person was reporting 6'2.25" (which would be 74.25, very close to the actual value of 74.3) but either (a) he mistakenly thought 6' translated into 80 inches and wrote 82.25" or (b) instead of recording the value in inches as directed, he wrote 6'2.25" and his handwritten "6'" was mistaken for "8" (a careful check of the original data often clarifies supposed outliers).

17. Since the 95% confidence interval $\bar{d} \pm t_{df,.025} \cdot s_d/\sqrt{n}$ is $0 < \mu_d < 1.2$, we know the following:

$$\bar{d} - t_{df,.025} \cdot s_d/\sqrt{n} = 0$$
$$\bar{d} = t_{df,.025} \cdot s_d/\sqrt{n}$$
$$\bar{d} - 0 = t_{df,.025} \cdot s_d/\sqrt{n}$$
$$(\bar{d} - 0)/(\cdot s_d/\sqrt{n}) = t_{df,.025}$$

Testing the claim $\mu_d > 0$ at the α level produces the following:

H_o: $\mu_d \leq 0$
H_1: $\mu_d > 0$
$\alpha = ?$
C.R. $t > t_{df,\alpha}$
calculations:

$$t_{\bar{d}} = (\bar{d} - \mu_{\bar{d}})/s_{\bar{d}}$$
$$= (\bar{d} - 0)/(\cdot s_d/\sqrt{n})$$
$$= t_{df,.025} \text{ [from the confidence interval above]}$$

conclusion:

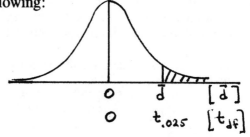

Reject H_o if the critical t is smaller than the calculated t -- i.e., whenever $t_{df,\alpha} < t_{df,.025}$.
Since $t_{df,\alpha} < t_{df,.025}$ occurs when $\alpha > .025$, the smallest possible value of α leading to the stipulated rejection of H_o is $\alpha = .025$.

8-4   Inferences about Two Proportions

NOTE: To be consistent with the notation of the previous sections, thereby reinforcing the patterns and concepts presented there, the manual uses the "usual" z formula written to apply to $\hat{p}_1-\hat{p}_2$'s

$$z_{\hat{p}_1-\hat{p}_2} = (\hat{p}_1-\hat{p}_2 - \mu_{\hat{p}_1-\hat{p}_2})/\sigma_{\hat{p}_1-\hat{p}_2}$$

with $\mu_{\hat{p}_1-\hat{p}_2} = p_1 - p_2$
and $\sigma_{\hat{p}_1-\hat{p}_2} = \sqrt{\bar{p}\bar{q}/n_1 + \bar{p}\bar{q}/n_2}$ [when H_o includes $p_1 = p_2$]
where $\bar{p} = (x_1 + x_2)/(n_1 + n_2)$

And so the formula for the z statistic may also be written as

$$z_{\hat{p}_1-\hat{p}_2} = ((\hat{p}_1-\hat{p}_2) - (p_1-p_2))/\sqrt{\bar{p}\bar{q}/n_1 + \bar{p}\bar{q}/n_2}$$

1. $x = \hat{p} \cdot n = .0425 \cdot 2750 = 116.875$, rounded to 117

3. $x = \hat{p} \cdot n = .289 \cdot 294 = 84.966$, rounded to 85

5. $\hat{p}_1 = x_1/n_1 = 10/20 = .500$
   $\hat{p}_2 = x_2/n_2 = 15/25 = .600$
   $\hat{p}_1-\hat{p}_2 = .500 - .600 = -.100$
   a. $\bar{p} = (x_1 + x_2)/(n_1 + n_2) = (10 + 15)/(20 + 25) = 25/45 = .556$
   b. $z_{\hat{p}_1-\hat{p}_2} = (\hat{p}_1-\hat{p}_2 - \mu_{\hat{p}_1-\hat{p}_2})/\sigma_{\hat{p}_1-\hat{p}_2}$
      $= (-.100 - 0)/\sqrt{(.556)(.444)/20 + (.556)(.444)/25}$
      $= -.100/.1491$
      $= -.671$
   c. $\pm z_{.025} = \pm 1.960$
   d. P-value $= 2 \cdot P(z < -.67) = 2 \cdot (.5000 - .2486) = 2 \cdot (.2514) = .5028$

NOTE: Since $\bar{p}$ is the weighted average of $\hat{p}_1$ and $\hat{p}_2$, it must always fall between those two values. If it does not, then an error has been made that must be corrected before proceeding. Calculation of $\sigma_{\hat{p}_1-\hat{p}_2} = \sqrt{\overline{pq}/n_1 + \overline{pq}/n_2}$ can be accomplished with no round-off loss on most calculators by calculating $\bar{p}$ and proceeding as follows: STORE 1-RECALL = * RECALL = STORE RECALL ÷ n_1 + RECALL ÷ n_2 = $\sqrt{}$. The quantity $\sigma_{\hat{p}_1-\hat{p}_2}$ may then be STORED for future use. Each calculator is different -- learn how your calculator works, and do the homework on the same calculator you will use for the exam. If you have any questions about performing/storing calculations on your calculator, check with your instructor or class assistant.

7. Let those who received the vaccine be group 1.
   original claim: $p_1-p_2 < 0$
   $\hat{p}_1 = x_1/n_1 = 14/1070 = .013$
   $\hat{p}_2 = x_2/n_2 = 95/532 = .179$
   $\hat{p}_1-\hat{p}_2 = .013 - .179 = -.165$
   $\bar{p} = (x_1 + x_2)/(n_1 + n_2)$
   $= (14 + 95)/(1070 + 532) = 109/1602 = .068$
   H_o: $p_1-p_2 \geq 0$
   H_1: $p_1-p_2 < 0$
   $\alpha = .05$ [assumed]
   C.R. $z < -z_{.05} = -1.645$
   calculations:

   $z_{\hat{p}_1-\hat{p}_2} = (\hat{p}_1-\hat{p}_2 - \mu_{\hat{p}_1-\hat{p}_2})/\sigma_{\hat{p}_1-\hat{p}_2}$
   $= (-.165 - 0)/\sqrt{(.068)(.932)/1070 + (.068)(.932)/532}$
   $= -.165/.0134$
   $= -12.388$
   conclusion:
   Reject H_o; there is sufficient evidence to conclude that $p_1-p_2 < 0$.
   P-value = $P(z < -12.39) = .5000 - .4999 = .0001$

9. Let those not wearing seat belts be group 1.
   original claim: $p_1-p_2 > 0$
   $\hat{p}_1 = x_1/n_1 = 50/290 = .1724$
   $\hat{p}_2 = x_2/n_2 = 16/123 = .1301$
   $\hat{p}_1-\hat{p}_2 = .1724 - .1301 = .0423$
   $\bar{p} = (x_1 + x_2)/(n_1 + n_2)$
   $= (50 + 16)/(290 + 123) = 66/413 = .160$
   H_o: $p_1-p_2 \leq 0$
   H_1: $p_1-p_2 > 0$
   $\alpha = .05$
   C.R. $z > z_{.05} = 1.645$
   calculations:

   $z_{\hat{p}_1-\hat{p}_2} = (\hat{p}_1-\hat{p}_2 - \mu_{\hat{p}_1-\hat{p}_2})/\sigma_{\hat{p}_1-\hat{p}_2}$
   $= (.0423 - 0)/\sqrt{(.160)(.840)/290 + (.160)(.840)/123}$
   $= .0423/.0394$
   $= 1.074$
   conclusion:
   Do not reject H_o; there is not sufficient evidence to conclude that $p_1-p_2 > 0$.
   P-value = $P(z > 1.07) = .5000 - .3577 = .1423$

Based on these results, no specific action should be taken.

NOTE: This test involves only children who were hospitalized. It is probably true that a much greater percentage of children wearing seat belts avoided having to be hospitalized in the first place.

11. Let those receiving the Salk vaccine be group 1.
    original claim: $p_1-p_2 < 0$
    $\hat{p}_1 = x_1/n_1 = 33/200,000 = .000165$
    $\hat{p}_2 = x_2/n_2 = 115/200,000 = .000575$
    $\hat{p}_1-\hat{p}_2 = .000165 - .000575 = -.000410$
    $\bar{p} = (x_1 + x_2)/(n_1 + n_2)$
    $\quad = (33 + 115)/(200,000 + 200,000)$
    $\quad = 148/400,000 = .00037$
    H_o: $p_1-p_2 \geq 0$
    H_1: $p_1-p_2 < 0$
    $\alpha = .01$
    C.R. $z < -z_{.01} = -2.327$
    calculations:

$$Z_{\hat{p}1-\hat{p}2} = (\hat{p}_1-\hat{p}_2 - \mu_{\hat{p}1-\hat{p}2})/\sigma_{\hat{p}1-\hat{p}2}$$
$$= (-.000410 - 0)/\sqrt{(.00037)(.99963)/200,000 + (.00037)(.99963)/200,000}$$
$$= -.000410/.0000608$$
$$= -6.7416 \quad \text{[from TI-83 Plus]}$$

conclusion:
    Reject H_o; there is sufficient evidence to conclude that $p_1-p_2 < 0$.
    P-value = $P(z < -6.74) = 7.88E-12 = .00000000000788$ [from TI-83 Plus]
    Yes; assuming the group assignments were made at random, the vaccine appears to be effective.

NOTE: The values .00165, .00575 and .00037 appear above in the intermediate steps of the problem and in scientific notation on the TI-83 Plus. Even problems with calculator and/or software display given in the text will typically be worked out in some detail in this manual. For better understanding, match the numbers appearing in the displays with those appearing in the intermediate steps.

13. Let those younger than 21 be group 1.
    $\hat{p}_1 = x_1/n_1 = .0425$
    $\hat{p}_2 = x_2/n_2 = .0455$
    $\hat{p}_1-\hat{p}_2 = .0425 - .0455 = -.0029$
    $(\hat{p}_1-\hat{p}_2) \pm z_{.025}\sqrt{\hat{p}_1\hat{q}_1/n_1 + \hat{p}_2\hat{q}_2/n_2}$
    $-.0029 \pm 1.960 \cdot \sqrt{(.0425)(.9575)/2750 + (.0455)(.9545)/2200}$
    $-.0029 \pm .0115$
    $-.0144 < p_1-p_2 < .0086$
    Yes; the interval contains zero, indicating no significant difference between the two rates of violent crimes.

15. Let those given the written survey be group 1.
    $\hat{p}_1 = x_1/n_1 = 67/850 = .079 \ (x_1 = .079 \cdot 850)$
    $\hat{p}_2 = x_2/n_2 = 105/850 = .124 \ (x_2 = .124 \cdot 850)$
    $\hat{p}_1-\hat{p}_2 = .079 - .124 = -.0447$
    $\bar{p} = (x_1 + x_2)/(n_1 + n_2)$
    $\quad = (67 + 105)/(850 + 850) = 172/1700 = .1012$

a. original claim: $p_1-p_2 = 0$
   H_o: $p_1-p_2 = 0$
   H_1: $p_1-p_2 \neq 0$
   $\alpha = .05$ [assumed]
   C.R. $z < -z_{.025} = -1.960$
         $z > z_{.025} = 1.960$
   calculations:

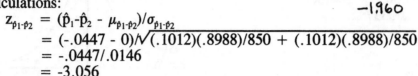

$$z_{\hat{p}_1-\hat{p}_2} = (\hat{p}_1-\hat{p}_2 - \mu_{\hat{p}_1-\hat{p}_2})/\sigma_{\hat{p}_1-\hat{p}_2}$$
$$= (-.0447 - 0)/\sqrt{(.1012)(.8988)/850 + (.1012)(.8988)/850}$$
$$= -.0447/.0146$$
$$= -3.056$$

conclusion:
    Reject H_o; there is sufficient evidence to reject the claim that $p_1 - p_2 = 0$ and to conclude that $p_1-p_2 \neq 0$ (in fact, that $p_1 - p_2 < 0$ -- i.e., that fewer students receiving the written test admit carrying a gun.)
    P-value $= 2 \cdot P(z < -3.06) = 2 \cdot (.5000 - .4989) = 2 \cdot (.0011) = .0022$
Yes; based on this result the difference between 7.9% and 12.4% is significant.

b. $(\hat{p}_1-\hat{p}_2) \pm z_{.005}\sqrt{\hat{p}_1\hat{q}_1/n_1 + \hat{p}_2\hat{q}_2/n_2}$
   $-.0447 \pm 2.575 \cdot \sqrt{(.079)(.921)/850 + (.124)(.876)/850}$
   $-.0447 \pm .0376$
   $-.0823 < p_1-p_2 < -.0071$

The interval does not contain zero, indicating a significant difference between the two response rates. We are 99% confident the interval from -8.3% to -0.7% contains the true difference between the population percentages.

17. Let the convicted arsonists be group 1.
    original claim: $p_1-p_2 > 0$
       $\hat{p}_1 = x_1/n_1 = 50/(50 + 43) = 50/93 = .538$
       $\hat{p}_2 = x_2/n_2 = 63/(63 + 144) = 63/207 = .304$
       $\hat{p}_1-\hat{p}_2 = .538 - .304 = .233$
       $\bar{p} = (x_1 + x_2)/(n_1 + n_2)$
          $= (50 + 63)/(93 + 207) = 113/300 = .377$
    H_o: $p_1-p_2 \leq 0$
    H_1: $p_1-p_2 > 0$
    $\alpha = .01$
    C.R. $z > z_{.01} = 2.327$
    calculations:

   $z_{\hat{p}_1-\hat{p}_2} = (\hat{p}_1-\hat{p}_2 - \mu_{\hat{p}_1-\hat{p}_2})/\sigma_{\hat{p}_1-\hat{p}_2}$
        $= (.233 - 0)/\sqrt{(.377)(.623)/93 + (.377)(.623)/207}$
        $= .233/.0605$
        $= 3.857$
conclusion:
    Reject H_o; there is sufficient evidence to conclude that $p_1-p_2 > 0$.
P-value $= P(z > 3.86) = .5000 - .4999 = .0001$

Yes; it does seem reasonable that drinking might be related to the type of crime. The kinds of problems and personalities associated with drinking may well be more likely to be associated with some crimes more than others.

19. Let those receiving warmth be group 1.
    original claim: $p_1 - p_2 < 0$
    $\hat{p}_1 = x_1/n_1 = 6/104 = .0577$
    $\hat{p}_2 = x_2/n_2 = 18/96 = .1875$
    $\hat{p}_1 - \hat{p}_2 = .0577 - .1875 = -.1298$
    $\bar{p} = (x_1 + x_2)/(n_1 + n_2)$
    $\quad = (6 + 18)/(104 + 96) = 24/200 = .12$
    $H_o: p_1 - p_2 \geq 0$
    $H_1: p_1 - p_2 < 0$
    $\alpha = .05$
    C.R. $z < -z_{.05} = -1.645$
    calculations:
    $z_{\hat{p}_1 - \hat{p}_2} = (\hat{p}_1 - \hat{p}_2 - \mu_{\hat{p}_1 - \hat{p}_2})/\sigma_{\hat{p}_1 - \hat{p}_2}$
    $\quad = (-.1298 - 0)/\sqrt{(.12)(.88)/104 + (.12)(.88)/96}$
    $\quad = -.1298/.0460$
    $\quad = -2.822$
    conclusion:
    Reject H_o; there is sufficient evidence to conclude that $p_1 - p_2 < 0$.
    P-value $= P(z < -2.82) = .5000 - .4976 = .0024$
    Yes; if these results are verified, surgical patients should be routinely warmed.

21. Let those using Viagra be group 1.
    original claim: $p_1 - p_2 = .10$
    $\hat{p}_1 = x_1/n_1 = .16$ [see note below]
    $\hat{p}_2 = x_2/n_2 = .04$ [see note below]
    $\hat{p}_1 - \hat{p}_2 = .16 - .04 = .12$
    $H_o: p_1 - p_2 = .10$
    $H_1: p_1 - p_2 \neq .10$
    $\alpha = .05$
    C.R. $z < -z_{.025} = -1.960$
    $\quad z > z_{.025} = 1.960$
    calculations:
    $z_{\hat{p}_1 - \hat{p}_2} = (\hat{p}_1 - \hat{p}_2 - \mu_{\hat{p}_1 - \hat{p}_2})/\sigma_{\hat{p}_1 - \hat{p}_2}$
    $\quad = (.12 - .10)/\sqrt{(.16)(.84)/734 + (.04)(.96)/725}$
    $\quad = .02/.0154$
    $\quad = 1.302$
    conclusion:
    Do not reject H_o; there is not sufficient evidence to reject the claim that $p_1 - p_2 = .10$.
    P-value $= 2 \cdot P(z > 1.30) = 2 \cdot (.5000 - .4032) = 2 \cdot (.0968) = .1936$
    NOTE: The values of x_1 and x_2 were not given. Since any x_1 between 114 and 121 inclusive
    and any x_2 between 26 and 32 inclusive round to the stated percents of 16% and 4%, no further
    accuracy is possible. The calculations proceed using only whole percents, and any attempt to
    identify more significant digits cannot be justified from the statement of the problem.

23. $E^2 = (z_{\alpha/2})^2(p_1q_1/n_1 + p_2q_2/n_2)$ squaring the original equation
    $E^2 = (z_{\alpha/2})^2(.25/n_1 + .25/n_2)$ setting the unknown proportions to .5
    $E^2 = (z_{\alpha/2})^2(.25/n + .25/n)$ requiring $n_1 = n_2 = n$
    $E^2 = (z_{\alpha/2})^2(.50/n)$ addition
    $n = (z_{\alpha/2})^2(.50)/E^2$ solving for n
    $\quad = (1.960)^2(.50)/(.03)^2$ for $\alpha = .05$ and $E = .03$
    $\quad = 2134.2$ rounded up to 2135

8-5 Comparing Variation in Two Samples

NOTE: The following conventions are used in this manual regarding the F test.
* The set of scores with the larger sample variance is designated group 1.
* Even though always designating the scores with the larger sample variance as group 1 makes lower critical values are unnecessary in two-tailed tests, the lower critical value will be calculated (using the method given in exercise #14) and included (in brackets) for completeness and for consistency with the other tests.
* The degrees of freedom for group 1 (numerator) and group 2 (denominator) will be given with the F as a superscript and subscript respectively.
* If the desired degrees of freedom does not appear in Table A-5, the closest entry will be used. If the desired degrees of freedom is exactly halfway between two tabled values, the conservative approach of using the smaller degrees of freedom is employed. Since any finite number is closer to 120 than ∞, 120 is used for all degrees of freedom larger than 120.
* Since all hypotheses in the text question the equality of σ_1^2 and σ_2^2, the calculation of F [which is statistically defined to be $F = (s_1^2/\sigma_1^2)/(s_2^2/\sigma_2^2)$] is shortened to $F = s_1^2/s_2^2$.

1. Let those receiving the treatment be group 1. original claim: $\sigma_1^2 \neq \sigma_2^2$
    $H_o: \sigma_1^2 = \sigma_2^2$
    $H_1: \sigma_1^2 \neq \sigma_2^2$
    $\alpha = .05$
    C.R. $F < F_{29,.975}^{15} = [.3783]$
    $\quad\quad F > F_{29,.025}^{15} = 2.3248$
    calculations:
    $\quad F = s_1^2/s_2^2$
    $\quad\quad = (150.0)^2/(75.0)^2$
    $\quad\quad = 4.0000$
    conclusion:

    Reject H_o; there is sufficient evidence to conclude that $\sigma_1^2 \neq \sigma_2^2$ (in fact, that $\sigma_1^2 > \sigma_2^2$).

3. Let the regular Coke be group 1. original claim: $\sigma_1^2 = \sigma_2^2$
    $H_o: \sigma_1^2 = \sigma_2^2$
    $H_1: \sigma_1^2 \neq \sigma_2^2$
    $\alpha = .05$
    C.R. $F < F_{35,.975}^{35} = [.4822]$
    $\quad\quad F > F_{35,.025}^{35} = 2.0739$
    calculations:
    $\quad F = s_1^2/s_2^2$
    $\quad\quad = (.007507)^2/(.004391)^2$
    $\quad\quad = 2.9228$
    conclusion:

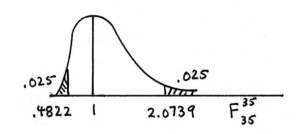

    Reject H_o; there is sufficient evidence to reject the claim that $\sigma_1^2 = \sigma_2^2$ and to conclude that $\sigma_1^2 \neq \sigma_2^2$ (in fact, that $\sigma_1^2 > \sigma_2^2$).

The significant difference in variability between the weights of regular and diet Coke is not necessarily due to differences in quality control efforts. Regular Coke may contain a relatively heavy ingredient (e.g., sugar) that is difficult to dispense with the same level of consistency as the other ingredients.

5. Let the stress sample be group 1. original claim: $\sigma_1 \neq \sigma_2$ [i.e., $\sigma_1^2 \neq \sigma_2^2$]
   H_o: $\sigma_1^2 = \sigma_2^2$
   H_1: $\sigma_1^2 \neq \sigma_2^2$
   $\alpha = .10$
   C.R. $F < F_{39,.95}^{39} = [.5907]$
       $F > F_{39,.05}^{39} = 1.6928$
   calculations:
       $F = s_1^2/s_2^2$
         $= (13.2)^2/(11.6)^2$
         $= 1.2949$
   conclusion:
       Do not reject H_o; there is not sufficient evidence to conclude that $\sigma_1^2 \neq \sigma_2^2$.

7. Let the students be group 1. original claim: $\sigma_1^2 > \sigma_2^2$
   H_o: $\sigma_1^2 \leq \sigma_2^2$
   H_1: $\sigma_1^2 > \sigma_2^2$
   $\alpha = .05$ [assumed]
   C.R. $F > F_{151,.05}^{216} = 1.3519$
   calculations:
       $F = s_1^2/s_2^2$
         $= (3.67)^2/(3.65)^2$
         $= 1.0110$
   conclusion:
       Do not reject H_o; there is not sufficient evidence to conclude that $\sigma_1^2 > \sigma_2^2$.

9. easy to difficult   difficult to easy
   $n = 25$ $n = 16$
   $\Sigma x = 677.88$ $\Sigma x = 507.65$
   $\Sigma x^2 = 19,509.3278$ $\Sigma x^2 = 16,379.0161$
   $\bar{x} = 27.115$ $\bar{x} = 31.728$
   $s^2 = 47.0198$ $s^2 = 18.1489$

   Let the easy to difficult sample be group 1. original claim: $\sigma_1^2 = \sigma_2^2$
   H_o: $\sigma_1^2 = \sigma_2^2$
   H_1: $\sigma_1^2 \neq \sigma_2^2$
   $\alpha = .05$
   C.R. $F < F_{15,.975}^{24} = [.4103]$
       $F > F_{15,.025}^{24} = 2.7006$
   calculations:
       $F = s_1^2/s_2^2$
         $= 47.0198/18.1489$
         $= 2.5908$
   conclusion:
       Do not reject H_o; there is not sufficient evidence to reject the claim that $\sigma_1^2 = \sigma_2^2$.

10. placebo   calcium
    $n = 13$ $n = 15$
    $\Sigma x = 1490.5$ $\Sigma x = 1740.4$
    $\Sigma x^2 = 171965.47$ $\Sigma x^2 = 202936.92$
    $\bar{x} = 114.65$ $\bar{x} = 116.03$
    $s^2 = 89.493$ $s^2 = 71.722$

Let those who took the placebo be group 1. original claim: $\sigma_1^2 = \sigma_2^2$
H_o: $\sigma_1^2 = \sigma_2^2$
H_1: $\sigma_1^2 \neq \sigma_2^2$
$\alpha = .10$
C.R. F $<$ $F_{14,.95}^{12} = [.3821]$
     F $>$ $F_{14,.05}^{12} = 2.5342$
calculations:
   F $= s_1^2/s_2^2$
     $= 89.493/71.722$
     $= 1.2478$

conclusion:
   Do not reject H_o; there is not sufficient evidence to reject the claim that $\sigma_1^2 = \sigma_2^2$.
P-value $= P(F_{14}^{12} > 1.2478) = .6852$ [from TI-83 Plus]

11. <u>Diet Coke</u>   <u>Diet Pepsi</u>
   n $= 36$ n $= 36$
   $\Sigma x = 28.2526$ $\Sigma x = 28.2189$
   $\Sigma x^2 = 22.17315828$ $\Sigma x^2 = 22.12028574$
   $\bar{x} = .78479$ $\bar{x} = .78386$
   $s^2 = .000019278$ $s^2 = .000019023$

Let the diet Coke be group 1. original claim: $\sigma_1 \neq \sigma_2$ [i.e., $\sigma_1^2 \neq \sigma_2^2$]
H_o: $\sigma_1^2 = \sigma_2^2$
H_1: $\sigma_1^2 \neq \sigma_2^2$
$\alpha = .05$
C.R. F $<$ $F_{35,.975}^{35} = [.4822]$
     F $>$ $F_{35,.025}^{35} = 2.0739$
calculations:
   F $= s_1^2/s_2^2$
     $= .000019278/.000019023$
     $= 1.0134$
conclusion:
   Do not reject H_o; there is not sufficient evidence to conclude that $\sigma_1^2 \neq \sigma_2^2$.

13. Refer to exercise #3.
The calculated statistic is $F_{35}^{35} = 2.9228$ in a two-tailed test.
Since $2.9228 > F_{35,.01}^{35} = 2.3860$, the largest tabled value, $P(F_{35}^{35} > 2.9228) < .01$.
P-value $= 2 \cdot P(F_{35}^{35} > 2.9228) < 2 \cdot (.01) = .02$ -- i.e., P-value $< .02$.

15. Let those who took the placebo be group 1.
   $F_L = F_{14,.975}^{12} = 1/F_{12,.025}^{14} = 1/3.1772 = .3147$
   $F_R = F_{14,.025}^{12} = 3.0502$
               $(s_1^2/s_2^2) \cdot (1/F_R) < \sigma_1^2/\sigma_2^2 < (s_1^2/s_2^2) \cdot (1/F_L)$
  $(89.493/71.722) \cdot (1/3.0502) < \sigma_1^2/\sigma_2^2 < (89.493/71.722) \cdot (1/.3147)$
                        $.409 < \sigma_1^2/\sigma_2^2 < 3.964$
NOTE: This $\alpha = .05$ confidence interval includes the ratio 1.00 (which represents $\sigma_1^2 = \sigma_2^2$),
thus agreeing with the $\alpha = .10$ test of hypothesis in exercise #10 which failed to reject
H_o: $\sigma_1^2 = \sigma_2^2$. If we could not reject the idea of equality in exercise #10 where we were willing
to make a false rejection 10% of the time, we would not expect to exclude the idea of equality
in exercise #15 where we were willing to make a false exclusion only 5% of the time.

8-6 Inferences about Two Means: Independent and Small samples

1. Let the treatment group be group 1.
   original claim: $\mu_1 - \mu_2 = 0$ [small samples and σ unknown, first test H_o: $\sigma_1^2 = \sigma_2^2$]
   Since $s_1 = .62$ and $s_2 = .61$, it is obvious we will not be able to reject H_o: $\sigma_1^2 = \sigma_2^2$.
   Although the directions for this exercise allow us to make that judgment without a formal test, we proceed for the sake of completeness [and those instructors who may require the test regardless].
   H_o: $\sigma_1^2 = \sigma_2^2$
   H_1: $\sigma_1^2 \neq \sigma_2^2$
   $\alpha = .05$
   C.R. $F < F_{24,.975}^{4} = [.1175]$
   $\quad\quad F > F_{24,.025}^{4} = 3.3794$
   calculations:
   $\quad F = s_1^2/s_2^2$
   $\quad\quad = (.62)^2/(.61)^2$
   $\quad\quad = 1.0331$

   conclusion:
   Do not reject H_o; there is not sufficient evidence to reject the claim that $\sigma_1^2 = \sigma_2^2$.
   Now proceed using s_p^2 for both s_1^2 and s_2^2.
   $\bar{x}_1 - \bar{x}_2 = 98.20 - 104.60 = -6.40$
   $s_p^2 = (df_1 \cdot s_1^2 + df_2 \cdot s_2^2)/(df_1 + df_2)$
   $\quad\quad = (4 \cdot .62^2 + 24 \cdot .61^2)/(4 + 24) = 10.4680/28 = .3739$
   H_o: $\mu_1 - \mu_2 = 0$
   H_1: $\mu_1 - \mu_2 \neq 0$
   $\alpha = .05$
   C.R. $t < -t_{28,.025} = -2.048$
   $\quad\quad t > t_{28,.025} = 2.048$
   calculations:
   $\quad t_{\bar{x}_1 - \bar{x}_2} = (\bar{x}_1 - \bar{x}_2 - \mu_{\bar{x}_1 - \bar{x}_2})/s_{\bar{x}_1 - \bar{x}_2}$
   $\quad\quad = (-6.40 - 0)/\sqrt{.3739/5 + .3739/25}$
   $\quad\quad = -6.40/.2995$
   $\quad\quad = -21.366$

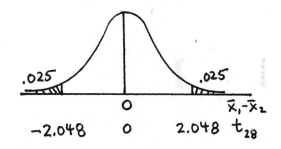

   conclusion:
   Reject H_o; there is sufficient evidence to reject the claim that $\mu_1 - \mu_2 = 0$ and to conclude that $\mu_1 - \mu_2 \neq 0$ (in fact, that $\mu_1 - \mu_2 < 0$).

3. Refer to exercise #1.
   $(\bar{x}_1 - \bar{x}_2) \pm t_{28,.025}\sqrt{s_p^2/n_1 + s_p^2/n_2}$
   $-6.40 \pm 2.048 \cdot \sqrt{.3739/5 + .3739/25}$
   $-6.40 \pm .61$
   $-7.01 < \mu_1 - \mu_2 < -5.79$

5. Let the men be group 1.
   original claim: $\mu_1-\mu_2 = 0$ [small samples and σ unknown, first test H_o: $\sigma_1^2 = \sigma_2^2$]
       Although the directions for this exercise allow one to proceed without performing this test,
   we proceed for the sake of completeness [and those instructors who may require the test
   regardless].

   H_o: $\sigma_1^2 = \sigma_2^2$
   H_1: $\sigma_1^2 \neq \sigma_2^2$
   $\alpha = .05$
   C.R. $F < F_{3,.975}^4 = [.1288]$
       $F > F_{3,.025}^4 = 15.101$
   calculations:
       $F = s_1^2/s_2^2$
         $= (2.747)^2/(1.698)^2$
         $= 2.6172$

   conclusion:
       Do not reject H_o; there is not sufficient evidence to reject the claim that $\sigma_1^2 = \sigma_2^2$.

   a. Assume $\sigma_1^2 = \sigma_2^2$ and proceed using s_p^2 for both s_1^2 and s_2^2.
       $\bar{x}_1-\bar{x}_2 = 69.920 - 64.725 = 5.195$
       $s_p^2 = (df_1 \cdot s_1^2 + df_2 \cdot s_2^2)/(df_1 + df_2)$
         $= (4 \cdot 2.747^2 + 3 \cdot 1.698^2)/(4 + 3) = 38.834/7 = 5.5477$
   H_o: $\mu_1-\mu_2 = 0$
   H_1: $\mu_1-\mu_2 \neq 0$
   $\alpha = .01$
   C.R. $t < -t_{7,.005} = -3.500$
       $t > t_{7,.005} = 3.500$
   calculations:
       $t_{\bar{x}_1-\bar{x}_2} = (\bar{x}_1-\bar{x}_2 - \mu_{\bar{x}_1-\bar{x}_2})/s_{\bar{x}_1-\bar{x}_2}$
         $= (5.195 - 0)/\sqrt{5.5477/5 + 5.5477/4}$
         $= 5.195/1.5800$
         $= 3.288$

   conclusion:
       Do not reject H_o; there is not sufficient evidence to reject the claim that $\mu_1-\mu_2 = 0$.

   b. Assume $\sigma_1^2 \neq \sigma_2^2$ and proceed using s_1^2 and s_2^2 and the smaller df (viz., $df_2 = 3$).
       $\bar{x}_1-\bar{x}_2 = 69.920 - 64.725 = 5.195$
   H_o: $\mu_1-\mu_2 = 0$
   H_1: $\mu_1-\mu_2 \neq 0$
   $\alpha = .01$
   C.R. $t < -t_{3,.005} = -5.841$
       $t > t_{3,.005} = 5.841$
   calculations:
       $t_{\bar{x}_1-\bar{x}_2} = (\bar{x}_1-\bar{x}_2 - \mu_{\bar{x}_1-\bar{x}_2})/s_{\bar{x}_1-\bar{x}_2}$
         $= (5.195 - 0)/\sqrt{(2.747)^2/5 + (1.698)^2/4}$
         $= 5.195/1.4933$
         $= 3.479$

   conclusion:
       Do not reject H_o; there is not sufficient evidence to reject the claim that $\mu_1-\mu_2 = 0$.

   c. Yes; both procedures lead to the same result. While there is not enough evidence to reject
   the claim that mean and women have the same mean height, that does not mean that the data
   support that claim.

   d. Based on the given sample data, we conclude that there is not enough evidence to reject the
   claim that men and women have the same mean height. The results do not agree with
   general knowledge about such heights, but the results do agree with the general knowledge
   about statistical tests -- that a specific conclusion should not be made on the basis of sample

sizes of $n_1 = 5$ and $n_2 = 4$ unless the evidence is overwhelming beyond all doubt, and these results are not.

7. Let the O-C patients be group 1.
   original claim: $\mu_1 - \mu_2 = 0$ [small samples and σ unknown, first test H_0: $\sigma_1^2 = \sigma_2^2$]
   H_0: $\sigma_1^2 = \sigma_2^2$
   H_1: $\sigma_1^2 \neq \sigma_2^2$
   $\alpha = .05$ [assumed]
   C.R. $F < F_{9,.975}^9 = [.2484]$
   $\quad\ \ F > F_{9,.025}^9 = 4.0260$
   calculations:
   $\quad F = s_1^2/s_2^2$
   $\quad\quad = (.08)^2/(.08)^2$
   $\quad\quad = 1.0000$
   conclusion:

   Do not reject H_0; there is not sufficient evidence to conclude that $\sigma_1^2 \neq \sigma_2^2$.
   Now proceed using s_p^2 for both s_1^2 and s_2^2
   $\overline{x}_1 - \overline{x}_2 = .34 - .45 = -.11$
   $s_p^2 = (df_1 \cdot s_1^2 + df_2 \cdot s_2^2)/(df_1 + df_2)$
   $\quad = (9 \cdot (.08)^2 + 9 \cdot (.08)^2)/(9 + 9) = .1152/18 = .0064$
   H_0: $\mu_1 - \mu_2 = 0$
   H_1: $\mu_1 - \mu_2 \neq 0$
   $\alpha = .01$
   C.R. $t < -t_{18,.005} = -2.878$
   $\quad\ \ t > t_{18,.005} = 2.878$
   calculations:
   $\quad t_{\overline{x}_1 - \overline{x}_2} = (\overline{x}_1 - \overline{x}_2 - \mu_{\overline{x}_1 - \overline{x}_2})/s_{\overline{x}_1 - \overline{x}_2}$
   $\quad\quad = (-.11 - 0)/\sqrt{.0064/10 + .0064/10}$
   $\quad\quad = -.11/.0358$
   $\quad\quad = -3.075$

   conclusion:
   Reject H_0; there is sufficient evidence to reject the claim that $\mu_1 - \mu_2 = 0$ and to conclude that $\mu_1 - \mu_2 \neq 0$ (in fact, $\mu_1 - \mu_2 < 0$).
   Based on this result, it appears that obsessive-compulsive disorders have a biological indicator.

9. Let the O-C patients be group 1.
   confidence interval: $\mu_1 - \mu_2$ [small samples and σ unknown, first test H_0: $\sigma_1^2 = \sigma_2^2$]
   Although the directions for this exercise allow one to proceed without performing this test, we proceed for the sake of completeness [and those instructors who may require the test regardless].
   H_0: $\sigma_1^2 = \sigma_2^2$
   H_1: $\sigma_1^2 \neq \sigma_2^2$
   $\alpha = .05$
   C.R. $F < F_{9,.975}^9 = [.2484]$
   $\quad\ \ F > F_{9,.025}^9 = 4.0260$
   calculations:
   $\quad F = s_1^2/s_2^2$
   $\quad\quad = (156.84)^2/(137.97)^2$
   $\quad\quad = 1.2922$
   conclusion:

   Do not reject H_0; there is not sufficient evidence to conclude that $\sigma_1^2 \neq \sigma_2^2$.

a. Assume $\sigma_1^2 \neq \sigma_2^2$ and proceed using s_p^2 for both s_1^2 and s_2^2.

$\bar{x}_1 - \bar{x}_2 = 1390.03 - 1268.41 = 121.62$

$s_p^2 = (df_1 \cdot s_1^2 + df_2 \cdot s_2^2)/(df_1 + df_2)$
$= (9 \cdot (156.84)^2 + 9 \cdot (137.97)^2)/(9 + 9) = 43634.5065/18 = 21817.3$

$(\bar{x}_1 - \bar{x}_2) \pm t_{18,.025}\sqrt{s_p^2/n_1 + s_p^2/n_2}$

$121.62 \pm 2.101 \cdot \sqrt{21817.3/10 + 21817.3/10}$

121.62 ± 138.78

$-17.16 < \mu_1 - \mu_2 < 260.40$

b. Assume $\sigma_1^2 \neq \sigma_2^2$ and proceed using s_1^2 and s_2^2 and the smaller df (viz., $df_1 = df_2 = 9$).

$\bar{x}_1 - \bar{x}_2 = 1390.03 - 1268.41 = 121.62$

$(\bar{x}_1 - \bar{x}_2) \pm t_{9,.025}\sqrt{s_1^2/n_1 + s_2^2/n_2}$

$121.62 \pm 2.262 \cdot \sqrt{(156.84)^2/10 + (137.97)^2/10}$

121.62 ± 149.42

$-27.80 < \mu_1 - \mu_2 < 271.04$

c. The confidence interval in part (b) is wider. This was not caused by the values of $s_{\bar{x}_1 - \bar{x}_2}$ (which were actually the same in this case -- and will be whenever $n_1 = n_2$), but by the larger value of t required because we were using fewer degrees of freedom. The results in part (a) are better -- because we are using all the data (and their combined df) to estimate only one unknown parameter (the common σ^2) instead of having to split the data (and the df) to estimate two unknown parameters (σ_1^2 and σ_2^2).

11. Let the orange ones be group 1.

original claim: $\mu_1 - \mu_2 = 0$ [small samples and σ unknown, first test $H_o: \sigma_1^2 = \sigma_2^2$]

$H_o: \sigma_1^2 = \sigma_2^2$

$H_1: \sigma_1^2 \neq \sigma_2^2$

$\alpha = .05$ [assumed]

C.R. $F < F_{20,.975}^7 = [.2239]$

$\quad\quad F > F_{20,.025}^7 = 3.0074$

calculations:

$F = s_1^2/s_2^2$

$\quad = (.0472)^2/(.0275)^2$

$\quad = 2.9459$

conclusion:

Do not reject H_o; there is not sufficient evidence to reject the claim that $\sigma_1^2 = \sigma_2^2$.

Now proceed using s_p^2 for both s_1^2 and s_2^2

$\bar{x}_1 - \bar{x}_2 = .9251 - .9097 = .0154$

$s_p^2 = (df_1 \cdot s_1^2 + df_2 \cdot s_2^2)/(df_1 + df_2)$
$\quad = (7 \cdot (.0472)^2 + 20 \cdot (.0275)^2)/(7 + 20) = .0307/27 = .00114$

$H_o: \mu_1 - \mu_2 = 0$

$H_1: \mu_1 - \mu_2 \neq 0$

$\alpha = .05$

C.R. $t < -t_{27,.025} = -2.052$

$\quad\quad t > t_{27,.025} = 2.052$

calculations:

$t_{\bar{x}_1 - \bar{x}_2} = (\bar{x}_1 - \bar{x}_2 - \mu_{\bar{x}_1 - \bar{x}_2})/s_{\bar{x}_1 - \bar{x}_2}$

$\quad = (.0154 - 0)/\sqrt{.00114/8 + .00114/21}$

$\quad = .0154/.0140$

$\quad = 1.099$

conclusion:

Do not reject H_o; there is not sufficient evidence to reject the claim that $\mu_1 - \mu_2 = 0$.

No; no corrective action is necessary.

13. filtered   non-filtered
   n = 21   n = 8
   $\Sigma x = 270$ $\Sigma x = 125$
   $\Sigma x^2 = 3660$ $\Sigma x^2 = 1963$
   $\bar{x} = 12.857$ $\bar{x} = 15.625$
   $s^2 = 9.429$ $s^2 = 1.411$

Let the filtered ones be group 1.

original claim: $\mu_1 - \mu_2 = 0$ [small samples and σ unknown, first test H_o: $\sigma_1^2 = \sigma_2^2$]

   H_o: $\sigma_1^2 = \sigma_2^2$
   H_1: $\sigma_1^2 \neq \sigma_2^2$
   $\alpha = .05$ [assumed]
   C.R. $F < F_{7,.975}^{20} = [.3325]$
       $F > F_{7,.025}^{20} = 4.4667$
   calculations:
      $F = s_1^2/s_2^2$
        $= 9.429/1.411$
        $= 6.6835$

   conclusion:

      Reject H_o; there is sufficient evidence to reject the claim that $\sigma_1^2 = \sigma_2^2$ and to conclude that $\sigma_1^2 \neq \sigma_2^2$ (in fact, that $\sigma_1^2 > \sigma_2^2$.).

Now proceed using s_1^2 and s_2^2 and the smaller degrees of freedom (viz., $df_2 = 7$).

   $\bar{x}_1 - \bar{x}_2 = 12.857 - 15.625 = -2.768$

H_o: $\mu_1 - \mu_2 = 0$
H_1: $\mu_1 - \mu_2 \neq 0$
$\alpha = .05$ [assumed]
C.R. $t < -t_{7,.025} = -2.365$
    $t > t_{7,.025} = 2.365$
calculations:
  $t_{\bar{x}_1 - \bar{x}_2} = (\bar{x}_1 - \bar{x}_2 - \mu_{\bar{x}_1 - \bar{x}_2})/s_{\bar{x}_1 - \bar{x}_2}$
      $= (-2.768 - 0)/\sqrt{9.429/21 + 1.411/8}$
      $= -2.768/.791$
      $= -3.500$

conclusion:

    Reject H_o; there is sufficient evidence to reject the claim that $\mu_1 - \mu_2 = 0$ and to conclude that $\mu_1 - \mu_2 \neq 0$ (in fact, that $\mu_1 - \mu_2 < 0$).

Based on this result and the examples of this section, it appears that cigarette filters are truly effective. The ultimate filter, which reduces harmful first-hand intake to zero, is to refrain from smoking. Even then, there is legitimate concern about second-hand smoke -- and cigarette filters probably do not reduce the harmful levels of second-hand smoke.

15. Let the peanut M&M's be group 1.

original claim: $\mu_1 - \mu_2 \neq 0$ [small samples and σ unknown, first test H_o: $\sigma_1^2 = \sigma_2^2$]

   H_o: $\sigma_1^2 = \sigma_2^2$
   H_1: $\sigma_1^2 \neq \sigma_2^2$
   $\alpha = .05$ [assumed]
   C.R. $F < F_{99,.975}^{20} = [.4638]$
      $F > F_{99,.025}^{20} = 1.8249$
   calculations:
      $F = s_1^2/s_2^2$
        $= (.3127)^2/(.0369)^2 = 71.8130$

   conclusion:

      Reject H_o; there is not sufficient evidence to reject the claim that $\sigma_1^2 = \sigma_2^2$ and to conclude that $\sigma_1^2 \neq \sigma_2^2$ (in fact, that $\sigma_1^2 > \sigma_2^2$).

Assume $\sigma_1^2 \neq \sigma_2^2$ and proceed using s_1^2 and s_2^2 and the smaller df (viz., $df_1 = 20$).
   $\bar{x}_1 - \bar{x}_2 = 2.4658 - .9147 = 1.5511$
H_o: $\mu_1 - \mu_2 = 0$
H_1: $\mu_1 - \mu_2 \neq 0$
$\alpha = .05$ [assumed]
C.R. $t < -t_{20,.025} = -2.086$
      $t > t_{20,.025} = 2.086$
calculations:

$t_{\bar{x}_1 - \bar{x}_2} = (\bar{x}_1 - \bar{x}_2 - \mu_{\bar{x}_1 - \bar{x}_2})/s_{\bar{x}_1 - \bar{x}_2}$
           $= (1.5511 - 0)/\sqrt{(.3127)^2/21 + (.0369)^2/100}$
           $= 1.5511/.06834$
           $= 22.698$
conclusion:
   Reject H_o; there is sufficient evidence to conclude that $\mu_1 - \mu_2 \neq 0$ (in fact, that $\mu_1 - \mu_2 > 0$).

17. Let the treatment sample be group 1.
    original claim: $\mu_1 - \mu_2 = 0$ [small samples and σ unknown, first test H_o: $\sigma_1^2 = \sigma_2^2$]
        H_o: $\sigma_1^2 = \sigma_2^2$
        H_1: $\sigma_1^2 \neq \sigma_2^2$
        $\alpha = .05$
        C.R. $F < F_{21,.975}^{21} = [.4124]$
             $F > F_{21,.025}^{21} = 2.4247$
        calculations:

            $F = s_1^2/s_2^2$
              $= (.015)^2/(.000)^2$
              $= \infty$
    conclusion:
        Reject H_o; there is sufficient evidence to reject the claim that $\sigma_1^2 = \sigma_2^2$ and to conclude that $\sigma_1^2 \neq \sigma_2^2$ (in fact, $\sigma_1^2 > \sigma_2^2$).
    Now proceed using s_1^2 and s_2^2 and the smaller degrees of freedom (viz., $df_1 = df_2 = 21$).
        $\bar{x}_1 - \bar{x}_2 = .049 - .000 = .049$
    H_o: $\mu_1 - \mu_2 = 0$
    H_1: $\mu_1 - \mu_2 \neq 0$
    $\alpha = .05$
    C.R. $t < -t_{21,.025} = -2.080$
         $t > t_{21,.025} = 2.080$
    calculations:

        $t_{\bar{x}_1 - \bar{x}_2} = (\bar{x}_1 - \bar{x}_2 - \mu_{\bar{x}_1 - \bar{x}_2})/s_{\bar{x}_1 - \bar{x}_2}$
              $= (.049 - 0)/\sqrt{(.015)^2/22 + (0)^2/22}$
              $= .049/.00319$
              $= 15.322$
conclusion:
    Reject H_o; there is sufficient evidence to reject the claim that $\mu_1 - \mu_2 = 0$ and to conclude that $\mu_1 - \mu_2 \neq 0$ (in fact, $\mu_1 - \mu_2 > 0$).

19. a. $A = s_1^2/n_1 = (3.7)^2/21 = .65190$
$B = s_2^2/n_2 = (1.7)^2/8 = .36125$
$df = (A + B)^2/(A^2/df_1 + B^2/df_2)$
  $= (1.01315)^2/(.42498/20 + .13050/7)$
  $= 1.02648/.039892$
  $= 25.7$

For the test of hypothesis:
  The C.R. changes from $t < -t_{7,.05} = -1.895$ to $t < -t_{26,.05} = -1.706$.
  The calculated test statistic of $t = -10.630$ is not affected.
  In this case, the conclusion remains the same.

For the confidence interval:
  $(\bar{x}_1-\bar{x}_2) \pm t_{25.7,.025}\sqrt{s_1^2/n_1 + s_2^2/n_2}$
  $-10.7 \pm 2.060 \cdot \sqrt{(3.7)^2/21 + (1.7)^2/8}$
  -10.7 ± 2.1
  $-12.8 < \mu_1-\mu_2 < -8.6$
  This is slightly narrower than the given confidence interval based on $df=7$.

b. The extreme outlier is the filtered king with 1 mg of tar. Removing this changes the
group 1 summary statistics to $n = 20$, $\Sigma x = 278$, $\Sigma x^2 = 3986$, $\bar{x} = 13.9$, $s = 2.53$.
  In the preliminary test of $H_o: \sigma_1^2 = \sigma_2^2$, the calculated ratio $F = (2.53)^2/(1.7)^2 = 2.2182$ is
  not in the C.R.
  Proceed using the pooled variance $s_p^2 = (df_1 \cdot s_1^2 + df_2 \cdot s_2^2)/(df_1 + df_2)$
    $= (19 \cdot (2.53)^2 + 7 \cdot (1.7)^2)/(19+7)$
    $= 142.03/26$
    $= 5.46$
  and $df = df_1 + df_2 = 19 + 7 = 26$.

For the test of hypothesis:
  The C.R. changes from $t < -t_{7,.05} = -1.895$ to $t < -t_{26,.05} = -1.706$.
  The calculated test statistic becomes $t = [(13.9 - 24.0) - 0]/\sqrt{5.46/20 + 5.46/8}$
    $= -10.1/.978$
    $= -10.330$.
  In this case, the conclusion remains the same.

For the confidence interval:
  $(\bar{x}_1-\bar{x}_2) \pm t_{26,.025}\sqrt{s_p^2/n_1 + s_p^2/n_2}$
  $-10.1 \pm 2.060 \cdot \sqrt{5.46/20 + 5.46/8}$
  -10.7 ± 2.0
  $-12.7 < \mu_1-\mu_2 < -8.7$
  This is slightly narrower than the given confidence interval using s_1, s_2 and $df = 7$.

Review Exercises

1. Let those who took Viagra be group 1.
   NOTE: Since there are a range of x_1 values and x_2 values that give $\hat{p}$'s which round to .07 and .02, and the original data is not given, the accuracy of this problem is limited. Use .07 and .02 when $\hat{p}$'s are called for. Use $x = \hat{p} \cdot n$ when x's are called for, even if that x is not a whole number. No other accuracy can justified from the given information.

   a. original claim: $p_1 - p_2 > 0$

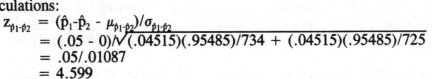

$$\hat{p}_1 = x_1/n_1 = 51.4/734 = .07 \quad [x_1 = .07 \cdot 734 = 51.4]$$
$$\hat{p}_2 = x_2/n_2 = 14.5/725 = .02 \quad [x_2 = .02 \cdot 725 = 14.5]$$
$$\hat{p}_1 - \hat{p}_2 = .07 - .02 = .05$$
$$\bar{p} = (x_1 + x_2)/(n_1 + n_2)$$
$$\quad = (51.4 + 14.5)/(734 + 725) = 65.9/1459 = .04515$$

$H_0: p_1 - p_2 \leq 0$
$H_1: p_1 - p_2 > 0$
$\alpha = .05$ [assumed]
C.R. $z > z_{.05} = 1.645$
calculations:

$$z_{\hat{p}_1 - \hat{p}_2} = (\hat{p}_1 - \hat{p}_2 - \mu_{\hat{p}_1 - \hat{p}_2})/\sigma_{\hat{p}_1 - \hat{p}_2}$$
$$= (.05 - 0)/\sqrt{(.04515)(.95485)/734 + (.04515)(.95485)/725}$$
$$= .05/.01087$$
$$= 4.599$$

conclusion:
   Reject H_0; there is sufficient evidence to conclude that $p_1 - p_2 > 0$.

   b. $(\hat{p}_1 - \hat{p}_2) \pm z_{.025}\sqrt{\hat{p}_1\hat{q}_1/n_1 + \hat{p}_2\hat{q}_2/n_2}$
   $.05 \pm 1.645 \cdot \sqrt{(.07)(.93)/734 + (.02)(.98)/725}$
   $.05 \pm .02$
   $.03 < p_1 - p_2 < .07$
   The interval does not contain zero, indicating a significant difference between the two dyspepsia rates.

2.

experimental	control
$n = 20$	$n = 16$
$\Sigma x = 1327.61$	$\Sigma x = 1660.28$
$\Sigma x^2 = 96477.3313$	$\Sigma x^2 = 181167.5442$
$\bar{x} = 66.3805$	$\bar{x} = 103.7675$
$s^2 = 439.469$	$s^2 = 592.296$

   Let the controls be group 1.

   a. original claim: $\sigma_1^2 = \sigma_2^2$
   $H_0: \sigma_1^2 = \sigma_2^2$
   $H_1: \sigma_1^2 \neq \sigma_2^2$
   $\alpha = .05$
   C.R. $F < F_{19,.975}^{15} = [.3629]$
   $\quad\quad F > F_{19,.025}^{15} = 2.6171$
   calculations:

   $F = s_1^2/s_2^2$
   $\quad = 592.296/439.469$
   $\quad = 1.3478$
   conclusion:
   Do not reject H_0; there is not sufficient evidence to reject the claim that $\sigma_1^2 = \sigma_2^2$.

b. original claim $\mu_1 - \mu_2 = 0$
   Now proceed using s_p^2 for both s_1^2 and s_2^2
   $\bar{x}_1 - \bar{x}_2 = 103.7675 - 66.3805 = 37.387$
   $s_p^2 = (df_1 \cdot s_1^2 + df_2 \cdot s_2^2)/(df_1 + df_2)$
   $= (15 \cdot 592.296 + 19 \cdot 439.469)/(15 + 19) = 17234.351/34 = 506.89$
   $H_o: \mu_1 - \mu_2 = 0$
   $H_1: \mu_1 - \mu_2 \neq 0$
   $\alpha = .05$
   C.R. $t < -t_{34,.025} = -1.960$
   $\quad t > t_{34,.025} = 1.960$
   calculations:

   $t_{\bar{x}_1-\bar{x}_2} = (\bar{x}_1 - \bar{x}_2 - \mu_{\bar{x}_1-\bar{x}_2})/s_{\bar{x}_1-\bar{x}_2}$
   $= (37.387 - 0)/\sqrt{506.89/16 + 506.89/20}$
   $= 37.387/7.552$
   $= 4.951$
   conclusion:
   Reject H_o; there is sufficient evidence to reject the claim that $\mu_1 - \mu_2 = 0$ and to conclude
   that $\mu_1 - \mu_2 \neq 0$ (in fact, that $\mu_1 - \mu_2 > 0$).

c. $(\bar{x}_1 - \bar{x}_2) \pm t_{34,.025}\sqrt{s_p^2/n_1 + s_p^2/n_2}$
   $37.387 \pm 1.960\sqrt{506.89/16 + 506.89/20}$
   37.387 ± 14.801
   $22.586 < \mu_1 - \mu_2 < 52.188$

3. Let those who saw someone else getting help be group 1.
   original claim: $p_1 - p_2 > 0$
   $\hat{p}_1 = x_1/n_1 = 58/2000 = .0290$ [$x_1 = .0290 \cdot 2000 = 58$]
   $\hat{p}_2 = x_2/n_2 = 35/2000 = .0175$ [$x_2 = .0175 \cdot 2000 = 35$]
   $\hat{p}_1 - \hat{p}_2 = .0290 - .0175 = .0115$
   $\bar{p} = (x_1 + x_2)/(n_1 + n_2)$
   $= (58 + 35)/(2000 + 2000) = 93/4000 = .02325$
   $H_o: p_1 - p_2 \leq 0$
   $H_1: p_1 - p_2 > 0$
   $\alpha = .05$
   C.R. $z > z_{.05} = 1.645$
   calculations:
   $z_{\hat{p}_1-\hat{p}_2} = (\hat{p}_1 - \hat{p}_2 - \mu_{\hat{p}_1-\hat{p}_2})/\sigma_{\hat{p}_1-\hat{p}_2}$
   $= (.0115 - 0)/\sqrt{(.02325)(.97675)/2000 + (.02325)(.97675)/2000}$
   $= .0115/.00477$
   $= 2.413$
   conclusion:
   Reject H_o; there is sufficient evidence to conclude that $p_1 - p_2 > 0$.
   P-value $= P(z > 2.41) = .5000 - .4920 = .0080$

4. Let the women be group 1. [n's > 30 and σ's known, use z]
   $\bar{x}_1 = 5$ hr 01 min $= 301$
   $\bar{x}_2 = 4$ hr 17 min $= 257$
   $\bar{x}_1 - \bar{x}_2 = 301 - 257 = 44$
   $(\bar{x}_1 - \bar{x}_2) \pm z_{.005}\sqrt{\sigma_1^2/n_1 + \sigma_2^2/n_2}$
   $44 \pm 2.575\sqrt{57^2/100 + 57^2/100}$
   44 ± 21
   $23 < \mu_1 - \mu_2 < 65$
   The confidence interval does not contain 0, suggesting that there is a significant difference
   between the two means -- in fact, that the mean is greater for the women.

5. original claim: $\mu_d = 0$ [$n \leq 30$ and σ_d unknown, use t]
   $d = x_1 - x_2$: 5 0 0 0 8 1 1 4 0 1
   $\quad n = 10$
   $\quad \Sigma d = 20 \qquad\qquad \bar{d} = 2.00$
   $\quad \Sigma d^2 = 108 \qquad\qquad s_d = 2.749$
   $H_o: \mu_d = 0$
   $H_1: \mu_d \neq 0$
   $\alpha = .05$ [assumed]
   C.R. $t < -t_{9,.025} = -2.262$
   $\qquad t > t_{9,.025} = 2.262$
   calculations:

   $\quad t_{\bar{d}} = (\bar{d} - \mu_{\bar{d}})/s_{\bar{d}}$
   $\qquad = (2.00 - 0)/(2.749/\sqrt{10})$
   $\qquad = 2.00/.869 = 2.301$
   conclusion:
   Reject H_o; there is sufficient evidence to reject the claim that $\mu_d = 0$ and to conclude that $\mu_d \neq 0$ (in fact, that $\mu_d > 0$ -- i.e., that training reduces the weights).

6. Refer to the notation and data summary for exercise #5.
   $\bar{d} \pm t_{9,.025} \cdot s_d/\sqrt{n}$
   $2.00 \pm 2.262 \cdot 2.749/\sqrt{10}$
   2.00 ± 1.97
   $0.0 < \mu_d < 4.0$

7. a. original claim: $\mu_1 - \mu_2 < 0$ [$n_1 > 30$ and $n_2 > 30$, use z (with s's for σ's)]
   $\quad \bar{x}_1 - \bar{x}_2 = 26,588 - 44,765 = -18,177$
   $\quad H_o: \mu_1 - \mu_2 \geq 0$
   $\quad H_1: \mu_1 - \mu_2 < 0$
   $\quad \alpha = .01$
   $\quad$ C.R. $z < -z_{.01} = -2.327$
   $\quad$ calculations:

   $\qquad z_{\bar{x}_1-\bar{x}_2} = (\bar{x}_1-\bar{x}_2 - \mu_{\bar{x}_1-\bar{x}_2})/\sigma_{\bar{x}_1-\bar{x}_2}$
   $\qquad\qquad = (-18,177 - 0)/\sqrt{(8441)^2/85 + (12,469)^2/120}$
   $\qquad\qquad = -18,177/1460.78$
   $\qquad\qquad = -12.443$
   $\quad$ conclusion:
   $\qquad$ Reject H_o; there is sufficient evidence to conclude that $\mu_1 - \mu_2 < 0$.
   $\quad$ P-value $= P(z < -12.44) = .5000 - .4999 = .0001$
   b. $(\bar{x}_1-\bar{x}_2) \pm z_{.005}\sqrt{\sigma_1^2/n_1 + \sigma_2^2/n_2}$
   $\quad -18,177 \pm 2.575\sqrt{(8441)^2/100 + (12,469)^2/100}$
   $\quad -18,177 \pm 3762$
   $\quad -21,939 < \mu_1 - \mu_2 < -14,415$

8. Let the multi-line sample be group 1.
   a. original claim: $\sigma_2 < \sigma_1$ [i.e., $\sigma_1^2 > \sigma_2^2$]
      $H_o: \sigma_1^2 \leq \sigma_2^2$
      $H_1: \sigma_1^2 > \sigma_2^2$
      $\alpha = .01$
      C.R. $F > F_{19,.01}^{24} = 2.9249$
      calculations:
      $$F = s_1^2/s_2^2$$
      $$= (3.619)^2/(1.841)^2$$
      $$= 3.8643$$

      conclusion:
         Reject H_o; there is sufficient evidence to conclude that $\sigma_1^2 > \sigma_2^2$.
   b. original claim: $\mu_1-\mu_2 = 0$
      Proceed using s_1^2 and s_2^2 and the smaller degrees of freedom (viz., $df_2 = 19$).
      $\bar{X}_1-\bar{X}_2 = 6.896 - 7.460 = -.564$
      $H_o: \mu_1-\mu_2 = 0$
      $H_1: \mu_1-\mu_2 \neq 0$
      $\alpha = .01$
      C.R. $t < -t_{19,.005} = -2.861$
      $\quad\quad t > t_{19,.005} = 2.861$
      calculations:
      $$t_{\bar{x}_1-\bar{x}_2} = (\bar{X}_1-\bar{X}_2 - \mu_{\bar{x}_1-\bar{x}_2})/s_{\bar{x}_1-\bar{x}_2}$$
      $$= (-.564 - 0)/\sqrt{(3.619)^2/25 + (1.841)^2/20}$$
      $$= -.564/.8326$$
      $$= -.677$$

      conclusion:
         Do not reject H_o; there is not sufficient evidence to reject the claim that $\mu_1-\mu_2 = 0$.

9. a. original claim $\mu_d > 0$ [$n \leq 30$ and σ_d unknown, use t]
      $d = x_{after} - x_{before}$: 90 25 17 16 65 10 5 9 29 6 30 46
      $n = 12$
      $\Sigma d = 348$ $\quad\quad\quad\quad \bar{d} = 29.0$
      $\Sigma d^2 = 17594$ $\quad\quad\quad s_d = 26.1$
      $H_o: \mu_d \leq 0$
      $H_1: \mu_d > 0$
      $\alpha = .05$
      C.R. $t > t_{11,.05} = 1.796$
      calculations:
      $$t_{\bar{d}} = (\bar{d} - \mu_{\bar{d}})/s_{\bar{d}}$$
      $$= (29.0 - 0)/(26.1/\sqrt{12})$$
      $$= 29.0/7.539$$
      $$= 3.847$$

      conclusion:
         Reject H_o; there is sufficient evidence to conclude that $\mu_d > 0$.
   b. $\bar{d} \pm t_{11,.025} \cdot s_{\bar{d}}$
      $29.0 \pm 2.201 \cdot 26.1/\sqrt{12}$
      29.0 ± 16.6
      $12.4 < \mu_d < 45.6$

10.
yellow	green
$n = 26$	$n = 7$
$\Sigma x = 23.849$	$\Sigma x = 6.486$
$\Sigma x^2 = 21.904543$	$\Sigma x^2 = 6.018734$
$\bar{x} = .9173$	$\bar{x} = .9266$
$s^2 = .001144$	$s^2 = .001499$

Let the green ones be group 1.

original claim: $\mu_1 - \mu_2 = 0$ [small samples and σ unknown, first test H_o: $\sigma_1^2 = \sigma_2^2$]

H_o: $\sigma_1^2 = \sigma_2^2$
H_1: $\sigma_1^2 \neq \sigma_2^2$
$\alpha = .05$ [assumed]
C.R. $F < F_{25,.975}^6 = [.1954]$
$\quad\quad F > F_{25,.025}^6 = 2.9685$
calculations:
$\quad F = s_1^2/s_2^2$
$\quad\quad = .001499/.001144 = 1.3105$

conclusion:
Do not reject H_o; there is not sufficient evidence to reject the claim that $\sigma_1^2 = \sigma_2^2$.

Now proceed using s_p^2 for both s_1^2 and s_2^2
$\bar{x}_1 - \bar{x}_2 = .9266 - .9173 = .0093$
$s_p^2 = (df_1 \cdot s_1^2 + df_2 \cdot s_2^2)/(df_1 + df_2)$
$\quad = (6 \cdot (.001499) + 25 \cdot (.001144))/(6 + 25) = .03758/31 = .001212$

H_o: $\mu_1 - \mu_2 = 0$
H_1: $\mu_1 - \mu_2 \neq 0$
$\alpha = .05$ [assumed]
C.R. $t < -t_{31,.025} = -1.960$
$\quad\quad t > t_{31,.025} = 1.960$
calculations:
$\quad t_{\bar{x}_1 - \bar{x}_2} = (\bar{x}_1 - \bar{x}_2 - \mu_{\bar{x}_1 - \bar{x}_2})/s_{\bar{x}_1 - \bar{x}_2}$
$\quad\quad = (.0093 - 0)/\sqrt{.001212/7 + .001212/26}$
$\quad\quad = .0093/.0148$
$\quad\quad = .627$

conclusion:
Do not reject H_o; there is not sufficient evidence to conclude that $\mu_1 - \mu_2 \neq 0$.
No. One cannot be 95% certain that the differences observed reflect anything other than the normal variability expected to occur when the process is in control as specified.

Cumulative Review Exercises

1. Refer to the summary table at the right.
   a. $P(Y) = 53/750$
   $\quad\quad = .0707$
   b. $P(M \text{ or } Y) = P(M) + P(Y) - P(M \text{ and } Y)$
   $\quad\quad\quad = 250/750 + 53/750 - 26/750$
   $\quad\quad\quad = 277/750 = .3693$
   c. $P(Y \mid M) = 26/250$
   $\quad\quad\quad = .104$
   d. $P(Y \mid F) = 27/500$
   $\quad\quad\quad = .054$

		TICKET?		
		Y	N	
SEX	M	26	224	250
	F	27	473	500
		53	697	750

e. Let the males be group 1.
   original claim: $p_1-p_2 > 0$
   $$\hat{p}_1 = x_1/n_1 = 26/250 = .104$$
   $$\hat{p}_2 = x_2/n_2 = 27/500 = .054$$
   $$\hat{p}_1-\hat{p}_2 = .104 - .054 = .050$$
   $$\bar{p} = (x_1 + x_2)/(n_1 + n_2)$$
   $$\quad = (26 + 27)/(250 + 500) = 53/750 = .0707$$
   H_o: $p_1-p_2 \le 0$
   H_1: $p_1-p_2 > 0$
   $\alpha = .05$
   C.R. $z > z_{.05} = 1.645$
   calculations:

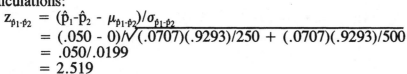

   $$z_{\hat{p}_1-\hat{p}_2} = (\hat{p}_1-\hat{p}_2 - \mu_{\hat{p}_1-\hat{p}_2})/\sigma_{\hat{p}_1-\hat{p}_2}$$
   $$= (.050 - 0)/\sqrt{(.0707)(.9293)/250 + (.0707)(.9293)/500}$$
   $$= .050/.0199$$
   $$= 2.519$$
   conclusion:
   Reject H_o; there is sufficient evidence to conclude that $p_1-p_2 > 0$.
   P-value = $P(z > 2.52) = .5000 - .4941 = .0059$
   No; we cannot conclude that men speed more, only that they are ticketed more. It is possible, for example, that men drive more.

2. There is a problem with the reported results, and no statistical analysis would be appropriate. Since there were 100 drivers in each group, and the number in each group owning a cell phone must be a whole number between 0 and 100 inclusive, the sample proportion for each group must be a whole percent. The reported values of 13.7% and 10.6% are not mathematical possibilities for the sample success rates of groups of 100.

3. Let the prepared students be group 1.
   a. original claim: $p_1 > .50$
   $$\hat{p}_1 = x_1/n_1 = 62/80 = .775$$
   H_o: $p_1 \le .50$
   H_1: $p_1 > .50$
   $\alpha = .05$
   C.R. $z > z_{.05} = 1.645$
   calculations:

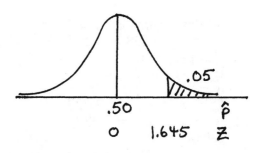

   $$z_{\hat{p}_1} = (\hat{p}_1 - \mu_{\hat{p}_1})/\sigma_{\hat{p}_1}$$
   $$= (.775 - .50)/\sqrt{(.50)(.50)/80}$$
   $$= .275/.0559$$
   $$= 4.919$$
   conclusion:
   Reject H_o; there is sufficient evidence to conclude that $p_1 > .50$.

b. original claim: $p_1-p_2 > 0$ [concerns p's, use z]

$\hat{p}_1 = x_1/n_1 = 62/80 = .775$

$\hat{p}_2 = x_2/n_2 = 23/50 = .460$

$\hat{p}_1-\hat{p}_2 = .775 - .460 = .315$

$\bar{p} = (x_1 + x_2)/(n_1 + n_2)$
$= (62 + 23)/(80 + 50) = 85/130 = .654$

H_0: $p_1-p_2 \leq 0$

H_1: $p_1-p_2 > 0$

$\alpha = .05$

C.R. $z > z_{.05} = 1.645$

calculations:

$z_{\hat{p}1-\hat{p}2} = (\hat{p}_1-\hat{p}_2 - \mu_{\hat{p}1-\hat{p}2})/\sigma_{\hat{p}1-\hat{p}2}$
$= (.315 - 0)/\sqrt{(.654)(.346)/80 + (.654)(.346)/50}$
$= .315/.0858$
$= 3.673$

conclusion:

Reject H_0; there is sufficient evidence to conclude that $p_1-p_2 > 0$.

4. Let the women be group 1.

a. original claim: $\sigma_1 = \sigma_2$ [i.e., $\sigma_1^2 = \sigma_2^2$]

H_0: $\sigma_1^2 = \sigma_2^2$

H_1: $\sigma_1^2 \neq \sigma_2^2$

$\alpha = .02$

C.R. $F < F^{67}_{85,.99} = [.5446]$

$F > F^{67}_{85,.01} = 1.8363$

calculations:

$F = s_1^2/s_2^2$
$= (114.16)^2/(97.23)^2$
$= 1.3786$

conclusion:

Do not reject H_0; there is not sufficient evidence to reject the claim that $\sigma_1^2 = \sigma_2^2$.

b. original claim: $\mu_1-\mu_2 = 0$ [$n_1 > 30$ and $n_2 > 30$, use z (with s's for σ's)]

$\bar{x}_1-\bar{x}_2 = 538.82 - 525.23 = 13.59$

H_0: $\mu_1-\mu_2 = 0$

H_1: $\mu_1-\mu_2 \neq 0$

$\alpha = .02$

C.R. $z < -z_{.01} = -2.327$

$z > z_{.01} = 2.327$

calculations:

$z_{\bar{x}1-\bar{x}2} = (\bar{x}_1-\bar{x}_2 - \mu_{\bar{x}1-\bar{x}2})/\sigma_{\bar{x}1-\bar{x}2}$
$= (13.59 - 0)/\sqrt{(114.16)^2/68 + (97.23)^2/86}$
$= 13.59/17.366$
$= .783$

conclusion:

Do not reject H_0; there is not sufficient evidence to reject the claim that $\mu_1-\mu_2 = 0$.

P-value $= 2 \cdot P(z > .78) = 2 \cdot (.5000 - .2823) = 2 \cdot (.2177) = .4354$

c. $\bar{x}_1 \pm z_{.01} \cdot \sigma_{\bar{x}1}$

$538.82 \pm 2.327 \cdot 114.16/\sqrt{68}$

538.82 ± 32.21

$506.61 < \mu < 571.03$

Chapter 9

Correlation and Regression

9-2 Correlation

1. a. From Table A-6, CV $= \pm.312$; therefore r $= .401$ indicates a significant (positive) linear correlation.
   b. The proportion of the variation in y that can be explained in terms of the variation in x is $r_2 = (.401)^2 = .1608$, or 16.08%.

NOTE: In addition to the value of n, calculation of r requires five sums: Σx, Σy, Σx^2, Σy^2 and Σxy. The next problem shows the chart prepared to find these sums. As the sums can usually be found conveniently using a calculator and without constructing the chart, subsequent problems typically give only the values of the sums and do not show a chart.

Also, calculation of r also involves three subcalculations.
(1) $n(\Sigma xy) - (\Sigma x)(\Sigma y)$ determines the sign of r. If large values of x are associated with large values of y, it will be positive. If large values of x are associated with small values of y, it will be negative. If not, a mistake has been made.
(2) $n(\Sigma x^2) - (\Sigma x)^2$ cannot be negative. If it is, a mistake has been made.
(3) $n(\Sigma y^2) - (\Sigma y)^2$ cannot be negative. If it is, a mistake has been made.

Finally, r must be between -1 and 1 inclusive. If not, a mistake has been made. If this or any of the previous mistakes occurs, stop immediately and find the error; continuing will be a fruitless waste of effort.

3.

x	y	xy	x^2	y^2
1	19	19	1	361
2	12	24	4	144
3	4	12	9	16
5	18	90	25	324
5	20	100	25	400
16	73	245	64	1245

$n(\Sigma xy) - (\Sigma x)(\Sigma y) = 5(245) - (16)(73) = 57$
$n(\Sigma x^2) - (\Sigma x)^2 = 5(64) - (16)^2 = 64$
$n(\Sigma y^2) - (\Sigma y)^2 = 5(1245) - (73)^2 = 896$

   a. According to the scatter diagram, there appears to be a significant "U-shaped" patten , but no <u>linear</u> relationship between x and y. Expect a value for r close to 0.
   b. $r = [n(\Sigma xy) - (\Sigma x)(\Sigma y)]/[\sqrt{n(\Sigma x^2) - (\Sigma x)^2} \cdot \sqrt{n(\Sigma y^2) - (\Sigma y)^2}]$
      $= 57/[\sqrt{64} \cdot \sqrt{896}]$
      $= .238$
      From Table A-6, CV $= \pm.878$; therefore r $= .238$ does not indicates a significant linear correlation. This agrees with the interpretation of the scatter diagram in part (a).

NOTE: In each of exercises 5-16, the first variable listed is given the designation x, and the second variable listed is given the designation y. In correlation problems, the designation of x and y is arbitrary -- so long as a person remains consistent after making the designation. For part (d) of each problem, the following summary statistics should be saved: n, Σx, Σy, Σx^2, Σy^2, Σxy.

5. a. weight (lbs)

$n = 9$
$\Sigma x = 632.0$
$\Sigma y = 1089$
$\Sigma x^2 = 44399.50$
$\Sigma y^2 = 132223$
$\Sigma xy = 76546.0$

b. $n(\Sigma xy) - (\Sigma x)(\Sigma y) = 9(76546) - (632)(1089) = 666$
   $n(\Sigma x^2) - (\Sigma x)^2 = 9(44399.5) - (632)^2 = 171.5$
   $n(\Sigma y^2) - (\Sigma y)^2 = 9(132223) - (1089)^2 = 4086$
   $r = [n(\Sigma xy) - (\Sigma x)(\Sigma y)]/[\sqrt{n(\Sigma x^2) - (\Sigma x)^2} \cdot \sqrt{n(\Sigma y^2) - (\Sigma y)^2}]$
   $= 666/[\sqrt{171.5} \cdot \sqrt{4086}\,]$
   $= .796$

c. $H_o: \rho = 0$
   $H_1: \rho \neq 0$
   $\alpha = .05$

   C.R. $r < -.666$ OR   C.R. $t < -t_{7,.025} = -2.365$
         $r > .666$ $t > t_{7,.025} = 2.365$
   calculations: calculations:
       $r = .796$ $t_r = (r - \mu_r)/s_r$
                               $= (.796 - 0)/\sqrt{(1-(.796)^2)/7}$
                               $= .796/.229$
                               $= 3.475$
   conclusion:
      Reject H_o; there is sufficient evidence to reject the claim that $\rho = 0$ and to conclude that $\rho \neq 0$ (in fact, that $\rho > 0$).

7. a. household size

$n = 8$
$\Sigma x = 14.60$
$\Sigma y = 26$
$\Sigma x^2 = 32.9632$
$\Sigma y^2 = 104$
$\Sigma xy = 56.80$

b. $n(\Sigma xy) - (\Sigma x)(\Sigma y) = 8(56.80) - (14.60)(26) = 74.80$
  $n(\Sigma x^2) - (\Sigma x)^2 = 8(32.9632) - (14.60)^2 = 50.5456$
  $n(\Sigma y^2) - (\Sigma y)^2 = 8(104) - (26)^2 = 156$
  $r = [n(\Sigma xy) - (\Sigma x)(\Sigma y)]/[\sqrt{n(\Sigma x^2) - (\Sigma x)^2} \cdot \sqrt{n(\Sigma y^2) - (\Sigma y)^2}]$
  $= 74.80/[\sqrt{50.5456} \cdot \sqrt{156}]$
  $= .842$

c. $H_o: \rho = 0$
  $H_1: \rho \neq 0$
  $\alpha = .05$
  C.R. $r < -.707$ OR   C.R. $t < -t_{6,.025} = -2.447$
  $\quad\quad r > .707$ $t > t_{6,.025} = 2.447$
  calculations: calculations:
  $\quad r = .842$ $t_r = (r - \mu_r)/s_r$
  $\quad\quad\quad\quad\quad\quad\quad\quad\quad = (.842 - 0)/\sqrt{(1-(.842)^2)/6}$
  $\quad\quad\quad\quad\quad\quad\quad\quad\quad = .842/.220$
  $\quad\quad\quad\quad\quad\quad\quad\quad\quad = 3.829$

conclusion:
  Reject H_o; there is sufficient evidence to reject the claim that $\rho = 0$ and to conclude that $\rho \neq 0$ (in fact, $\rho > 0$).
  Yes; there does appear to be a linear relationship between the amount of discarded plastic and household size.

9. a. BAC (%)

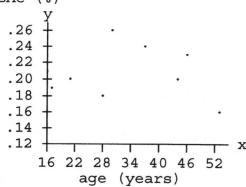

$n = 8$
$\Sigma x = 276.6$
$\Sigma y = 1.66$
$\Sigma x^2 = 10680.48$
$\Sigma y^2 = .3522$
$\Sigma xy = 57.191$

b. $n(\Sigma xy) - (\Sigma x)(\Sigma y) = 8(57.191) - (276.6)(1.66) = -1.628$
  $n(\Sigma x^2) - (\Sigma x)^2 = 8(10680.48) - (276.6)^2 = 8936.28$
  $n(\Sigma y^2) - (\Sigma y)^2 = 8(.3522) - (1.66)^2 = .062$
  $r = [n(\Sigma xy) - (\Sigma x)(\Sigma y)]/[\sqrt{n(\Sigma x^2) - (\Sigma x)^2} \cdot \sqrt{n(\Sigma y^2) - (\Sigma y)^2}]$
  $= -1.628/[\sqrt{8936.28} \cdot \sqrt{.062}]$
  $= -.069$

c. H_o: $\rho = 0$
   H_1: $\rho \neq 0$
   $\alpha = .05$
   C.R. r < -.707   OR   C.R. t < $-t_{6,.025}$ = -2.447
       r > .707   t > $t_{6,.025}$ = 2.447
   calculations: calculations:
       r = -.069 $t_r = (r - \mu_r)/s_r$
                                        $= (-.069 - 0)/\sqrt{(1-(-.069)^2)/6}$
                                        $= -.069/.407$
                                        $= -.170$

conclusion:
   Do not reject H_o; there is not sufficient evidence to reject the claim that $\rho = 0$.
   No; based on this result, the BAC level does not seem to be related to the age of the person tested.

11. A. Interval-duration relationship?
    a. interval (minutes)

n = 50
Σx = 10832
Σy = 4033
Σx^2 = 2513280
Σy^2 = 332331
Σxy = 903488

duration (seconds)

   b. $n(\Sigma xy) - (\Sigma x)(\Sigma y) = 50(903488) - (10832)(4033) = 1488944$
      $n(\Sigma x^2) - (\Sigma x)^2 = 50(2513280) - (10832)^2 = 8331776$
      $n(\Sigma y^2) - (\Sigma y)^2 = 50(332331) - (4033)^2 = 351461$
      $r = [n(\Sigma xy) - (\Sigma x)(\Sigma y)]/[\sqrt{n(\Sigma x^2) - (\Sigma x)^2} \cdot \sqrt{n(\Sigma y^2) - (\Sigma y)^2}]$
        $= 1488944/[\sqrt{8331776} \cdot \sqrt{351461}]$
        $= .870$

   c. H_o: $\rho = 0$
      H_1: $\rho \neq 0$
      $\alpha = .05$
      C.R. r < -.279   OR   C.R. t < $-t_{48,.025}$ = -1.960
          r > .279   t > $t_{48,.025}$ = 1.960
      calculations: calculations:
          r = .870 $t_r = (r - \mu_r)/s_r$
                                          $= (.870 - 0)/\sqrt{(1-(.870)^2)/48}$
                                          $= .870/.0711$
                                          $= 12.231$

   conclusion:
      Reject H_o; there is sufficient evidence to reject the claim that $\rho = 0$ and to conclude that
      $\rho \neq 0$ (in fact, $\rho > 0$).
   Yes; there is a significant positive linear correlation, suggesting that the interval after an
   eruption is related to the duration of the eruption. NOTE: The longer the duration of an
   eruption, the more pressure has been released and the longer it will take the geyser to build
   back up for another eruption. In fact, the park rangers use the duration of one eruption to
   predict the time of the next eruption.

B. Interval-height relationship?
a. interval (minutes)

n = 50
Σx = 6904
Σy = 4033
Σx^2 = 961150
Σy^2 = 332331
Σxy = 556804

* = 2 occurrences

b. $n(\Sigma xy) - (\Sigma x)(\Sigma y) = 50(556804) - (6904)(4033) = -3632$
$n(\Sigma x^2) - (\Sigma x)^2 = 50(961150) - (6904)^2 = 392284$
$n(\Sigma y^2) - (\Sigma y)^2 = 50(332331) - (4033)^2 = 351461$
$r = [n(\Sigma xy) - (\Sigma x)(\Sigma y)]/[\sqrt{n(\Sigma x^2) - (\Sigma x)^2} \cdot \sqrt{n(\Sigma y^2) - (\Sigma y)^2}]$
$= -3632/[\sqrt{392284} \cdot \sqrt{351461}]$
$= -.00978$

c. $H_o: \rho = 0$
$H_1: \rho \neq 0$
$\alpha = .05$
C.R. r < -.279   OR   C.R. $t < -t_{48,.025} = -1.960$
    r > .279 $t > t_{48,.025} = 1.960$
calculations: calculations:
    r = -.00978 $t_r = (r - \mu_r)/s_r$
                            $= (-.00978 - 0)/\sqrt{(1-(-.00978)^2)/48}$
                            $= -.00978/.144$
                            $= -.068$
conclusion:
    Do not reject H_o; there is not sufficient evidence to reject the claim that $\rho = 0$.
No; there is a not significant linear correlation, suggesting that the interval after an
eruption is not so related to the height of the eruption.

C. Duration is the more relevant predictor of the interval until the next eruption -- because it
has a significant correlation with interval, while height does not.

3. A. Price-weight relationship? (NOTE: For convenience, we measure the price y in $1000's)
a. price ($1000's)

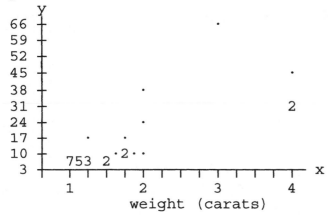

n = 30
Σx = 50.65
Σy = 433.23
Σx^2 = 109.74
Σy^2 = 12378.86
Σxy = 1027.03

2 = 2 occurrences
3 = 3 occurrences
etc.

b. $n(\Sigma xy) - (\Sigma x)(\Sigma y) = 30(1027.03) - (50.65)(433.23) = 8867.80$
$n(\Sigma x^2) - (\Sigma x)^2 = 30(109.74) - (50.65)^2 = 726.78$
$n(\Sigma y^2) - (\Sigma y)^2 = 30(12378.86) - (433.23)^2 = 183677.57$
$r = [n(\Sigma xy) - (\Sigma x)(\Sigma y)]/[\sqrt{n(\Sigma x^2) - (\Sigma x)^2} \cdot \sqrt{n(\Sigma y^2) - (\Sigma y)^2}]$
$\quad = 8867.80/[\sqrt{726.78} \cdot \sqrt{183677.57}]$
$\quad = .767$

c. $H_o: \rho = 0$
$H_1: \rho \neq 0$
$\alpha = .05$

C.R. $r < -.361$ 　　**OR**　　 C.R. $t < -t_{28,.025} = -2.048$
　　　$r > .361$ 　　　　　　　　　　　$t > t_{28,.025} = 2.048$

calculations: 　　　　　　　　　calculations:
　　$r = .767$ 　　　　　　　　　　$t_r = (r - \mu_r)/s_r$
　　　　　　　　　　　　　　　　　　$= (.767 - 0)/\sqrt{(1-(.767)^2)/28}$
　　　　　　　　　　　　　　　　　　$= .767/.121$
　　　　　　　　　　　　　　　　　　$= 6.325$

conclusion:
Reject H_o; there is sufficient evidence to reject the claim that $\rho = 0$ and to conclude that $\rho \neq 0$ (in fact, that $\rho > 0$).

Yes; based on this result there is a significant positive linear correlation between the price of a diamond and its weight in carats.

B. Price-color relationship? (NOTE: For convenience, we measure the price y in $1000's)
a. price ($1000's)

$n = 30$
$\Sigma x = 140$
$\Sigma y = 433.23$
$\Sigma x^2 = 782$
$\Sigma y^2 = 12378.86$
$\Sigma xy = 1630.03$

2 = 2 occurrences
3 = 3 occurrences
etc.

b. $n(\Sigma xy) - (\Sigma x)(\Sigma y) = 30(1630.03) - (140)(433.23) = -11751.30$
$n(\Sigma x^2) - (\Sigma x)^2 = 30(782) - (140)^2 = 3860$
$n(\Sigma y^2) - (\Sigma y)^2 = 30(12378.86) - (433.23)^2 = 183677.57$
$r = [n(\Sigma xy) - (\Sigma x)(\Sigma y)]/[\sqrt{n(\Sigma x^2) - (\Sigma x)^2} \cdot \sqrt{n(\Sigma y^2) - (\Sigma y)^2}]$
$\quad = -11751.30/[\sqrt{3860} \cdot \sqrt{183677.57}]$
$\quad = -.441$

c. H_o: $\rho = 0$
   H_1: $\rho \neq 0$
   $\alpha = .05$
   C.R. r < −.361   OR   C.R. t < −$t_{28,.025}$ = −2.048
        r > .361   t > $t_{28,.025}$ = 2.048
   calculations: calculations:
        r = −.441 $t_r = (r - \mu_r)/s_r$
                                  $= (-.441 - 0)/\sqrt{(1-(-.441)^2)/28}$
                                  $= -.441/.170$
                                  $= -2.600$

   conclusion:
      Reject H_o; there is sufficient evidence to reject the claim that $\rho = 0$ and to conclude that
      $\rho \neq 0$ (in fact, that $\rho < 0$).
   Yes; based on this result there is a significant negative linear correlation between the price of
   a diamond and its color -- i.e., the lower the color number, the higher the price.

C. When determining the value of a diamond, weight is a more important than color. The
   weight of a diamond is more highly correlated (i.e., has an r value farther from 0) with the
   selling price than is the color.

15. A. Gross-budget relationship?
   a. gross (million $'s)

```
  y
600 +                                        .
    +                                            n = 36
500 +                                            Σx = 1931.375
    +                                            Σy  = 4966.978
400 +                                            Σx² = 173580.583
    +                                            Σy² = 1153595.699
300 +      .      .                              Σxy = 338719.616
    +   ..                  .
200 +.              .          .        .        2 = 2 occurrences
    +.           .        .                      3 = 3 occurrences
    +  .      .      .
100 +..    .                 .
    + 2 .      .   ..     2  . .
    +3       .      .       2 . .
    +--+--+--+--+--+--+--+--+--+-- x
    0    40   80   120  160  200
        budget (million $'s)
```

   b. n(Σxy) − (Σx)(Σy) = 36(338719.616) − (1931.375)(4966.978) = 2600808.825
      n(Σx²) − (Σx)² = 36(173580.583) − (1931.375)² = 2518691.489
      n(Σy²) − (Σy)² = 36(1153595.699) − (4966.978)² = 16858571.148
      r = [n(Σxy) − (Σx)(Σy)]/[$\sqrt{n(Σx^2) - (Σx)^2}$ ·$\sqrt{n(Σy^2) - (Σy)^2}$]
        = 2600808.825/[$\sqrt{2518691.489}$ ·$\sqrt{16858571.148}$]
        = .399

c. H_o: $\rho = 0$
   H_1: $\rho \neq 0$
   $\alpha = .05$
   C.R. r < -.335   OR   C.R. t < $-t_{34,.025}$ = -1.960
        r > .335   t > $t_{34,.025}$ = 1.960
   calculations: calculations:
        r = .399 $t_r = (r - \mu_r)/s_r$
                                        $= (.399 - 0)/\sqrt{(1-(.399)^2)/34}$
                                        $= .399/.157$
                                        $= 2.538$

conclusion:
   Reject H_o; there is sufficient evidence to reject the claim that $\rho = 0$ and to conclude that $\rho \neq 0$ (in fact, that $\rho > 0$).

Yes, there is sufficient evidence to say there is a significant linear correlation between these two variables.

B. Gross-rating relationship?
a. gross (million $'s)

n = 36
Σx = 267.2
Σy = 4966.978
Σx^2 = 2017.40
Σy^2 = 1153595.699
Σxy = 37995.901

* = 2 occurrences

b. $n(\Sigma xy) - (\Sigma x)(\Sigma y) = 36(37995.901) - (267.2)(4966.978) = 40675.914$
   $n(\Sigma x^2) - (\Sigma x)^2 = 36(2017.40) - (267.2)^2 = 1230.56$
   $n(\Sigma y^2) - (\Sigma y)^2 = 36(1153595.699) - (4966.978)^2 = 16858571.148$
   $r = [n(\Sigma xy) - (\Sigma x)(\Sigma y)]/[\sqrt{n(\Sigma x^2) - (\Sigma x)^2} \cdot \sqrt{n(\Sigma y^2) - (\Sigma y)^2}]$
     $= 40675.914/[\sqrt{1230.56} \cdot \sqrt{16858571.148}]$
     $= .282$

c. H_o: $\rho = 0$
   H_1: $\rho \neq 0$
   $\alpha = .05$
   C.R. r < -.335 OR   C.R. t < $-t_{34,.025}$ = -1.960
        r > .335   t > $t_{34,.025}$ = 1.960
   calculations: calculations:
        r = .282 $t_r = (r - \mu_r)/s_r$
                                      $= (.282 - 0)/\sqrt{(1-(.282)^2)/34}$
                                      $= .282/.165$
                                      $= 1.717$

conclusion:
   Do not reject H_o; there is not sufficient evidence to reject the claim that $\rho = 0$.

No, there is not sufficient evidence to say there is a significant linear correlation between these two variables.

C. Based on the preceding results, invest in the big budget movies. There is a stronger correlation between budget and gross than between viewer rating and gross. It appears that the publicity associated with big budgets attracts more patrons than do high ratings by other viewers.

A linear correlation coefficient very close to zero indicates <u>no</u> significant linear correlation and no tendencies can be inferred.

A linear correlation coefficient very close to zero indicates no significant <u>linear</u> correlation, but there may some other type of relationship between the variables.

There are n=13 data points.
From the scatterplot, it appears that the correlation is about +.75.

H_o: $\rho = 0$
H_1: $\rho \neq 0$
$\alpha = .05$

| C.R. $r < -.553$ | OR | C.R. $t < -t_{11,.025} = -2.201$ |
| $r > .553$ | | $t > t_{11,.025} = 2.201$ |

calculations: calculations:

$r = .75$ $t_r = (r - \mu_r)/s_r$

$$= (.75 - 0)/\sqrt{(1-(.75)^2)/11}$$
$$= .75/.199$$
$$= 3.761$$

conclusion:
    Reject H_o; there is sufficient evidence to reject the claim that $\rho = 0$ and to conclude that $\rho \neq 0$ (in fact, that $\rho > 0$).

Yes; based on this result there is a significant positive linear correlation between the mortality rate and the rate of infections acquired in intensive care units.

a. For $\pm t_{48,.025} = \pm 1.960$, the critical values are $r = \pm 1.960/\sqrt{(\pm 1.960)^2 + 48} = \pm.272$.

b. For $\pm t_{73,.05} = \pm 1.645$, the critical values are $r = \pm 1.645/\sqrt{(\pm 1.645)^2 + 73} = \pm.189$.

c. For $-t_{18,.05} = -1.734$, the critical value is $r = -1.734/\sqrt{(-1.734)^2 + 18} = -.378$.

d. For $t_{8,.05} = 1.860$, the critical value is $r = 1.860/\sqrt{(1.860)^2 + 8} = .549$.

e. For $t_{10,.01} = 2.764$, the critical value is $r = 2.764/\sqrt{(2.764)^2 + 10} = .658$.

No; correlation cannot be used to address the issue of whether male statistics students exaggerate their heights. The two concepts are not related. A high positive correlation near $r = 1.000$, for example, indicates either that the reported and actual heights tended to agree or that the reported heights were over-stated (or under-stated) by a consistent linear pattern (e.g., each student over-stating by approximately 1.5").

NOTE: The underlying principle is that a high correlation indicates that one value can be predicted from the other -- not that the values are equal. A related and often incorrectly analyzed scenario is that a high correlation between grades on test 1 and test 2 does not mean that each student received approximately the same grade on the two tests -- only that there is some linear relationship that allows us to predict one score if we know the other.

9-3 Regression

NOTE: For exercises 1-20, the exact summary statistics (i.e., without any rounding) are given on the right. While the intermediate calculations on the left are presented rounded to various degrees of accuracy, the entire unrounded values were preserved in the calculator until the very end.

1. $\bar{x} = 3.00$
   $\bar{y} = 8.00$
   $b_1 = [n(\Sigma xy) - (\Sigma x)(\Sigma y)]/[n(\Sigma x^2) - (\Sigma x)^2]$
       $= [4(104) - (12)(32)]/[4(44) - (12)^2]$
       $= 32/32$
       $= 1.00$
   $b_0 = \bar{y} - b_1\bar{x}$
       $= 8.00 - (1.00)(3.00)$
       $= 5.00$
   $\hat{y} = b_0 + b_1x$
       $= 5.00 + 1.00x$

   $n = 4$
   $\Sigma x = 12$
   $\Sigma y = 32$
   $\Sigma x^2 = 44$
   $\Sigma y^2 = 264$
   $\Sigma xy = 104$

3. $\bar{x} = 3.20$
   $\bar{y} = 14.60$
   $b_1 = [n(\Sigma xy) - (\Sigma x)(\Sigma y)]/[n(\Sigma x^2) - (\Sigma x)^2]$
       $= [5(245) - (16)(73)]/[5(64) - (16)^2]$
       $= 57/64$
       $= 0.89$
   $b_0 = \bar{y} - b_1\bar{x}$
       $= 14.60 - (0.89)(3.20)$
       $= 11.75$
   $\hat{y} = b_0 + b_1x$
       $= 11.75 + 0.89x$

   $n = 5$
   $\Sigma x = 16$
   $\Sigma y = 73$
   $\Sigma x^2 = 64$
   $\Sigma y^2 = 1245$
   $\Sigma xy = 245$

NOTE: In the exercises that follow, this manual uses the full accuracy of b_0 and b_1 when calculating $\hat{y}$. Using only the rounded values as stated in the equation produces slightly different answers.

5. $\bar{x} = 70.22$
   $\bar{y} = 121.00$
   $b_1 = [n(\Sigma xy) - (\Sigma x)(\Sigma y)]/[n(\Sigma x^2) - (\Sigma x)^2]$
       $= 666.0/171.50$
       $= 3.88$
   $b_0 = \bar{y} - b_1\bar{x}$
       $= 121.00 - (3.88)(70.22)$
       $= -151.70$
   $\hat{y} = b_0 + b_1x$
       $= -151.7 + 3.88x$
   $\hat{y}_{69} = -151.7 + 3.88(69)$
        $= 116.3$ lbs

   $n = 9$
   $\Sigma x = 632.0$
   $\Sigma y = 1089$
   $\Sigma x^2 = 44399.50$
   $\Sigma y^2 = 132223$
   $\Sigma xy = 76546.0$

7. $\bar{x} = 1.825$

$\bar{y} = 3.25$

$b_1 = [n(\Sigma xy) - (\Sigma x)(\Sigma y)]/[n(\Sigma x^2) - (\Sigma x)^2]$

$\quad = 74.80/50.5456$

$\quad = 1.480$

$b_0 = \bar{y} - b_1\bar{x}$

$\quad = 3.25 - (1.480)(1.825)$

$\quad = .549$

$\hat{y} = b_0 + b_1x$

$\quad = .549 + 1.480x$

$\hat{y}_{.50} = .549 + 1.480(.50)$

$\quad = 1.3$ people

$n = 8$

$\Sigma x = 14.60$

$\Sigma y = 26$

$\Sigma x^2 = 32.9632$

$\Sigma y^2 = 104$

$\Sigma xy = 56.80$

9. $\bar{x} = 34.575$

$\bar{y} = .2075$

$b_1 = [n(\Sigma xy) - (\Sigma x)(\Sigma y)]/[n(\Sigma x^2) - (\Sigma x)^2]$

$\quad = -1.628/8936.28$

$\quad = -.000182$

$b_0 = \bar{y} - b_1\bar{x}$

$\quad = .2075 - (-.000182)(34.575)$

$\quad = .214$

$\hat{y} = b_0 + b_1x$

$\quad = .214 - .000182x$

$\hat{y}_{21} = \bar{y}$ [no significant correlation]

$\quad = .2075\%$

$n = 8$

$\Sigma x = 276.6$

$\Sigma y = 1.66$

$\Sigma x^2 = 10680.48$

$\Sigma y^2 = .3522$

$\Sigma xy = 57.191$

11. a. interval/duration relationship

$\bar{x} = 216.64$

$\bar{y} = 80.66$

$b_1 = [n(\Sigma xy) - (\Sigma x)(\Sigma y)]/[n(\Sigma x^2) - (\Sigma x)^2]$

$\quad = 1488944/8331776$

$\quad = .179$

$b_0 = \bar{y} - b_1\bar{x}$

$\quad = 80.66 - (.179)(216.64)$

$\quad = 41.9$

$\hat{y} = b_0 + b_1x$

$\quad = 41.9 + .179x$

$\hat{y}_{210} = 41.9 + .179(210)$

$\quad = 79.5$ minutes

$n = 50$

$\Sigma x = 10832$

$\Sigma y = 4033$

$\Sigma x^2 = 2513280$

$\Sigma y^2 = 332331$

$\Sigma xy = 903488$

b. interval/height relationship

$\bar{x} = 138.08$

$\bar{y} = 80.66$

$b_1 = [n(\Sigma xy) - (\Sigma x)(\Sigma y)]/[n(\Sigma x^2) - (\Sigma x)^2]$

$\quad = -3632/392284$

$\quad = -.00926$

$b_0 = \bar{y} - b_1\bar{x}$

$\quad = 80.66 - (-.00926)(138.08)$

$\quad = 81.9$

$\hat{y} = b_0 + b_1x$

$\quad = 81.9 - .00926x$

$\hat{y}_{275} = \bar{y}$ [no significant correlation]

$\quad = 80.7$ minutes

$n = 50$

$\Sigma x = 6904$

$\Sigma y = 4033$

$\Sigma x^2 = 961150$

$\Sigma y^2 = 332331$

$\Sigma xy = 556804$

c. The predicted time in part (a) is better, since interval and duration are significantly correlated. Since interval and height are not significantly correlated, the predicted time in part (b) did not even use the height data.

13. a. price/weight relationship [NOTE: For convenience, we measure the price y in $1000's.)

$\bar{x} = 1.688$ $n = 30$

$\bar{y} = 14.441$ $\Sigma x = 50.65$

$b_1 = [n(\Sigma xy) - (\Sigma x)(\Sigma y)]/[n(\Sigma x^2) - (\Sigma x)^2]$ $\Sigma y = 433.23$

     $= 8867.80/726.78$ $\Sigma x^2 = 109.74$

     $= 12.202$ $\Sigma y^2 = 12378.86$

$b_0 = \bar{y} - b_1\bar{x}$ $\Sigma xy = 1027.03$

     $= 14.441 - 12.202(1.688)$

     $= -6.159$

$\hat{y} = b_0 + b_1x$

     $= -6.159 + 12.202x$

$\hat{y}_{1.5} = -6.159 + 12.202(1.5)$

     $= 12.143$ (i.e., $12,143)

b. price/color relationship (NOTE: For convenience, we measure the price y in $100's.)

$\bar{x} = 4.667$ $n = 50$

$\bar{y} = 14.441$ $\Sigma x = 140$

$b_1 = [n(\Sigma xy) - (\Sigma x)(\Sigma y)]/[n(\Sigma x^2) - (\Sigma x)^2]$ $\Sigma y = 433.23$

     $= -11751.3/3860$ $\Sigma x^2 = 782$

     $= -3.044$ $\Sigma y^2 = 12378.86$

$b_0 = \bar{y} - b_1\bar{x}$ $\Sigma xy = 1630.03$

     $= 14.441 - (-3.044)(4.667)$

     $= 28.648$

$\hat{y} = b_0 + b_1x$

     $= 28.648 + (-3.044)x$

$\hat{y}_3 = 28.648 - 3.044(3)$

     $= 19.515$ (i.e., $19,515)

c. The predicted time in part (a) is better, since there is a stronger correlation between price and weight than between price and color.

15. a. gross/budget relationship

$\bar{x} = 53.649$ $n = 36$

$\bar{y} = 137.972$ $\Sigma x = 1931.375$

$b_1 = [n(\Sigma xy) - (\Sigma x)(\Sigma y)]/[n(\Sigma x^2) - (\Sigma x)^2]$ $\Sigma y = 4966.978$

     $= 2600808.8/2518691.5$ $\Sigma x^2 = 173580.583$

     $= 1.033$ $\Sigma y^2 = 1153595.699$

$b_0 = \bar{y} - b_1\bar{x}$ $\Sigma xy = 338719.616$

     $= 137.972 - 1.033(53.649)$

     $= 82.573$

$\hat{y} = b_0 + b_1x$

     $= 82.573 + 1.033x$

$\hat{y}_{15} = 82.573 + 1.033(15)$

     $= 98.06$ (i.e., $98,060,000)

b. gross/rating relationship

$\bar{x} = 7.422$

$\bar{y} = 137.972$

$b_1 = [n(\Sigma xy) - (\Sigma x)(\Sigma y)]/[n(\Sigma x^2) - (\Sigma x)^2]$

$\quad = 40675.91/1230.56$

$\quad = 33.055$

$b_0 = \bar{y} - b_1\bar{x}$

$\quad = 137.972 - 33.055(7.422)$

$\quad = -107.368$

$\hat{y} = b_0 + b_1x$

$\quad = -107.368 + 33.055x$

$\hat{y}_7 = \bar{y}$ [no significant correlation]

$\quad = 137.97$ (i.e., \$137,970,000)

$n = 30$

$\Sigma x = 267.2$

$\Sigma y = 4966.978$

$\Sigma x^2 = 2017.40$

$\Sigma y^2 = 1153595.699$

$\Sigma xy = 37995.901$

c. The predicted time in part (a) is better, since gross and budget are significantly correlated. Since gross and rating are not significantly correlated, the predicted gross in part (b) did not even use the rating data.

7. The .05 critical values for r are taken from Table A-6.

a. CV $= \pm.632$; r $= .931$ is significant

use $\hat{y} = 4.00 + 2.00x$

$\quad \hat{y}_{3.00} = 4.00 + 2.00(3.00)$

$\quad\quad = 10.00$

b. CV $= \pm.220$; r $= -.033$ is not significant

use $\hat{y} = \bar{y}$

$\quad \hat{y}_{3.00} = \bar{y}$

$\quad\quad = 2.50$

9. a. Yes; the point is an outlier. It is far away from the other data points; \$400 is considerably larger than the other bills, and \$50 is considerably larger than the other tips.

b. No; the point is not an influential one. The original regression line $\hat{y} = -.347 + .149x$ predicts $\hat{y}_{400} = -.347 + .149(400) = \$59.25$ -- which is relatively close to \$50.00 and suggests that the point will not greatly affect the regression equation. The following check confirms this.

$\bar{x} = 120.26$

$\bar{y} = 16.23$

$b_1 = [n(\Sigma xy) - (\Sigma x)(\Sigma y)]/[n(\Sigma x^2) - (\Sigma x)^2]$

$\quad = 81545.01/668584.39$

$\quad = .1220$

$b_0 = \bar{y} - b_1\bar{x}$

$\quad = 16.23 - .1220(120.26)$

$\quad = 1.55$

$\hat{y} = b_0 + b_1x$

$\quad = 1.55 + .1220x$

$n = 7$

$\Sigma x = 841.84$

$\Sigma y = 113.58$

$\Sigma x^2 = 196754.1416$

$\Sigma y^2 = 3309.5364$

$\Sigma xy = 25308.7436$

. original data

$n=5$

$\Sigma x = 4,234,178$

$\Sigma y = 576$

$\Sigma x^2 = 3,595,324,583,102$

$\Sigma y^2 = 67552$

$\Sigma xy = 491,173,342$

original data divided by 1000

$n = 5$

$\Sigma x = 4,234.178$

$\Sigma y = 576$

$\Sigma x^2 = 3,595,324.583102$

$\Sigma y^2 = 67552$

$\Sigma xy = 491,173.342$

$\bar{x} = 846835.6$ $\bar{x} = 846.8356$

$\bar{y} = 115.2$ $\bar{y} = 115.2$

$n\Sigma xy - (\Sigma x)(\Sigma y) = 16,980,182$ $n\Sigma xy - (\Sigma x)(\Sigma y) = 16,980.182$

$n\Sigma x^2 - (\Sigma x)^2 = 48,459,579,826$ $n\Sigma x^2 - (\Sigma x)^2 = 48,459.579826$

$b_1 = 16,980,182/48,459,579,826$ $b_1 = 16,980.182/48,459.579826$

$\quad = .0003504$ $\quad = .3504$

$b_0 = \bar{y} - b_1\bar{x}$ $b_0 = \bar{y} - b_1\bar{x}$

$\quad = 115.2 - .0003504(846835.6)$ $\quad = 115.2 - .3504(846.8356)$

$\quad = -181.53$ $\quad = -181.53$

$\hat{y} = b_0 + b_1x$ $\hat{y} = b_0 + b_1x$

$\quad = -181.53 + .0003504x$ $\quad = -181.53 + .3504x$

Dividing each x by 1000 multiplies b_1, the coefficient of x in the regression equation, by 1000; multiplying the x coefficient by 1000 and dividing x by 1000 will "cancel out" and all predictions remain the same.

Dividing each y by 1000 divides both b_1 and b_0 by 1000; consistent with the new "units" for y, all predictions will also turn out divided by 1000.

23.   original data $n = 4$

x	y
2.0	12.0
2.5	18.7
4.2	53.0
10.0	225.0

$\Sigma x = 18.7$

$\Sigma y = 308.7$

$\Sigma x^2 = 127.89$

$\Sigma y^2 = 53927.69$

$\Sigma xy = 2543.35$

$b_1 = [n(\Sigma xy) - (\Sigma x)(\Sigma y)]/[n(\Sigma x^2) - (\Sigma x)^2]$

$\quad = 4400.71/161.87$

$\quad = 27.2$

$b_0 = \bar{y} - b_1\bar{x}$

$\quad = (308.7/4) - (27.2)(18.7/4)$

$\quad = -49.9$

$\hat{y} = b_0 + b_1x$

$\quad = -49.9 + 27.2x$

$r = [n(\Sigma xy) - (\Sigma x)(\Sigma y)]/[\sqrt{n(\Sigma x^2) - (\Sigma x)^2} \cdot \sqrt{n(\Sigma y^2) - (\Sigma y)^2}]$

$\quad = 4400.71/[\sqrt{161.87} \cdot \sqrt{120415.07}]$

$\quad = .9968$

Based on the value of the associated correlations (.9968 > .9631), the above equation using the original data seems to fit the data better than the following equation using ln(x) instead of x.

NOTE: Both x and y (perhaps, especially y) seem to be growing exponentially. A wiser choice for a transformation might be to use both ln(x) for x and ln(y) for y (or, perhaps, only ln(y) for y).

using ln(x) for x $n = 4$

x	y
.693	12.0
.916	18.7
1.435	53.0
2.303	225.0

$\Sigma x = 5.347$

$\Sigma y = 308.7$

$\Sigma x^2 = 8.681$

$\Sigma y^2 = 53927.69$

$\Sigma xy = 619.594$

$b_1 = [n(\Sigma xy) - (\Sigma x)(\Sigma y)]/[n(\Sigma x^2) - (\Sigma x)^2]$

$\quad = 827.7220/6.134071$

$\quad = 134.9$

$b_0 = \bar{y} - b_1\bar{x}$

$\quad = (308.7/4) - (134.9)(5.347/4)$

$\quad = 103.2$

$\hat{y} = b_o + b_1 \cdot \ln(x)$ [since the "x" is really $\ln(x)$]
  $= -103.2 + 134.9 \cdot \ln(x)$

$r = [n(\Sigma xy) - (\Sigma x)(\Sigma y)]/[\sqrt{n(\Sigma x^2) - (\Sigma x)^2} \cdot \sqrt{n(\Sigma y^2) - (\Sigma y)^2}]$
  $= 827.7220/[\sqrt{6.134071} \cdot \sqrt{120415.07}]$
  $= .9631$

-4 Variation and Prediction Intervals

1. The coefficient of determination is $r^2 = (.3)^2 = .09$.
   The portion of the total variation explained by the regression line is $r^2 = .09 = 9\%$.

3. The coefficient of determination is $r^2 = (-.327)^2 = .107$.
   The portion of the total variation explained by the regression line is $r^2 = .107 = 10.7\%$.

5. Since $r^2 = .928$, $r = +.963$ (positive because $b_1 = 12.5$ is positive).
   From Table A-6, the critical values necessary for significance are $\pm.279$.
   Since $.963 > .279$, we conclude there is a significant positive linear correlation between chest
      sizes of bears and their weights.

7. 262.74 from the "fit" value on the Minitab display
   NOTE: $\hat{y} = -264.48 + 12.5444x$
   $ \hat{y}_{50} = -264.48 + 12.5444(50)$
   $ = 362.74$

9. The predicted values were calculated using the regression line $\hat{y} = 2 + 3x$.

x	y	$\hat{y}$	$\bar{y}$	$\hat{y}-\bar{y}$	$(\hat{y}-\bar{y})^2$	$y-\hat{y}$	$(y-\hat{y})^2$	$y-\bar{y}$	$(y-\bar{y})^2$
1	5	5	12.2	-7.2	51.84	0	0	-7.2	51.84
2	8	8	12.2	-4.2	17.64	0	0	-4.2	17.64
3	11	11	12.2	-1.2	1.44	0	0	-1.2	1.44
5	17	17	12.2	4.8	23.04	0	0	4.8	23.04
6	20	20	12.2	7.8	60.84	0	0	7.8	60.84
17	61	61	61.0	0	154.80	0	0	0	154.80

a. The explained variation is $\Sigma(\hat{y}-\bar{y})^2 = 154.80$

b. The unexplained variation is $\Sigma(y-\hat{y})^2 = 0$

c. The total variation is $\Sigma(y-\bar{y})^2 = 154.80$

d. $r^2 = \Sigma(\hat{y}-\bar{y})^2/\Sigma(y-\bar{y})^2 = 154.80/154.80 = 1.00$

e. $s_e^2 = \Sigma(y-\hat{y})^2/(n-2) = 0/3 = 0$
   $s_e = 0$

NOTE: A table such as the one in the preceding problem organizes the work and provides all the
values needed to discuss variation. In such a table, the following must always be true and can be
used as a check before proceeding.
   * $\Sigma y = \Sigma \hat{y} = \Sigma \bar{y}$
   * $\Sigma(\hat{y}-\bar{y}) = \Sigma(y-\hat{y}) = \Sigma(y-\bar{y}) = 0$
   * $\Sigma(y-\bar{y})^2 + \Sigma(y-\hat{y})^2 = \Sigma(y-\bar{y})^2$

11. The predicted values were calculated using the regression line $\hat{y} = .549270 + 1.47985x$.

x	y	$\hat{y}$	$\bar{y}$	$\hat{y}-\bar{y}$	$(\hat{y}-\bar{y})^2$	$y-\hat{y}$	$(y-\hat{y})^2$	$y-\bar{y}$	$(y-\bar{y})^2$
.27	2	.949	3.25	-2.301	5.295	1.051	1.105	-1.25	1.5625
1.41	3	2.636	3.25	-.614	.377	.364	.133	-.25	.0625
2.19	3	3.790	3.25	.540	.292	-.790	.624	-.25	.0625
2.83	6	4.737	3.25	1.487	2.212	1.263	1.595	2.75	7.5625
2.19	4	3.790	3.25	.540	.292	.210	.044	.75	.5625
1.81	2	3.228	3.25	-.022	.000	-1.228	1.507	-1.25	1.5625
.85	1	1.807	3.25	-1.443	2.082	-.807	.651	-2.25	5.0625
3.05	5	5.063	3.25	1.813	3.286	-.063	.004	1.75	3.0625
14.60	26	26.000	26.00		0 13.837		0 5.663		0 19.5000

a. The explained variation is $\Sigma(\hat{y}-\bar{y})^2 = 13.837$

b. The unexplained variation is $\Sigma(y-\hat{y})^2 = 5.663$

c. The total variation is $\Sigma(y-\bar{y})^2 = 19.500$

d. $r^2 = \Sigma(\hat{y}-\bar{y})^2/\Sigma(y-\bar{y})^2 = 13.837/19.500 = .7096$

e. $s_e^2 = \Sigma(y-\hat{y})^2/(n-2) = 5.663/6 = .9438$
   $s_e = .9715$

13. a. $\hat{y} = 2 + 3x$
   $\hat{y}_4 = 2 + 3(4)$
   $= 14$

   b. $\hat{y} \pm t_{n-2,\alpha/2}s_e\sqrt{1 + 1/n + n(x_o-\bar{x})^2/[n\Sigma x^2-(\Sigma x)^2]}$
   $\hat{y} \pm 0,$ since $s_e = 0$
   The prediction "interval" in this case shrinks to a single point. Since $r^2 = 1.00$ (i.e., 100% of the variability in the y's can be explained by the regression), a perfect prediction can be made. For a practical example of such a situation, consider the regression line $\hat{y} = 1.399x$ for predicting the amount of money y due for purchasing x gallons of gasoline at $1.399 per gallon -- the "prediction" will be exactly correct every time because of the perfect correlation between the number of gallons purchased and the amount of money due.

15. a. $\hat{y} = .549270 + 1.47985x$
   $\hat{y}_{2.50} = .549270 + 1.47985(2.50) = 4.25$

   b. preliminary calculations
   $n = 8$
   $\Sigma x = 14.60$ $\bar{x} = 14.60/8 = 1.825$
   $\Sigma x^2 = 32.9632$ $n\Sigma x^2-(\Sigma x)^2 = 8(32.9632)-(14.60)^2 = 50.5456$
   $\hat{y} \pm t_{n-2,\alpha/2}s_e\sqrt{1 + 1/n + n(x_o-\bar{x})^2/[n\Sigma x^2-(\Sigma x)^2]}$
   $\hat{y}_{2.50} \pm t_{6,.025}(.9715)\sqrt{1 + 1/8 + 8(2.50-1.825)^2/[50.5456]}$
   $4.25 \pm (2.447)(.9715)\sqrt{1.19711}$
   4.25 ± 2.60
   $1.65 < y_{2.50} < 6.85$ [Fractional values may represent part-time occupancy.]

Exercises 17-20 refer to the chapter problem of Table 9-1. They use the following, which are calculated and/or discussed at various places in the text,

$n = 6$

$\Sigma x = 441.84$ $\hat{y} = -.347279 + .148614x$

$\Sigma x^2 = 36754.1416$ $s_e = 3.26584$

and the values obtained below.

$\bar{x} = (\Sigma x)/n = 441.84/6 = 73.64$

$n\Sigma x^2-(\Sigma x)^2 = 6(36754.1416) - (441.84)^2 = 25302.2604$

17. $\hat{y}_{50} = -.347279 + .148614(50)$

$= 7.083$

$\hat{y} \pm t_{n-2,\alpha/2}s_e\sqrt{1 + 1/n + n(x_o-\bar{x})^2/[n\Sigma x^2-(\Sigma x)^2]}$

$\hat{y}_{50} \pm t_{4,.025}(3.26584)\sqrt{1 + 1/6 + 6(50-73.64)^2/[25302.2604]}$

$7.083 \pm (2.776)(3.26584)\sqrt{1.29919}$

7.083 ± 10.334

$-3.25 < y_{50} < 17.42$

$0 \le y_{50} < 17.42$

19. $\hat{y}_{80} = -.347279 + .148614(80)$

$= 11.542$

$\hat{y} \pm t_{n-2,\alpha/2}s_e\sqrt{1 + 1/n + n(x_o-\bar{x})^2/[n\Sigma x^2-(\Sigma x)^2]}$

$\hat{y}_{80} \pm t_{4,.01}(3.26584)\sqrt{1 + 1/6 + 6(80-73.64)^2/[25302.2604]}$

$11.542 \pm (3.747)(3.26584)\sqrt{1.17629}$

11.542 ± 13.272

$-1.73 < y_{80} < 24.81$

$0 \le y_{80}\ 24.81$

21. This exercise uses the following values from the chapter problem of Table 9-1, which are calculated and/or discussed at various places in the text,

$n = 6$ $\Sigma x = 441.84$ $b_o = -.347279$

$\Sigma x^2 = 36754.1416$ $b_1 = .148614$

and the values obtained below. $s_e = 3.26584$

$\bar{x} = (\Sigma x)/n = 441.84/6 = 73.64$

$\Sigma x^2-(\Sigma x)^2/n = 36754.1416 - (441.84)^2/6 = 4217.0434$

a. $b_o \pm t_{n-2,\alpha/2}s_e\sqrt{1/n + \bar{x}^2/[\Sigma x^2-(\Sigma x)^2/n]}$

$-.347279 \pm t_{4,.025}(3.26584)\sqrt{1/6 + (73.64)^2/[4217.0434]}$

$-.347279 \pm (2.776)(3.26584)\sqrt{1.45260}$

$-.347279 \pm 10.92667$

$-11.27 < \beta_o < 10.60$

b. $b_1 \pm t_{n-2,\alpha/2}s_e/\sqrt{\Sigma x^2-(\Sigma x)^2/n}$

$.148614 \pm t_{4,.025}(3.26584)/\sqrt{4217.0434}$

$.148614 \pm (2.776)(3.26584)/\sqrt{4217.0434}$

$.148614 \pm .139608$

$.009 < \beta_1 < .288$

From these results, we conclude that a sample of size n=6 is too small to make a very accurate prediction about the amount of a tip. The "lower limit" for a tip (i.e., the amount of a tip for a bill of $0.00, as indicated by β_o) could be as high as $10.60. At the very least, since the confidence interval for β_1 is completely (albeit barely) above 0.00, we can state that there is a positive correlation between the tip and the amount of the bill.

23. a. Since $s_e^2 = \Sigma(y-\hat{y})^2/(n-2)$, then $(n-2)s_e^2 = \Sigma(y-\hat{y})^2$.
   The unexplained variation is equal to $(n-2)s_e^2$.

   b.
   $$r^2 = \text{(explained variation)/(total variation)}$$
   $$\text{(total variation)} \cdot r^2 = \text{(explained variation)}$$
   $$[\text{(explained variation)} + \text{(unexplained variation)}] \cdot r^2 = \text{(explained variation)}$$
   $$\text{(explained variation)} \cdot r^2 + \text{(unexplained variation)} \cdot r^2 = \text{(explained variation)}$$
   $$\text{(unexplained variation)} \cdot r^2 = \text{(explained variation)} - \text{(explained variation)} \cdot r^2$$
   $$\text{(unexplained variation)} \cdot r^2 = \text{(explained variation)}(1 - r^2)$$
   $$\text{(unexplained variation)} \cdot r^2/(1 - r^2) = \text{(explained variation)}$$

   The explained variation is equal to (unexplained variation)$\cdot r^2/(1 - r^2)$.

   c. If $r^2 = .900$, then $r = \pm.949$.
   Since the regression line has a negative slope (i.e., $b_1 = -2$), we choose the negative root.
   The linear correlation coefficient, therefore, is $r = -.949$.

9-5 Multiple Regression

1. WEIGHT $= -285 - 1.38\cdot(\text{AGE}) - 11.2\cdot(\text{HEADWDTH}) + 28.6\cdot(\text{NECK})$
   $\hat{y} = -285 - 1.38x_1 - 11.2x_2 + 28.6x_3$

3. Yes, since the overall P-value of .002 is less than .05.

5. a. WEIGHT $= -274 + .426\cdot(\text{LENGTH}) + 12.1\cdot(\text{CHEST})$
   $\hat{y} = -274 + .426x_1 + 12.1x_2$
   b. $R^2 = 92.8\% = .928$
   adjusted $R^2 = 92.5\% = .925$
   P-value $= .000$
   c. Yes, since the overall P-value of .000 [actually, .000$^+$] is less than .05.

7. a. WEIGHT $= -235 + .403\cdot(\text{AGE}) + 5.11\cdot(\text{NECK}) - .555\cdot(\text{LENGTH}) + 9.19\cdot(\text{CHEST})$
   $\hat{y} = -235 + .403x_1 + 5.11x_2 - .555x_3 + 9.19x_4$
   b. $R^2 = 94.2\% = .942$
   adjusted $R^2 = 93.8\% = .938$
   P-value $= .000$
   c. Yes, since the overall P-value of .000 [actually, .000$^+$] is less than .05.

9. a. VIEWER $= 7.74 - .00585\cdot(\text{BUDGET})$
   $\hat{y} = 7.74 - .00585x_1$
   b. $R^2 = 7.0\% = .070$
   adjusted $R^2 = 4.3\% = .043$
   P-value $= .119$
   c. No, since the overall P-value of .119 is greater than .05.

11. a. VIEWER $= 7.41 - .00992\cdot(\text{BUDGET}) + .00394\cdot(\text{GROSS})$
    $\hat{y} = 7.41 - .00922x_1 + .00394x_2$
    b. $R^2 = 24.9\% = .249$
    adjusted $R^2 = 20.4\% = .204$
    P-value $= .009$
    c. Yes, since the overall P-value of .009 is less than .05.

3. a. PRICE = -6158 + 12201·(CARAT)
   $\hat{y} = -6158 + 12201x_1$
   b. $R^2 = 58.9\% = .589$
   adjusted $R^2 = 57.4\% = .574$
   P-value = .000
   c. Yes, since the overall P-value of .000 [actually, .000⁺] is less than .05.

5. a. PRICE = 8221 + 12297·(CARAT) - 3116·(COLOR)
   $\hat{y} = 8221 + 12297x_1 - 3116x_2$
   b. $R^2 = 79.3\% = .793$
   adjusted $R^2 = 77.8\% = .778$
   P-value = .000
   c. Yes, since the overall P-value of .000 [actually, .000⁺] is less than .05.

7. To find the best variables for a regression to predict NICOTINE, consider the correlations. Arranged in one table, the correlations are as follows.

	TAR	NICOTINE	CO
TAR	1.000		
NICOTINE	0.961	1.000	
CO	0.934	0.863	1.000

The variable having the highest correlation (0.961) with NICOTINE is TAR. This positive correlation indicates there tends to be more nicotine in the cigarettes containing more tar.

The variable having the second highest correlation (0.934) with NICOTINE is CO. Since the correlation (0.863) between TAR and CO is a high one, those two variables can be accurately predicted from each other are giving duplicate information. They should not both be included.

The best multiple regression equation with nicotine as the independent variable, therefore, is probably the simple linear regression
   NICOTINE = .1540 + .0651·(TAR)
   $R^2 = 92.4\% = .924$
   adjusted $R^2 = 92.1\% = .921$
   overall P-value = .000

The adjusted R^2 and overall P-value indicate that this is a suitable multiple regression equation for making predictions on the amount of nicotine in a cigarette. Yes; this equation is also good for predicting the amount of nicotine given the amount of tar and carbon monoxide -- as discussed above, the amount of carbon monoxide would probably add little and can simply be ignored without loss of any significant useful information.

9. To find the best variables for a regression to predict selling price SELL, consider the correlations. Arranged in one table, the correlations are as follows.

	SELL	LIST	AREA	ROOM	BEDR	BATH	AGEH	ACRE	TAXH
SELL	1.000								
LIST	0.997	1.000							
AREA	0.879	0.892	1.000						
ROOM	0.560	0.571	0.751	1.000					
BEDR	0.335	0.320	0.476	0.657	1.000				
BATH	0.640	0.640	0.668	0.555	0.458	1.000			
AGEH	-0.147	-0.130	0.125	0.371	0.141	-0.023	1.000		
ACRE	0.169	0.167	0.177	0.282	0.037	0.301	0.304	1.000	
TAXH	0.899	0.907	0.810	0.513	0.314	0.583	-0.189	0.060	1.000

The variable having the highest correlation (0.997) with SELL is LIST. This positive correlation indicates that the houses with the higher selling prices tend to be those with the higher listing price. This would be the best single variable for predicting selling price. The best additional predictor variable to add would be one having a low correlation with LIST (so that it is not merely duplicating the information provided by LIST). This suggests that AGEH or ACRE would be appropriate variables to consider next. But since the seller and/or realtor undoubtedly considered each of the other variables when determining LIST, there is a sense in which the information of the other variables is already incorporated into LIST.

The best multiple regression equation with corn production as the independent variable, therefore, is probably

SELL = 7.26 + .914 · (LIST)
R^2 = 99.5% = .995
adjusted R^2 = 99.5% = .995
overall P-value = .000

The adjusted R^2 and overall P-value indicate that this is a suitable multiple regression equation for making predictions on the selling price. Since the adjusted R^2 is so close to 1.000, there is little room for improvement and no additional variable could add any significant improvement.

9-6 Modeling

1. The graph appears to be that of a quadratic function.
    Try a regression of the form $y = ax^2 + bx + c$.
    This produces $y = 2x^2 - 12x + 18$ with $R^2 = 100\%$, a perfect fit.

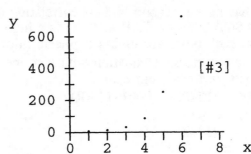

3. The graph appears to be that of an exponential function.
    The equation has the form $y = a \cdot b^x$
    Taking the natural log of both sides
       $\ln y = \ln (a \cdot b^x)$
            $= \ln a + x \cdot (\ln b)$
    Try a regression of the form $\ln y = c + dx$
    This produces $\ln y = -.00009 + 1.1x$ with $R^2 \approx 100\%$, a "perfect" fit.
       solving for the original parameters

$$\ln a = -.0009 \qquad \ln b = 1.1$$
$$a = e^{-.00009} \qquad b = e^{1.1}$$
$$= 1.00 \qquad = 3.00$$

       which yields the exponential equation
$$y = (1.00) \cdot (3.00)^x$$

5. Code the years as 0,1,2,... for 1980,1981,1982,...
    The graph is irregular, but appears to be a declining function.
    Try an exponential function with a base less than 1.00.
    The equation has the form $y = a \cdot b^x$
    Taking the natural log of both sides
       $\ln y = \ln (a \cdot b^x)$
            $= \ln a + x \cdot (\ln b)$

Try a regression of the form $\ln y = c + dx$

This produces $\ln y = 13.88 - .1201x$ with $R^2 = 72.6\%$, a reasonable fit for the data.

solving for the original parameters: $\ln a = 13.88$ $\ln b = -.1201$

$$a = e^{13.88}$$ $$b = e^{-.1201}$$
$$= 1,067,000$$ $$= .8868$$

which yields the exponential equation: $y = (1,067,000) \cdot (.8868)^x$

For 1998, the prediction is: $\hat{y}_{18} = (1,067,000) \cdot (.887)^{18}$
$$= 122,800$$

. The graph appears to be that of a logarithmic function.

The equation has the form $y = a + b \cdot (\ln x)$

This produces $y = 1253 + 1970 \cdot (\ln x)$ with $R^2 = 96.5\%$, an excellent fit.

The graph could also be a power function with an exponent less than 1.00.

The equation has the form $y = ax^b$

Taking the natural log of both sides

$$\ln y = \ln (a \cdot x^b)$$
$$= \ln a + b \cdot (\ln x)$$

This produces $\ln y = 7.601 + .3999 \cdot (\ln x)$ with $R^2 \approx 100\%$, a "perfect" fit.

solving for the original parameter

$$\ln a = 7.601$$
$$a = e^{7.601}$$
$$= 2000$$

which yields the power equation
$$y = 2000 \cdot x^{.3999}$$

The power function gives a slightly better fit than the logarithmic function.

9. First we must find the regression equation for the linear model.

$\bar{x} = 5.5$ $n = 10$

$\bar{y} = 83.3$ $\Sigma x = 55$

$b_1 = [n(\Sigma xy) - (\Sigma x)(\Sigma y)]/[n(\Sigma x^2) - (\Sigma x)^2]$ $\Sigma y = 833$

$\quad = [10(6585) - (55)(833)]/[10(385) - (55)^2]$ $\Sigma x^2 = 385$

$\quad = 20035/825$ $\Sigma y^2 = 121881$

$\quad = 24.285$ $\Sigma xy = 6585$

$b_o = \bar{y} - b_1\bar{x}$

$\quad = 83.3 - (24.284)(5.5)$

$\quad = -50.267$

$\hat{y} = b_o + b_1 x$

$\quad = -50.267 + 24.285x$

The linear predicted values were calculated using $\hat{y} = -50.227 + 24.285x$.

The quadratic predicted values were calculated using $\hat{q} = 2.6742x^2 - 5.132x + 8.567$.

NOTE: The above quadratic equation uses more accuracy than the one given in this section. While the values given in the table below are rounded to the tenths, more accuracy was used to carry out the calculations.

x	y	$\hat{y}$	$\hat{q}$	$(y-\hat{y})^2$	$(q-\hat{q})^2$
1	5	-26.0	6.1	959.9	1.2
2	10	-1.7	9.0	136.8	1.0
3	17	22.6	17.2	31.2	.1
4	31	46.9	30.8	252.0	.0
5	50	71.2	49.8	447.7	.1
6	76	95.4	74.0	378.0	3.8
7	106	119.7	103.7	188.5	5.4
8	132	144.0	138.7	144.3	44.4
9	179	168.3	179.0	114.5	0.0
10	227	192.6	224.7	1184.5	5.4
55	833	833.0	833.0	3837.4	61.4

a. The sum of the squares of the residuals (i.e., the unexplained variation) resulting from the linear model is $\Sigma(y-\hat{y})^2 = 3837.4$.

b. The sum of the squares of the residuals (i.e., the unexplained variation) resulting from the quadratic model is $\Sigma(q-\hat{q})^2 = 61.4$.

c. According to the sum of squares criterion, the quadratic model for these data is far superior to the linear one.

NOTE: $\Sigma(y-\hat{y}) = \Sigma(q-\hat{q}) = 0$. This demonstrates there is no unique model for which the sum of the residuals is zero. In general, the goal is to find the model for which the sum of the squares of the residuals is a minimum.

Review Exercises

1. Let x be the price and y be the consumption.

$n = 10$ $n(\Sigma xy) - (\Sigma x)(\Sigma y) = 10(4.74330) - (13.66)(3.468)$

$\Sigma x = 13.66$ $= .06012$

$\Sigma y = 3.468$ $n(\Sigma x^2) - (\Sigma x)^2 = 10(18.6694) - (13.66)^2$

$\Sigma x^2 = 18.6694$ $= .0984$

$\Sigma y^2 = 1.234888$ $n(\Sigma y^2) - (\Sigma y)^2 = 10(1.234888) - (3.468)^2$

$\Sigma xy = 4.74330$ $= .321856$

$r = [n(\Sigma xy) - (\Sigma x)(\Sigma y)]/[\sqrt{n(\Sigma x^2) - (\Sigma x)^2} \cdot \sqrt{n(\Sigma y^2) - (\Sigma y)^2}]$
$\quad = [.06012]/[\sqrt{.0984} \cdot \sqrt{.321856}]$
$\quad = .338$

a. $H_o: \rho = 0$
$\quad H_1: \rho \neq 0$
$\quad \alpha = .05$
C.R. $r < -.632$ OR   C.R. $t < -t_{8,.025} = -2.306$
$\qquad r > .632$ $t > t_{8,.025} = 2.306$
calculations: calculations:
$\qquad r = .338$ $t_r = (r - \mu_r)/s_r$
$\qquad\qquad\qquad\qquad\qquad\qquad = (.338 - 0)/\sqrt{(1-(.338)^2)/8}$
$\qquad\qquad\qquad\qquad\qquad\qquad = .338/.333$
$\qquad\qquad\qquad\qquad\qquad\qquad = 1.015$

conclusion:
    Do not reject H_o; there is not sufficient evidence to reject the claim that $\rho = 0$.

b. $r^2 = (.338)^2 = .114 = 11.4\%$

c. $b_1 = [n(\Sigma xy) - (\Sigma x)(\Sigma y)]/[n(\Sigma x^2) - (\Sigma x)^2]$
$\qquad = .06012/.0984$
$\qquad = .611$

$\quad b_o = \bar{y} - b_1\bar{x}$
$\qquad = (3.468/10) - (.611)(13.66/10)$
$\qquad = -.488$

$\quad \hat{y} = b_o + b_1 x$
$\qquad = -.488 + .611x$

d. $\hat{y} = -.488 + .611x$
$\quad \hat{y}_{1.38} = \bar{y}$ [no significant correlation]
$\qquad = .3468$ pints per capita per week

2. Let x be the income and y be the consumption.
     $n = 10$ $n(\Sigma xy) - (\Sigma x)(\Sigma y) = 10(1230.996) - (3548)(3.468)$
     $\Sigma x = 3548$ $= 5.496$
     $\Sigma y = 3.468$ $n(\Sigma x^2) - (\Sigma x)^2 = 10(1259524) - (3548)^2$
     $\Sigma x^2 = 1259524$ $= 6936$
     $\Sigma y^2 = 1.234888$ $n(\Sigma y^2) - (\Sigma y)^2 = 10(1.234888) - (3.468)^2$
     $\Sigma xy = 1230.996$ $= .321856$

$r = [n(\Sigma xy) - (\Sigma x)(\Sigma y)]/[\sqrt{n(\Sigma x^2) - (\Sigma x)^2} \cdot \sqrt{n(\Sigma y^2) - (\Sigma y)^2}]$
$\quad = [5.496]/[\sqrt{6936} \cdot \sqrt{.321856}]$
$\quad = .116$

a. $H_o: \rho = 0$
$\quad H_1: \rho \neq 0$
$\quad \alpha = .05$
C.R. $r < -.632$ OR   C.R. $t < -t_{8,.025} = -2.306$
$\qquad r > .632$ $t > t_{8,.025} = 2.306$
calculations: calculations:
$\qquad r = .116$ $t_r = (r - \mu_r)/s_r$
$\qquad\qquad\qquad\qquad\qquad\qquad = (.116 - 0)/\sqrt{(1-(.116)^2)/8}$
$\qquad\qquad\qquad\qquad\qquad\qquad = .116/.351$
$\qquad\qquad\qquad\qquad\qquad\qquad = .331$

conclusion:
    Do not reject H_o; there is not sufficient evidence to reject the claim that $\rho = 0$.

b. $r^2 = (.116)^2 = .013 = 1.3\%$

c. $b_1 = [n(\Sigma xy) - (\Sigma x)(\Sigma y)]/[n(\Sigma x^2) - (\Sigma x)^2]$
$= 5.496/6936$
$= .000792$

$b_0 = \bar{y} - b_1\bar{x}$
$= (3.468/10) - (.000792)(3548/10)$
$= .0657$

$\hat{y} = b_0 + b_1x$
$= .0657 + .000792x$

d. $\hat{y} = .0657 + .000792x$
$\hat{y}_{365} = \bar{y}$ [no significant correlation]
$= .3468$ pints per capita per week

3. Let x be the temperature and y be the consumption.

$n = 10$ $n(\Sigma xy) - (\Sigma x)(\Sigma y) = 10(189.038) - (526)(3.468)$
$\Sigma x = 526$ $= 66.212$
$\Sigma y = 3.468$ $n(\Sigma x^2) - (\Sigma x)^2 = 10(29926) - (526)^2$
$\Sigma x^2 = 29926$ $= 22584$
$\Sigma y^2 = 1.234888$ $n(\Sigma y^2) - (\Sigma y)^2 = 10(1.234888) - (3.468)^2$
$\Sigma xy = 189.038$ $= .321856$

$r = [n(\Sigma xy) - (\Sigma x)(\Sigma y)]/[\sqrt{n(\Sigma x^2) - (\Sigma x)^2} \cdot \sqrt{n(\Sigma y^2) - (\Sigma y)^2}]$
$= [66.212]/[\sqrt{22584} \cdot \sqrt{.321856}]$
$= .777$

a. $H_0: \rho = 0$
$H_1: \rho \neq 0$
$\alpha = .05$
C.R. $r < -.632$ OR   C.R. $t < -t_{8,.025} = -2.306$
$r > .632$ $t > t_{8,.025} = 2.306$
calculations: calculations:
$r = .777$ $t_r = (r - \mu_r)/s_r$
$= (.777 - 0)/\sqrt{(1-(.777)^2)/8}$
$= .777/.223$
$= 3.486$

conclusion:
Reject H_0; there is sufficient evidence to reject the claim that $\rho = 0$ and to conclude that $\rho \neq 0$ (in fact, $\rho > 0$)

b. $r^2 = (.777)^2 = .604 = 60.4\%$

c. $b_1 = [n(\Sigma xy) - (\Sigma x)(\Sigma y)]/[n(\Sigma x^2) - (\Sigma x)^2]$
$= 66.212/22584$
$= .00293$

$b_0 = \bar{y} - b_1\bar{x}$
$= (3.468/10) - (.00293)(526/10)$
$= .193$

$\hat{y} = b_0 + b_1x$
$= .193 + .00293x$

d. $\hat{y} = .193 + .00293x$
$\hat{y}_{32} = .193 + .00293(32)$
$= .2864$ pints per capita per week

4. Using Minitab and the given notation,
$\hat{y} = -.053 + .747x_1 - .00220x_2 + .00303x_3$
$R^2 = 72.6\% = .726$
adjusted $R^2 = 58.9\% = .589$
overall P-value $= .040$

Yes; since the overall P-value of .040 is less than .05, the regression equation can be used to predict ice cream consumption. Individually, x_1 and x_2 were not significantly related to consumption; only x_3 could be used to predict consumption. This suggests that variables x_1 and x_2 might not be making a worthwhile contribution to the regression, and that the equation from exercise #3 (using x_3 alone) might actually be a better and more efficient predictive equation. To check this, calculate the adjusted R^2 for exercise #3.

$$\begin{aligned}
\text{adjusted } R^2 &= 1 - \{(n-1)/[n-(k+1)]\} \cdot (1-R^2) \\
&= 1 - \{9/[10-(2)]\} \cdot (1-R^2) \\
&= 1 - (9/8) \cdot (1-.777^2) \\
&= .554
\end{aligned}$$

Since the adjusted R^2 of .589 is higher than the adjusted R^2 of .554 from exercise #3, however, there is a well-defined statistical sense in which this multiple regression equation is the best of the equations considered in these exercises.

5. Let x be the precipitation and y be the corn production.

$n = 10$	$n(\Sigma xy) - (\Sigma x)(\Sigma y) = 10(483531.0) - (339.1)(14059)$
$\Sigma x = 339.1$	$= 67903.1$
$\Sigma y = 14059$	$n(\Sigma x^2) - (\Sigma x)^2 = 10(11896.05) - (339.1)^2$
$\Sigma x^2 = 11896.05$	$= 3971.69$
$\Sigma y^2 = 20770345$	$n(\Sigma y^2) - (\Sigma y)^2 = 10(20770345) - (14059)^2$
$\Sigma xy = 483531.0$	$= 10047969$

$$\begin{aligned}
r &= [n(\Sigma xy) - (\Sigma x)(\Sigma y)]/[\sqrt{n(\Sigma x^2) - (\Sigma x)^2} \cdot \sqrt{n(\Sigma y^2) - (\Sigma y)^2}] \\
&= [67903.1]/[\sqrt{3971.69} \cdot \sqrt{10047969}] \\
&= .340
\end{aligned}$$

a. $H_o: \rho = 0$
$H_1: \rho \neq 0$
$\alpha = .05$

C.R. $r < -.632$ OR   C.R. $t < -t_{8,.025} = -2.306$
     $r > .632$ $t > t_{8,.025} = 2.306$

calculations: calculations:
    $r = .340$ $t_r = (r - \mu_r)/s_r$
                       $= (.340 - 0)/\sqrt{(1-(.340)^2)/8}$
                       $= .340/.333$
                       $= 1.022$

conclusion:
    Do not reject H_o; there is not sufficient evidence to reject the claim that $\rho = 0$.

b. $$\begin{aligned}
b_1 &= [n(\Sigma xy) - (\Sigma x)(\Sigma y)]/[n(\Sigma x^2) - (\Sigma x)^2] \\
&= 67903.1/3971.69 \\
&= 17.10
\end{aligned}$$
$$\begin{aligned}
b_o &= \bar{y} - b_1\bar{x} \\
&= 1405.9 - (17.10)(33.91) \\
&= 826.1
\end{aligned}$$
$$\begin{aligned}
\hat{y} &= b_o + b_1 x \\
&= 826.1 + 17.10x
\end{aligned}$$

c. $\hat{y} = 826.1 + 17.10x$
   $\hat{y}_{29.3} = \bar{y}$ [no significant correlation]
        $= 1405.9$ million bushels

6. Let x be the temperature and y be the corn production.

$n = 10$ $\qquad$ $n(\Sigma xy) - (\Sigma x)(\Sigma y) = 10(684765.06) - (487.61)(14059)$
$\Sigma x = 487.61$ $\qquad\qquad\qquad = -7658.39$
$\Sigma y = 14059$ $\qquad$ $n(\Sigma x^2) - (\Sigma x)^2 = 10(23799.9197) - (487.61)^2$
$\Sigma x^2 = 23799.9197$ $\qquad\qquad\qquad = 235.6849$
$\Sigma y^2 = 20770345$ $\qquad$ $n(\Sigma y^2) - (\Sigma y)^2 = 10(20770345) - (14059)^2$
$\Sigma xy = 68476.06$ $\qquad\qquad\qquad = 10047969$

$r = [n(\Sigma xy) - (\Sigma x)(\Sigma y)]/[\sqrt{n(\Sigma x^2) - (\Sigma x)^2} \cdot \sqrt{n(\Sigma y^2) - (\Sigma y)^2}]$
$\quad = [-7658.39/[\sqrt{235.6849} \cdot \sqrt{10047969}\,]$
$\quad = -.157$

a. $H_o: \rho = 0$
$H_1: \rho \neq 0$
$\alpha = .05$
C.R. $r < -.632$ $\qquad$ OR $\qquad$ C.R. $t < -t_{8,.025} = -2.306$
$\qquad r > .632$ $\qquad\qquad\qquad\qquad t > t_{8,.025} = 2.306$
calculations: $\qquad\qquad\qquad$ calculations:
$\quad r = -.157$ $\qquad\qquad\qquad\qquad t_r = (r - \mu_r)/s_r$
$\qquad\qquad\qquad\qquad\qquad\qquad = (-.157 - 0)/\sqrt{(1-(-.157)^2)/8}$
$\qquad\qquad\qquad\qquad\qquad\qquad = -.157/.122$
$\qquad\qquad\qquad\qquad\qquad\qquad = -1.291$

conclusion:
Do not reject H_o; there is not sufficient evidence to reject the claim that $\rho = 0$.

b. $b_1 = [n(\Sigma xy) - (\Sigma x)(\Sigma y)]/[n(\Sigma x^2) - (\Sigma x)^2]$
$\quad = -7658.39/235.6849 = -32.49$
$b_o = \bar{y} - b_1\bar{x}$
$\quad = 1405.9 - (-32.49)(48.761) = 2990.3$
$\hat{y} = b_o + b_1 x$
$\quad = 2990.3 - 32.49x$

c. $\hat{y} = 2990.3 - 32.49x$
$\hat{y}_{48.86} = \bar{y}$ [no significant correlation]
$\qquad = 1405.9$ million bushels

7. Let x be the acreage and y be the corn production.

$n = 10$ $\qquad$ $n(\Sigma xy) - (\Sigma x)(\Sigma y) = 10(172561550) - (119550)(14059)$
$\Sigma x = 119550$ $\qquad\qquad\qquad = 44862050$
$\Sigma y = 14059$ $\qquad$ $n(\Sigma x^2) - (\Sigma x)^2 = 10(1454397500) - (119550)^2$
$\Sigma x^2 = 1454397500$ $\qquad\qquad\qquad = 251772500$
$\Sigma y^2 = 20770345$ $\qquad$ $n(\Sigma y^2) - (\Sigma y)^2 = 10(20770345) - (14059)^2$
$\Sigma xy = 172561550$ $\qquad\qquad\qquad = 10047969$

$r = [n(\Sigma xy) - (\Sigma x)(\Sigma y)]/[\sqrt{n(\Sigma x^2) - (\Sigma x)^2} \cdot \sqrt{n(\Sigma y^2) - (\Sigma y)^2}]$
$\quad = [44862050/[\sqrt{251772500} \cdot \sqrt{10047969}\,]$
$\quad = .892$

a. $H_o: \rho = 0$
$H_1: \rho \neq 0$
$\alpha = .05$
C.R. $r < -.632$ $\qquad$ OR $\qquad$ C.R. $t < -t_{8,.025} = -2.306$
$\qquad r > .632$ $\qquad\qquad\qquad\qquad t > t_{8,.025} = 2.306$
calculations: $\qquad\qquad\qquad$ calculations:
$\quad r = .892$ $\qquad\qquad\qquad\qquad t_r = (r - \mu_r)/s_r$
$\qquad\qquad\qquad\qquad\qquad\qquad = (.892 - 0)/\sqrt{(1-(.892)^2)/8}$
$\qquad\qquad\qquad\qquad\qquad\qquad = .892/.160 = 5.579$

conclusion:
Reject H_o; there is sufficient evidence to reject the claim that $\rho = 0$ and conclude that $\rho \neq 0$ (in fact, that $\rho > 0$).

b. $b_1 = [n(\Sigma xy) - (\Sigma x)(\Sigma y)]/[n(\Sigma x^2) - (\Sigma x)^2]$
   $= 44862050/251772500$
   $= .1782$

   $b_0 = \bar{y} - b_1\bar{x}$
   $= 1405.9 - (.1782)(11955.0)$
   $= -724.3$

   $\hat{y} = b_0 + b_1 x$
   $= -724.3 + .1782x$

c. $\hat{y} = -724.3 + .1782x$
   $\hat{y}_{13300} = -724.3 + .1782(13300)$
   $= 1645.8$ million bushels

8. Using Minitab and the given notation,
   $\hat{y} = -3145 + 10.1x_1 + 40.4x_2 + .187x_3$
   $R^2 = 87.4\% = .874$
   adjusted $R^2 = 81.1\% = .811$
   overall P-value $= .004$

   Yes; since the overall P-value of .004 is less than .05, the regression equation can be used to predict corn production. Individually, x_1 and x_2 were not significantly related to production; only x_3 could be used to predict production. This suggests that variables x_1 and x_2 might not be making a worthwhile contribution to the regression, and that the equation from exercise #7 (using x_3 alone) might actually be a better and more efficient predictive equation. To check this, calculate the adjusted R^2 for exercise #7.

   adjusted $R^2 = 1 - \{(n-1)/[n-(k+1)]\} \cdot (1-R^2)$
   $= 1 - \{9/[10-(2)]\} \cdot (1-R^2)$
   $= 1 - (9/8) \cdot (1-.892^2)$
   $= .770$

   Since the adjusted R^2 of .874 is higher than the adjusted R^2 of .770 from exercise #7, however, there is a well-defined statistical sense in which this multiple regression equation is the best of the equations considered in these exercises.

Cumulative Review Exercises

1. concerns μ: n > 30, use z [with s for σ]
   summary statistics: n = 50, $\Sigma x = 4033$, $\Sigma x^2 = 332331$, $\bar{x} = 80.66$,   s = 11.98

   a. original claim $\mu > 66$
      $H_0: \mu \leq 66$
      $H_1: \mu > 66$
      $\alpha = .05$ [assumed]
      C.R. z > $z_{.05} = 1.645$
      calculations:
      $z_{\bar{x}} = (\bar{x} - \mu)/\sigma_{\bar{x}}$
      $= (80.66 - 66)/(11.98/\sqrt{50}$
      $= 14.66/1.694$
      $= 8.655$

      conclusion: reject H_0; there is sufficient evidence to conclude that $\mu > 66$.

   b. $\bar{x} \pm z_{.025} \cdot \sigma_{\bar{x}}$
      $80.66 \pm 1.96 \cdot 11.98/\sqrt{50}$
      80.66 ± 3.32
      $77.3 < \mu < 84.0$

2. Let x and y be as given.

n = 12 $n(\Sigma xy) - (\Sigma x)(\Sigma y) = 12(122836) - (1189)(1234)$
$\Sigma x = 1189$ $= 6806$
$\Sigma y = 1234$ $n(\Sigma x^2) - (\Sigma x)^2 = 12(118599) - (1189)^2$
$\Sigma x^2 = 118599$ $= 9467$
$\Sigma y^2 = 127724$ $n(\Sigma y^2) - (\Sigma y)^2 = 12(127724) - (1234)^2$
$\Sigma xy = 122836$ $= 9932$

$r = [n(\Sigma xy) - (\Sigma x)(\Sigma y)]/[\sqrt{n(\Sigma x^2) - (\Sigma x)^2} \cdot \sqrt{n(\Sigma y^2) - (\Sigma y)^2}]$
$= [6806]/[\sqrt{9467}\ \sqrt{9932}\ ]$
$= .702$

a. $\bar{x} = 1189/12 = 99.1$
$s_x = \sqrt{9467/(12 \cdot 11)} = 8.47$

b. $\bar{y} = 1234/12 = 102.8$
$s_y = \sqrt{9932/(12 \cdot 11)} = 8.67$

c. No; there does not appear to be a difference between the means of the two populations. In exploring the relationship between the IQ's of twins, such a two sample approach is not appropriate because it completely ignores the pairings of the scores.

d. Correlation would be appropriate for answering the question "is there a (linear) relationship?"

H_0: $\rho = 0$
H_1: $\rho \neq 0$
$\alpha = .05$ [assumed]

C.R. r < -.576   OR   C.R. $t < -t_{10,.025} = -2.228$
    r > .576 $t > t_{10,.025} = 2.228$

calculations: calculations:
    r = .702 $t_r = (r - \mu_r)/s_r$
                                     $= (.702 - 0)/\sqrt{(1-(.702)^2)/10}$
                                     $= .702/.225$
                                     $= 3.116$

conclusion:
Reject H_0; there is sufficient evidence to reject the claim that $\rho = 0$ and to conclude that $\rho \neq 0$ (in fact, $\rho > 0$).

Ordinarily, the conclusion would be that about $R^2 = (.702)^2 = .493 = 49.3\%$ of the variation in x can be explained in terms of y (and vice-versa). In this context that means that about 49.3% of the variation among the IQ's in one group can be explained in terms of the IQ's of their twins (i.e, in terms of heredity). In simplest terms, it seems intelligence is about ½ due to heredity and ½ due to environment.

NOTE: The study in exercise #2 contains at least two interesting subtleties.
(1) Correlation addresses only whether there is a relationship between the IQ's, and not whether the IQ's are close to each other. If each older twin had an IQ 20 points higher than the corresponding younger twin, there would be a perfect correlation -- but the twins would not be similar in IQ at all. Beware of the misconception that correlation implies similarity.
(2) Within each pair, the older twin was designated x. That was an arbitrary decision to produce an objective rule, and there is no biological basis for putting all the older twins in one group and the younger twins in another. The x-y designation within pairs could just as properly have been made alphabetically, randomly, or by another rule. But the rule affects the results. If it happens to designate all the twins with the higher IQ as x, the correlation rises to r = .810. If it happens to designate the twins with the higher IQ as x in the first six pairs and as y in the last six, the correlation falls to r = .633.

One technique which addresses both of the above issues (i.e., it tests for similarity and does not depend upon an x-y designation at all) involves comparing the variation within pairs to the overall variation in IQ's. If there is significantly less variability between twins than there is variability in the general population (or between non-identical twins raised apart), then there is a significant relationship (i.e., in the sense of similarity) between the IQ's of identical twins. One possible approach is given below.

pair	x_1	x_2	$\overline{x}$	s^2
1	107	111	109.0	8.0
2	96	97	96.5	0.5
3	103	116	109.5	84.5
4	90	107	98.5	144.5
5	96	99	97.5	4.5
6	113	111	112.0	2.0
7	86	85	85.5	0.5
8	99	108	103.5	40.5
9	109	102	105.5	24.5
10	105	105	105.0	0.0
11	96	100	98.0	8.0
12	89	93	91.0	8.0

To estimate the overall s^2 use the variance for all 24 IQ scores taken as a group: $s_1^2 = 73.955$.

To estimate the s^2 between twins, use the mean of the 12 s^2 values above: $s_2^2 = 27.125$.

Now use the F statistic to test the claim that $s_1^2 > s_2^2$.

3. a. No; testing for a correlation will not determine whether the position has an effect on the measured values. Adding the 3.00 to each sitting value, for example, would certainly create a position effect. Such a change would affect the regression equation for predicting one value from the other, but it would not affect the correlation. In general, any linear transformation (e.g., measuring in different units) will not effect the correlation -- which is based on the relationship between the standardized scores for each set of scores.

b. The appropriate test would be the μ_d test of the previous chapter -- using the differences between the paired scores.

original claim $\mu_d = 0$ [$n \leq 30$ and σ_d unknown, use t]

d = $x_{sitting} - x_{supine}$: .03 -.64 -.35 .11 .17 .47 .49 .40 .29 .44

n = 10

$\Sigma d = 1.41$ $\qquad$ $\overline{d} = .141$

$\Sigma d^2 = 1.4727$ $\qquad$ $s_d = .376$

$H_o: \mu_d = 0$

$H_1: \mu_d \neq 0$

$\alpha = .05$

C.R. $t < -t_{9,.025} = -2.262$

$\quad t > t_{9,.025} = 2.262$

calculations:

$t_{\overline{d}} = (\overline{d} - \mu_d)/s_{\overline{d}}$

$\quad = (.141 - 0)/(.376/\sqrt{10})$

$\quad = .141/.119$

$\quad = 1.185$

conclusion:

Do not reject H_o; there is not sufficient evidence to reject the claim that $\mu_d = 0$.

Chapter 10

Multinomial Experiments and Contingency Tables

10-2 Multinomial Experiments: Goodness-of-Fit

1. a. $\chi^2_{37,.10} = 51.805$
   b. $\chi^2_{37,.10} = 51.085 > 38.232 > 29.051 = \chi^2_{37,.90}$
   $.10 < \text{P-value} < .90$
   c. There is not enough evidence to reject the claim that the 38 results are equally likely.

NOTE: In multinomial problems, always verify that $\Sigma E = \Sigma O$ before proceeding. If these sums are not equal, then an error has been made and further calculations have no meaning.

3. H_o: $p_{LF} = p_{RF} = p_{LR} = p_{RR} = .25$
   H_1: at least one of the proportions is different from .25
   $\alpha = .05$
   C.R. $\chi^2 > \chi^2_{3,.05} = 7.815$
   calculations:

tire	O	E	$(O-E)^2/E$
LF	11	10	.100
RF	15	10	2.500
LR	8	10	.400
RR	6	10	1.600
	40	40	4.600

$\chi^2 = \Sigma[(O-E)^2/E]$
$= 4.600$

conclusion:
   Do not reject H_o; there is not sufficient evidence to conclude that at least one of the proportions is different from .25.
While we cannot be 95% certain that there is a tendency to pick one tire more than any other, it might be worthwhile to take a larger sample to see if the trend toward picking front tires becomes significant.

5. H_o: $p_{Sun} = p_{Mon} = p_{Tue} = ... = p_{Sat} = 1/7$
   H_1: at least one of the proportions is different from 1/7
   $\alpha = .05$ [assumed]
   C.R. $\chi^2 > \chi^2_{6,.05} = 12.592$
   calculations:

day	O	E	$(O-E)^2/E$
Sun	40	30.857	2.709
Mon	24	30.857	1.524
Tue	25	30.857	1.112
Wed	28	30.857	.265
Thu	29	30.857	.112
Fri	32	30.857	.042
Sat	38	30.857	1.653
	216	216.000	7.417

$\chi^2 = \Sigma[(O-E)^2/E]$
$= 7.417$

conclusion:
   Do not reject H_o; there is not sufficient evidence to conclude that at least one of the proportions is different from 1/7.
There is not enough evidence to reject with 95% confidence the daily drinker theory in favor of theory of the casual drinker who binges on Friday and Saturday.

7. H_o: $p_{Mon} = .30$, $p_{Tue} = .15$, $p_{Wed} = .15$, $p_{Thu} = .20$, $p_{Fri} = .20$
   H_1: at least one of the proportions is different from what is claimed
   $\alpha = .05$
   C.R. $\chi^2 > \chi^2_{4,.05} = 9.488$
   calculations:

day	O	E	$(O-E)^2/E$
Mon	31	44.10	3.891
Tue	42	22.05	18.050
Wed	18	22.05	.744
Thu	25	29.40	.659
Fri	31	29.40	.087
	147	147.00	23.431

$\chi^2 = \Sigma[(O-E)^2/E]$
$= 23.431$

conclusion:
   Reject H_o; there is sufficient evidence to conclude that at least one of the proportions is different from what is claimed.

Rejection of this claim may indirectly help to correct the accident problem. It may indicate that the safety expert is wrong and that a better such person should be employed. It may indicate that the safety expert is correct but that Tuesday (which makes an unusually large contribution to $\Sigma[(O-E)^2/E]$) involves circumstances unique to the plant being studied. At the very least, rejection of the hypothesis indicates that medical staffing should not be based on the proportions claimed.

9. H_o: $p_0 = p_1 = p_2 = ... = p_9 = .10$
   H_1: at least one of the proportions is different from .10
   $\alpha = .05$
   C.R. $\chi^2 > \chi^2_{9,.05} = 16.919$
   calculations:

digit	O	E	$(O-E)^2/E$
0	8	10	.400
1	8	10	.400
2	12	10	.400
3	11	10	.100
4	10	10	.000
5	8	10	.400
6	9	10	.100
7	8	10	.400
8	12	10	.400
9	14	10	1.600
	100	100	4.200

$\chi^2 = \Sigma[(O-E)^2/E]$
$= 4.200$

conclusion:
   Do not reject H_o; there is not sufficient evidence to conclude that at least one of the proportions is different from .10.

11. H_o: $p_1 = .16$, $p_2 = .44$, $p_3 = .27$, $p_4 = .13$
   H_1: at least one of the proportions is different from the license proportions
   $\alpha = .05$
   C.R. $\chi^2 > \chi^2_{3,.05} = 7.815$
   calculations:

group	O	E	$(O-E)^2/E$
1: < 25	36	14.08	34.125
2: 25-44	21	38.72	8.109
3: 45-64	12	23.76	5.821
4: > 64	19	11.44	4.996
	88	88.00	53.051

$\chi^2 = \Sigma[(O-E)^2/E]$
$= 53.051$

conclusion:
   Reject H_o; there is sufficient evidence to conclude that at least one of the proportions is different from the license proportions.

Yes; the "under 25" group appears to have a disproportionate number of crashes. It would be fairer, but much more difficult, to base the E values on the proportion of miles driven and not on the proportion of licenses possessed.

13. H_o: $p_1 = p_2 = p_3 = \ldots = p_8 = 1/8$
    H_1: at least one of the proportions is different from 1/8
    $\alpha = .05$ [assumed]
    C.R. $\chi^2 > \chi^2_{7,.05} = 14.067$
    calculations:

start	O	E	$(O-E)^2/E$
1	29	18	6.722
2	19	18	.056
3	18	18	.000
4	25	18	2.722
5	17	18	.056
6	10	18	3.556
7	15	18	.500
8	11	18	2.722
	144	144	16.333

$\chi^2 = \Sigma[(O-E)^2/E]$
$= 16.333$

conclusion:
    Reject H_o; there is sufficient evidence to conclude that at least one of the proportions is different from 1/8 (i.e., that the probabilities of winning in the different starting positions are not all the same).

15. H_o: $p_0 = p_1 = p_2 = \ldots = p_9 = .10$
    H_1: at least one of the proportions is different from .10
    $\alpha = .05$ [assumed]
    C.R. $\chi^2 > \chi^2_{9,.05} = 16.919$
    calculations:

digit	O	E	$(O-E)^2/E$
0	33	5.2	148.623
1	10	5.2	4.431
2	3	5.2	.931
3	2	5.2	1.969
4	0	5.2	5.200
5	1	5.2	3.392
6	0	5.2	5.200
7	1	5.2	3.392
8	0	5.2	5.200
9	2	5.2	1.969
	52	52.0	180.308

$\chi^2 = \Sigma[(O-E)^2/E]$
$= 180.308$

conclusion:
    Reject H_o; there is sufficient evidence to conclude that at least one of the proportions is different from .10.

No; the reasoning applied in the section example that leads to the conclusion that the values are obtained from estimation rather than actual measurement does not apply here. Since there are so many days with 0.0 precipitation, one would expect a large number of 0's.

7. H_o: $p_{LF} = p_{RF} = p_{LR} = p_{RR} = .25$
   H_1: at least one of the proportions is different from .25
   $\alpha = .05$
   C.R. $\chi^2 > \chi^2_{3,.05} = 7.815$
   calculations:

tire	O	E	$(O-E)^2/E$
LF	11	23.5	6.649
RF	15	23.5	3.074
LR	8	23.5	10.223
RR	60	23.5	56.691
	94	94.0	76.638

$$\chi^2 = \Sigma[(O-E)^2/E]$$
$$= 76.638$$

conclusion:
   Reject H_o; there is sufficient evidence to conclude that at least one of the proportions is different from .25.

In general, an "outlier" is a point that is different from the normal pattern. Since that (i.e., "difference from the normal pattern") is essentially what the goodness-of-fit test is designed to detect, an outlier will tend to increase the probability of rejecting H_o.

9. NOTE: Both outcomes having the same expected frequency is equivalent to $p_1 = p_2 = .5$.
   a. H_o: $p_1 = p_2 = .5$
      H_1: at least one of the proportions is different from .5
      $\alpha = .05$
      C.R. $\chi^2 > \chi^2_{1,.05} = 3.841$
      calculations

type	O	E	O-E	$(O-E)^2$	$(O-E)^2/E$
A	f_1	$(f_1+f_2)/2$	$(f_1-f_2)/2$	$(f_1-f_2)^2/4$	$[(f_1-f_2)^2/4]/[(f_1+f_2)/2]$
B	f_2	$(f_1+f_2)/2$	$(f_2-f_1)/2$	$*(f_1-f_2)^2/4$	$[(f_1-f_2)^2/4]/[(f_1+f_2)/2]$
	f_1+f_2	f_1+f_2			$[(f_1-f_2)^2/2]/[(f_1+f_2)/2]$

*NOTE: $(f_2-f_1)^2 = (f_1-f_2)^2$
$$\chi^2 = \Sigma[(O-E)^2/E]$$
$$= [(f_1-f_2)^2/2]/[(f_1+f_2)/2]$$
$$= (f_1-f_2)^2/(f_1+f_2)$$

   b. H_o: $p = .5$
      H_1: $p \neq .5$
      $\alpha = .05$
      C.R. $z < -z_{.025} = -1.960$
            $z > z_{.025} = 1.960$
      calculations:

$$z_{\hat{p}} = (\hat{p} - \mu_{\hat{p}})/\sigma_{\hat{p}}$$
$$= [f_1/(f_1+f_2) - .5]/\sqrt{(.5)(.5)/(f_1+f_2)}$$
$$= [.5(f_1-f_2)/(f_1+f_2)]/\sqrt{(.5)(.5)/(f_1+f_2)}$$
$$= [(f_1-f_2)/(f_1+f_2)]/\sqrt{1/(f_1+f_2)}$$
$$= (f_1-f_2)/\sqrt{f_1+f_2}$$

Note that $z^2 = \chi^2$ since $[(f_1-f_2)/\sqrt{f_1+f_2}]^2 = (f_1-f_2)^2/(f_1+f_2)$ and $(\pm 1.960)^2 = 3.841$.

21. NOTE: Usually the hypothesized p_i's are given and the formula $E_i = np_i$ is used to find the individual expected values. Here, the individual expected values are given. To state a null hypothesis, either use the formula "in reverse" to solve for the hypothesized p_i's or use the term "goodness of fit" without mentioning specific probabilities.

H_o: there is goodness of fit to the Poisson distribution
H_1: there is not goodness of fit to the Poisson distribution
$\alpha = .05$
C.R. $\chi^2 > \chi^2_{2,.05} = 5.991$
calculations:

homicides	O	E	$(O-E)^2/E$
0	268	265.6	.022
1	79	84.4	.345
2+	18	15.0*	.600
	365	365.0	.967

$\chi^2 = \Sigma[(O-E)^2/E]$
$= .967$

conclusion:

Do not reject H_o; there is not sufficient evidence to conclude that there is not goodness of fit to the Poisson distribution.

* Combine the last three cells so that $E_i \geq 5$ for each i. Make $E_{2+} = 15.0$ so that $\Sigma E_i = 365$. This adjusts for the .1 "lost" in the original table due to round-off.

10-3   Contingency Tables: Independence and Homogeneity

NOTE: For each row and each column it must be true that $\Sigma O = \Sigma E$. After the marginal row and column totals are calculated, both the row totals and the column totals must sum to produce the same grand total. If either of the preceding is not true, then an error has been made and further calculations have no meaning. In addition, the following are true for all χ^2 contingency table analyses in this manual.

* The E values for each cell are given in parentheses below the O values.
* The addends used to calculate the χ^2 test statistic follow the physical arrangement of the cells in the original contingency table. This practice makes it easier to monitor the large number of inter-mediate steps involved and helps to prevent errors caused by missing or double-counting cells.
* The accompanying chi-square illustration follows the "usual" shape as pictured with Table A-6, even though that shape is not correct for df=1 or df=2.

1. H_o: drug treatment and oral reaction are independent
   H_1: drug treatment and oral reaction are related
   $\alpha = .05$
   C.R. $\chi^2 > \chi^2_{1,.05} = 3.841$
   calculations:

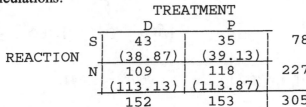

		TREATMENT		
		D	P	
REACTION	S	43 (38.87)	35 (39.13)	78
	N	109 (113.13)	118 (113.87)	227
		152	153	305

$\chi^2 = \Sigma[(O-E)^2/E]$
$= .4383 + .4355$
$\quad .1506 + .1496$
$= 1.174$

conclusion:
Do not reject H_o; there is not sufficient evidence to conclude that the drug treatment

and the oral reaction are related. NOTE: The P-value from Minitab can also be used to make the decision. Since .279 > .05, do not reject H_0.

A person thinking about using Nicorette might still want to be concerned about mouth soreness. The direction of the data is toward those using the real drug being more likely to experience soreness. While we cannot be 95% sure that there $\underline{is}$ soreness associated with Nicorette, neither can we be sure that there is $\underline{not}$ such soreness.

3. H_o: gender and opinion are independent
H_1: gender and opinion are related
$\alpha = .01$
C.R. $\chi^2 > \chi^2_{1,.01} = 6.635$
calculations:

OPINION

		Y	N	
	M	391	425	816
		(457.95)	(358.05)	
GENDER	F	480	256	736
		(413.05)	(322.95)	
		871	681	1552

$\chi^2 = \Sigma[(O-E)^2/E]$
$= 9.7873 + 12.5180$
$\quad 10.8511 + 13.8787$
$= 47.035$

conclusion:
   Reject H_o; there is sufficient evidence to reject the claim that gender and opinion are independent and to conclude that these variables are related.

Since more men than women are charged with drunk driving, the men would have more to lose.

. H_o: there is homogeneity of proportions across gender of interviewer
H_1: there is not homogeneity of proportions
$\alpha = .01$
C.R. $\chi^2 > \chi^2_{1,.01} = 6.635$
calculations:

INTERVIEWER

		M	F	
FEMALE	A	512	336	848
		(565.3)	(282.7)	
RESPONSE	D	288	64	352
		(234.7)	(117.3)	
		800	400	1200

$\chi^2 = \Sigma[(O-E)^2/E]$
$= 5.031 + 10.063$
$\quad 12.121 + 24.242$
$= 51.458$

conclusion:
   Reject H_o; there is sufficient evidence to conclude that there is not homogeneity of proportions across gender of interviewer -- i.e., the proportion of agree/disagree responses vary according to the gender of the interviewer.

7. NOTE: We assume the sample sizes of 200 men and 300 women were predetermined, making this exercise "homogeneity of proportions analysis" rather than "independence of variables."
H_o: there is homogeneity of proportions across gender of respondent
H_1: there is not homogeneity of proportions
$\alpha = .05$
C.R. $\chi^2 > \chi^2_{2,.05} = 5.991$
calculations:

		CONFIDENCE			
		GREAT	SOME	LITTLE	
RESPONDENT	M	115	56	29	200
		(116)	(60)	(24)	
	F	175	94	31	300
		(174)	(90)	(36)	
		290	150	60	500

$\chi^2 = \Sigma[(O-E)^2/E]$
$\quad = .009 + .267 + 1.042$
$\quad\quad .006 + .178 + .694$
$\quad = 2.195$

conclusion:
Do not reject H_o; there is not sufficient evidence to conclude that there is not homogeneity of proportions across gender of respondent -- i.e., the proportion of persons in the given confidence categories does not vary significantly according to the gender of the respondent.

9. H_o: distribution of planes and airline company are independent
H_1: distribution of planes and airline company are related
$\alpha = .01$
C.R. $\chi^2 > \chi^2_{2,.01} = 9.210$
calculations:

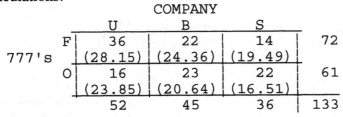

		COMPANY			
		U	B	S	
777's	F	36	22	14	72
		(28.15)	(24.36)	(19.49)	
	O	16	23	22	61
		(23.85)	(20.64)	(16.51)	
		52	45	36	133

$\chi^2 = \Sigma[(O-E)^2/E]$
$\quad = 2.1888 + .2288 + 1.5458$
$\quad\quad 2.5835 + .2701 + 1.8246$
$\quad = 8.642$

conclusion:
Do not reject H_o; there is not sufficient evidence to reject the claim that the distribution of planes is independent of the airline company.
NOTE: This is not evidence to support the given claim, but merely a statement that there is not enough evidence to reject the claim.

1. H_o: type of crime and criminal/victim connection are independent
   H_1: type of crime and criminal/victim connection are related
   $\alpha = .05$
   C.R. $\chi^2 > \chi^2_{2,.05} = 5.991$
   calculations:

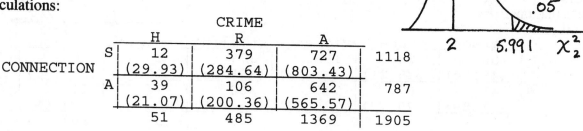

CRIME

		H	R	A	
	S	12	379	727	1118
CONNECTION		(29.93)	(284.64)	(803.43)	
	A	39	106	642	787
		(21.07)	(200.36)	(565.57)	
		51	485	1369	1905

$$\chi^2 = \Sigma[(O-E)^2/E]$$
$$= 10.7418 + 31.2847 + 7.2715$$
$$15.2600 + 44.4425 + 10.3298$$
$$= 119.330$$

conclusion:
   Reject H_o; there is sufficient evidence to conclude that the type of crime and the criminal/victim connection are related.

3. H_o: sentence and plea are independent
   H_1: sentence and plea are related
   $\alpha = .05$
   C.R. $\chi^2 > \chi^2_{1,.05} = 3.841$
   calculations:

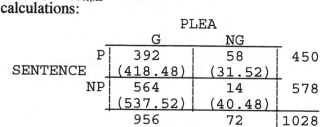

PLEA

		G	NG	
	P	392	58	450
SENTENCE		(418.48)	(31.52)	
	NP	564	14	578
		(537.52)	(40.48)	
		956	72	1028

$$\chi^2 = \Sigma[(O-E)^2/E]$$
$$= 1.6759 + 22.2518$$
$$1.3047 + 17.3241$$
$$= 42.557$$

conclusion:
   Reject H_o; there is sufficient evidence to conclude that a person's sentence and his original plea are related.

Yes; assuming that those who are really guilty will indeed be convicted with a trial, these results suggest that a guilty plea should be encouraged. But the study reported only those who plead not guilty and were convicted in trials. Suppose there were also guilty 50 persons who plead not guilty and were acquitted. Including them in the no prison category makes the table

PLEA

		G	NG	
	P	392	58	450
SENTENCE		(399.07)	(50.93)	
	NP	564	64	628
		(556.93)	(71.07)	
		956	112	1078

$$\chi^2 = \Sigma[(O-E)^2/E]$$
$$= .125 + .982$$
$$.090 + .704$$

15. H_o: alcohol use and type of crime are independent
    H_1: alcohol use and type of crime are related
    $\alpha = .05$ [assumed]
    C.R. $\chi^2 > \chi^2_{5,.05} = 11.071$
    calculations:

		arson	rape	violence	stealing	coining	fraud	χ^2_5
	Y	50	88	155	379	18	63	753
USE		(49.11)	(79.21)	(139.93)	(358.55)	(16.90)	(109.31)	
	N	43	62	110	300	14	144	673
		(43.89)	(70.79)	(125.07)	(320.45)	(15.10)	(97.69)	
		93	150	265	679	32	207	1426

with CRIME spanning the arson–fraud columns.

$$\chi^2 = \Sigma[(O-E)^2/E]$$
$$= .016 + 0.976 + 1.622 + 1.167 + .072 + 19.617$$
$$\quad .018 + 1.092 + 1.815 + 1.306 + .080 + 21.949$$
$$= 49.731$$

conclusion:
   Reject H_o; there is sufficient evidence to conclude that alcohol use and type of crime are related.

Fraud seems to be different from the other crimes in that it is more like to be committed by someone who abstains from alcohol.

17. H_o: gender and smoking are independent
    H_1: gender and smoking are related
    $\alpha = .05$ [assumed]
    C.R. $\chi^2 > \chi^2_{1,.05} = 3.841$
    calculations:

		SMOKING Y	N	
	M	15	42	57
GENDER		(14.25)	(42.75)	
	F	10	33	43
		(41.24)	(32.25)	
		25	75	100

$$\chi^2 = \Sigma[(O-E)^2/E]$$
$$= .039 + .013$$
$$\quad .052 + .017$$
$$= .122$$

conclusion:
   Do not reject H_o; there is not sufficient evidence to conclude that gender and smoking are related.

19. H_o: drug treatment and oral reaction are independent
    H_1: drug treatment and oral reaction are related
    $\alpha = .05$
    C.R. $\chi^2 > \chi^2_{1,.05} = 3.841$
    calculations:

		TREATMENT D	P	
	S	43	35	78
REACTION		(38.87)	(39.13)	
	N	109	118	227
		(113.13)	(113.87)	
		152	153	305

$$\chi^2 = \Sigma[(\mid O\text{-}E \mid -.5)^2/E]$$
$$= .3386 + .3364$$
$$.1163 + .1156$$
$$= .907$$

conclusion:

Do not reject H_o; there is not sufficient evidence to conclude that the drug treatment and the oral reaction are related.

Without the correction for continuity [see the solution for exercise #1 for the details] the calculated test statistic is 1.174. Since $(\mid O\text{-}E \mid -.5)^2 < (O\text{-}E)^2$ whenever $\mid O\text{-}E \mid > .25$, Yates' correction generally lowers the calculated test statistic.

Review Exercises

1. H_o: NRA membership and storage method are independent
H_1: NRA membership and storage method are related
$\alpha = .01$
C.R. $\chi^2 > \chi^2_{1,.01} = 6.635$
calculations:

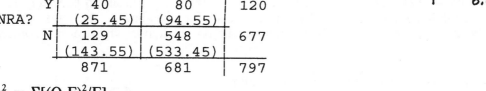

UNSAFE STORAGE?

		Y	N	
NRA?	Y	40 (25.45)	80 (94.55)	120
	N	129 (143.55)	548 (533.45)	677
		871	681	797

$$\chi^2 = \Sigma[(O\text{-}E)^2/E]$$
$$= 8.3251 + 2.2404$$
$$1.4756 + .3971$$
$$= 12.438$$

conclusion:

Reject H_o; there is sufficient evidence to reject the claim that NRA membership and gun storage method are independent and to conclude that these variables are related.

The largest discrepancy (in the Y-Y cell) indicates that NRA members tend to store their guns loaded and unlocked at a higher rate than non-NRA members. Had the results been in the opposite direction, it would be logical to credit gun safety programs within the NRA. While the above results make it difficult to identify reasonable explanatory factors, some possibilities may be: (1) NRA members tend to live in households where they suppose the absence of minors doesn't call for safe storage. (2) NRA members tend to view so-called safe storage as infringing on the right to bear arms.

2. H_o: $p_1 = p_2 = p_3 = ... = p_6 = 1/6$
H_1: at least one of the proportions is different from 1/6
$\alpha = .05$
C.R. $\chi^2 > \chi^2_{5,.05} = 11.071$
calculations:

die	O	E	$(O\text{-}E)^2/E$
1	2	5	1.800
2	8	5	1.800
3	4	5	.200
4	3	5	.800
5	7	5	.800
6	6	5	.200
	30	30	5.600

$$\chi^2 = \Sigma[(O\text{-}E)^2/E]$$
$$= 5.600$$

conclusion:
    Do not reject H_o; there is not sufficient evidence to conclude that at least one of the proportions is different from 1/6.

3. H_o: $p_{Mon} = p_{Tue} = p_{Wed} = \ldots = p_{Sun} = 1/7$
    H_1: at least one of the proportions is different from 1/7
    $\alpha = .05$
    C.R. $\chi^2 > \chi^2_{6,.05} = 12.592$
    calculations:

day	O	E	$(O-E)^2/E$
Mon	74	66.286	.8978
Tue	60	66.286	.5961
Wed	66	66.286	.0012
Thu	71	66.286	.3353
Fri	51	66.286	3.5249
Sat	66	66.286	.0012
Sun	76	66.286	1.4236
	464	464.000	6.7801

$\chi^2 = \Sigma[(O-E)^2/E]$
$= 6.780$

conclusion:
    Do not reject H_o; there is not sufficient evidence to conclude that at least one of the proportions is different from 1/7.
    The data are very close [calculated $\chi^2 = 6.78$, $E(\chi^2) = 6.00$ if H_o is true] to precisely the amount of random fluctuation we expect if there are no differences among the days, and they do not support the theory that more gunfire deaths occur on weekends when more people are at home.

4. H_o: there is homogeneity of proportions across airlines
    H_1: there is not homogeneity of proportions
    $\alpha = .05$
    C.R. $\chi^2 > \chi^2_{2,.05} = 5.991$
    calculations:

		AIRLINE			
		USAir	Americ	Delta	
ON TIME?	Y	80 (77.67)	77 (77.67)	76 (77.67)	233
	N	20 (22.23)	23 (23.33)	24 (22.33)	67
		100	100	100	300

$\chi^2 = \Sigma[(O-E)^2/E]$
$\quad = .070 + .006 + .036$
$\qquad .244 + .020 + .124$
$\quad = .500$

conclusion:
    Do not reject H_o; there is not sufficient evidence to conclude that there is not homogeneity of proportions.

. H_o: headache occurrence and drug usage are independent
H_1: headache occurrence and drug usage are related
$\alpha = .05$
C.R. $\chi^2 > \chi^2_{2,.05} = 5.991$
calculations:

		DRUG			
		Seldane	Placebo	Control	
HEADACHE?	Y	49	49	24	122
		(45.99)	(39.16)	(36.86)	
	N	732	616	602	1950
		(735.01)	(625.84)	(589.14)	
		781	665	626	2072

$$\chi^2 = \Sigma[(O-E)^2/E]$$
$$= .1976 + 2.4752 + 4.4862$$
$$\quad .0124 + .1549 + .2807$$
$$= 7.607$$

conclusion:
    Reject H_o; there is sufficient evidence to conclude that headache occurrence and drug usage are related.

Careful examination of the table indicates that while Seldane users have more headaches than those taking no medication (6.3% vs. 3.8%), they have fewer headaches than those taking a placebo (6.3% vs. 7.4%). This suggests that Seldane usage is associated with but not necessarily responsible for the headaches.

Cumulative Review Exercises

. scores in order: 66 75 77 80 82 84 89 94
$n = 8$
$\Sigma x = 647$
$\Sigma x^2 = 52847$
$\bar{x} = 647/8 = 80.9$
$\tilde{x} = (80 + 82)/2 = 81.0$
$R = 94 - 66 = 28$
$s^2 = [8(52847) - (647)^2]/[8(7)] = 74.41$
$s = 8.6$
five-number summary:
    $x_1 = 66$
    $P_{25} = (75 + 77)/2 = 76.0$ [$(.25)8 = 2$, a whole number, use $x_{2.5}$]
    $P_{50} = \tilde{x} = 81.0$
    $P_{75} = (84 + 89)/2 = 86.5$ [$(.75)8 = 6$, a whole number, use $x_{6.5}$]
    $x_8 = 94$

. Consider the table at the right.
a. $P(C) = 176/647 = .272$
b. $P(M) = 303/647 = .468$
c. $P(M \text{ or } C) = P(M) + P(C) - P(M \text{ and } C)$
        $= (303/647) + (176/647) - (82/647)$
        $= 397/647$
        $= .614$
d. $P(F_1 \text{ and } F_2) = P(F_1) \cdot P(F_2 \mid F_1)$
        $= (344/647) \cdot (343/646)$
        $= .282$

	A	B	C	D	
M	66	80	82	75	303
F	77	89	94	84	344
	143	169	176	159	647

3. H_o: gender and selection are independent
   H_1: gender and selection are related
   $\alpha = .05$ [assumed]
   C.R. $\chi^2 > \chi^2_{3,.05} = 7.815$
   calculations:

		SELECTION				
		A	B	C	D	
	Y	66	80	82	75	303
GENDER		(66.97)	(79.15)	(82.42)	(74.46)	
	N	77	89	94	84	344
		(76.03)	(89.85)	(93.58)	(84.54)	
		143	169	176	159	647

$$\chi^2 = \Sigma[(O-E)^2/E]$$
$$= .014 + .009 + .002 + .004$$
$$ .012 + .008 + .002 + .003$$
$$= .055$$

conclusion:
Do not reject H_o; there is not sufficient evidence to reject the claim that gender and selection are independent.

4. $n = 4$ $n(\Sigma xy) - (\Sigma x)(\Sigma y) = 4(26210) - (303)(344) = 608$
   $\Sigma x = 303$ $n(\Sigma x^2) - (\Sigma x)^2 = 4(23105) - (303)^2 = 611$
   $\Sigma y = 344$ $n(\Sigma y^2) - (\Sigma y)^2 = 4(29742) - (344)^2 = 632$
   $\Sigma x^2 = 23105$
   $\Sigma y^2 = 29742$
   $\Sigma xy = 26210$
   $r = [n(\Sigma xy) - (\Sigma x)(\Sigma y)]/[\sqrt{n(\Sigma x^2) - (\Sigma x)^2} \cdot \sqrt{n(\Sigma y^2) - (\Sigma y)^2}]$
   $= 608/[\sqrt{611} \cdot \sqrt{632}] = .978$
   original claim: $\rho \neq 0$

   H_o: $\rho = 0$
   H_1: $\rho \neq 0$
   $\alpha = .05$ [assumed]

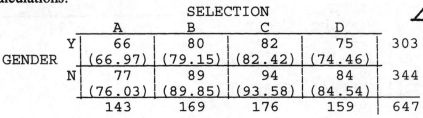

   C.R. $r < -.950$ OR C.R. $t < -t_{2,.025} = -4.303$
    $r > .950$ $t > t_{2,.025} = 4.303$
   calculations: calculations:
    $r = .978$ $t_r = (r - \mu_r)/s_r$
   $= (.978 - 0)/\sqrt{(1-(.978)^2)/2}$
   $.978/.146$
   $ = 6.696$

   conclusion:
   Reject H_o; there is sufficient evidence to conclude that $\rho \neq 0$ (in fact, $\rho > 0$).

5. original claim: $\mu_d > 0$ [n ≤ 30 and σ_d unknown, use t]
   d = y - x: 11 9 12 9
   n = 4
   $\Sigma d = 41$ $\bar{d} = 10.25$
   $\Sigma d^2 = 427$ $s_d = 1.50$
   H_o: $\mu_d \leq 0$
   H_1: $\mu_d > 0$
   $\alpha = .05$ [assumed]
   C.R. t > $t_{3,.05} = 2.353$
   calculations:
   $t_{\bar{d}} = (\bar{d} - \mu_{\bar{d}})/s_{\bar{d}}$
   $\phantom{t_{\bar{d}}} = (10.25 - 0)/(1.50/\sqrt{4})$
   $\phantom{t_{\bar{d}}} = 10.25/.750$
   $\phantom{t_{\bar{d}}} = 13.667$

   conclusion:
   Reject H_o; there is sufficient evidence to conclude that $\mu_d > 0$.

6. summary statistics
   males: n = 4 $\Sigma x = 303$ $\Sigma x^2 = 23105$ $\bar{x} = 75.75$ $s^2 = 50.917$
   females: n = 4 $\Sigma x = 344$ $\Sigma x^2 = 29742$ $\bar{x} = 86.00$ $s^2 = 52.667$
   Let the females be group 1.
   original claim: $\mu_1 - \mu_2 = 0$ [small samples and σ unknown, first test H_o: $\sigma_1^2 = \sigma_2^2$]
   H_o: $\sigma_1^2 = \sigma_2^2$
   H_1: $\sigma_1^2 \neq \sigma_2^2$
   $\alpha = .05$ [assumed]
   C.R. F < $F^3_{3,.975} = [.0648]$
    F > $F^3_{3,.025} = 15.439$
   calculations:
   $F = s_1^2/s_2^2$
   $ = 52.667/50.917$
   $ = 1.0344$

   conclusion:
   Do not reject H_o; there is not sufficient evidence to conclude that $\sigma_1^2 \neq \sigma_2^2$.
   Now proceed using s_p^2 for both s_1^2 and s_2^2
   $\bar{x}_1 - \bar{x}_2 = 86.00 - 75.75 = 10.25$
   $s_p^2 = (df_1 \cdot s_1^2 + df_2 \cdot s_2^2)/(df_1 + df_2)$
   $ = (3 \cdot 52.667 + 3 \cdot 50.917)/(3 + 3) = 310.75/6 = 51.792$
   H_o: $\mu_1 - \mu_2 = 0$
   H_1: $\mu_1 - \mu_2 \neq 0$
   $\alpha = .05$ [assumed]
   C.R. t < $-t_{6,.025} = -2.447$
    t > $t_{6,.025} = 2.447$
   calculations:
   $t_{\bar{x}_1-\bar{x}_2} = (\bar{x}_1-\bar{x}_2 - \mu_{\bar{x}_1-\bar{x}_2})/s_{\bar{x}_1-\bar{x}_2}$
   $\phantom{t_{\bar{x}_1-\bar{x}_2}} = (10.25 - 0)/\sqrt{51.792/4 + 51.792/4}$
   $\phantom{t_{\bar{x}_1-\bar{x}_2}} = 10.25/5.089$
   $\phantom{t_{\bar{x}_1-\bar{x}_2}} = 2.014$
   conclusion:
   Do not reject H_o; there is not sufficient evidence to reject the claim that $\mu_1 - \mu_2 = 0$.

Chapter 11

Analysis of Variance

11-2 One-Way ANOVA

1. a. $n = $ (total df $+ 1) = 20 + 1 = 21$
   b. $F_{15}^5 = 3.33$
   c. $F_{15,.05}^5 = 2.9013$
   d. P-value $= .032$
   e. The decision can be made using either the F statistic or the P-value.
   Using F: since $3.33 > 2.9013$,
   reject $H_o: \mu_1 = \mu_2 = ... = \mu_6$ and conclude at least one μ_i is different.
   Using P-value: since $.032 < .05$,
   reject $H_o: \mu_1 = \mu_2 = ... = \mu_6$ and conclude at least one μ_i is different.
   The two methods will always agree with each other.

3. a. $k = 3$, $\Sigma \bar{x}_i = 294.32$, $\Sigma \bar{x}_i^2 = 28875.1$
   $$\bar{\bar{x}} = \Sigma \bar{x}_i / k$$
   $$= 294.32/3$$
   $$= 98.107$$
   $$s_{\bar{x}}^2 = \Sigma(\bar{x}_i - \bar{\bar{x}})^2/(k-1)$$
   $$= [(97.94-98.107)^2 + (98.58-98.107)^2 + (97.8-98.107)^2]/2$$
   $$= .3459/2$$
   $$= .1729$$
   $$s_{\bar{x}}^2 = [k \cdot \Sigma \bar{x}_i^2 - (\Sigma \bar{x}_i)^2]/[k(k-1)]$$
   $$= [3 \cdot 28875.1 - (294.32)^2]/[3(2)]$$
   $$= 1.0376/6$$
   $$= .1729$$

   NOTE: Either of the above formulas can be used to calculate $s_{\bar{x}}^2$. While the latter (short-cut) formula was generally preferred in the earlier chapters, we choose to use the former (definition) formula here. Since the ANOVA calculations involve so many steps, using the definition formula helps to keep the calculations more organized and reinforces the concepts involved -- especially when the n_i values are not equal.
   $$ns_{\bar{x}}^2 = 5 \cdot .1729 = .8647$$
   b. $s_p^2 = \Sigma df_i s_i^2 / \Sigma df_i = [4(.568)^2 + 4(.701)^2 + 4(.752)^2]/12 = .4598$ [NOTE: The manual used the rounded s's provided. Using the original raw data yields a slightly different value.]
   c. $F = ns_{\bar{x}}^2/s_p^2 = .8647/.4598 = 1.8804$
   d. C.V. is $F = 3.8853$ [C.R. is $F > F_{12,.05}^2 = 3.8853$]
   e. Using F: since $1.8804 < 3.8853$, fail to reject $H_o: \mu_{18-20} = \mu_{21-29} = \mu_{30+}$.
   NOTE: Exercise 3 was worked in complete detail, showing even work done on the calculator without having to be written down. Subsequent exercises are worked showing intermediate steps, but without writing down detail for routine work done on the calculator. While the manual typically shows only three decimal places for the intermediate steps, all decimal places were carried in the calculator. DO NOT ROUND OFF INTERMEDIATE ANSWERS. SAVE calculated values that will be used again. See your instructor or class assistant if you need help using your calculator accurately and efficiently.

NOTE: This section is calculation-oriented. Do not get so involved with the formulas that you miss concepts. This manual arranges the calculations to promote both computational efficiency and understanding of the underlying principles. The following notation is used in this section.

k = the number of groups
n_i = the number of scores in group i (where i = 1,2,...,k)
$\bar{x}_i$ = the mean of group i
s_i^2 = the variance of group i
$\bar{\bar{x}}$ = $\Sigma n_i \bar{x}_i / \Sigma n_i$ = the (weighted) mean of the group means
    = the overall mean of all the scores in all the groups
    = $\Sigma \bar{x}_i / k$ = simplified form when each group has equal size n
s_B^2 = the variance between the groups
    = $\Sigma n_i (\bar{x}_i - \bar{\bar{x}})^2 / (k-1)$
    = $n\Sigma(\bar{x}_i - \bar{\bar{x}})^2 / (k-1)$ = $ns_{\bar{x}}^2$ = simplified form when each group has equal size n
s_p^2 = the variance within the groups
    = $\Sigma df_i s_i^2 / \Sigma df_i$ = the (weighted) mean of the group variances
        = the formula for s_p^2 from the two-sample problem generalized to k samples
    = $\Sigma s_i^2 / k$ = simplified form when each group has equal size n
numerator df = k-1
denominator df = Σdf_i
        = k(n-1) = simplified form when each group has equal size n
$F = s_B^2 / s_p^2$ = (variance between groups)/(variance within groups)

5. Since each group has equal size n, the simplified forms can be used.
The following preliminary values are identified and/or calculated.

	subcompact	compact	midsize	full size
n	5	5	5	5
Σx	252	265	244	230
Σx^2	12880	14131	11952	10782
$\bar{x}$	50.4	53.0	48.8	46.0
s^2	44.8	21.5	11.2	50.5

$k = 4$ $\qquad\qquad$ $\bar{\bar{x}} = \Sigma \bar{x}_i / k$
$n = 5$ $\qquad\qquad\qquad$ $= 49.55$
$s_{\bar{x}}^2 = \Sigma(\bar{x}_i - \bar{\bar{x}})^2 / (k-1)$ $\quad$ $s_p^2 = \Sigma s_i^2 / k$
$\qquad = 8.597$ $\qquad\qquad\qquad = 32.000$

$H_o: \mu_1 = \mu_2 = \mu_3 = \mu_4$
H_1: at least one mean is different
$\alpha = .05$
C.R. $F > F_{16,.05}^3 = 3.2389$
calculations:
$\quad F = ns_{\bar{x}}^2 / s_p^2$
$\qquad = 5(8.597)/32.000$
$\qquad = 1.3422$

conclusion:
    Do not reject H_o; there is not sufficient evidence to conclude that at least one mean is different.

7. Since each group has equal size n, the simplified forms can be used.
The following preliminary values are identified and/or calculated.

	4000 BC	1850 BC	150 AD
n	9	9	9
Σx	1194	1210	1243
Σx^2	158544	162768	171853
$\bar{x}$	132.67	134.44	138.11
s^2	17.500	11.278	22.611

$k = 3$ $\bar{\bar{x}} = \Sigma\bar{x}_i/k$
$n = 9$ $= 135.07$
$s_{\bar{x}}^2 = \Sigma(\bar{x}_i-\bar{\bar{x}})^2/(k-1)$ $s_p^2 = \Sigma s_i^2/k$
     $= 15.416/2 = 7.708$ $= 51.389/3 = 17.130$

H_o: $\mu_1 = \mu_2 = \mu_3$
H_1: at least one mean is different
$\alpha = .05$
C.R. $F > F_{24,.05}^2 = 3.4028$
calculations:
    $F = ns_{\bar{x}}^2/s_p^2$
      $= 9(7.708)/17.130$
      $= 4.0498$

conclusion:
    Reject H_o; there is sufficient evidence to conclude that at least one mean is different.

9. The following preliminary values are identified and/or calculated.

	1	2	3	4	5	total
n	11	11	10	9	7	48
Σx	36.7	39.6	32.5	27.6	25.5	161.9
Σx^2	124.49	144.68	106.73	85.22	94.65	555.77
$\bar{x}$	3.3364	3.6000	3.2500	3.0667	3.6429	3.3729
s^2	.2045	.2120	.1228	.0725	.2929	

$\bar{\bar{x}} = \Sigma n_i\bar{x}_i/\Sigma n_i$
   $= [11(3.3364) + 11(3.6000) + 10(3.2500) + 9(3.0667) + 7(3.6429)]/48$
   $= 161.9/48$
   $= 3.3729$
NOTE: This must always agree with the $\bar{x}$ in the "total" column.
$\Sigma n_i(\bar{x}_i-\bar{\bar{x}})^2 = 11(3.3364-3.3729)^2 + 11(3.6000-3.3729)^2 + 10(3.2500-3.3729)^2$
                     $+ 9(3.0667-3.3729)^2 + 7(3.6429-3.3729)^2$
       $= 2.087$
$\Sigma df_i s_i^2 = 10(.2045) + 10(.2120) + 9(.1228) + 8(.0725) + 6(.2929)$
       $= 7.608$
$s_B^2 = \Sigma n_i(\bar{x}_i-\bar{\bar{x}})^2/(k-1)$
     $= 2.087/4$
     $= .5217$
$s_p^2 = \Sigma df_i s_i^2/\Sigma df_i$
     $= 7.606/43$
     $= .1769$

H_o: $\mu_1 = \mu_2 = \mu_3 = \mu_4 = \mu_5$
H_1: at least one mean is different
$\alpha = .05$
C.R. $F > F^4_{43,.05} = 2.6060$
calculations:
  $F = s_B^2/s_p^2$
    $= .5217/.1769$
    $= 2.9491$

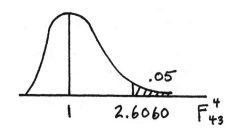

conclusion:
  Reject H_o; there is sufficient evidence to conclude that at least one mean is different.
NOTE: An ANOVA table may be completed as follows:

source	SS	df	MS	F
Trt	2.087	4	.522	2.9491
Error	7.608	43	.177	
Total	9.694	47		

$F = MS_{Trt}/MS_{Err}$
  $= .522/.177$
  $= 2.9491$

(1) Enter $SS_{Trt} = \Sigma n_i(\bar{x}_i - \bar{\bar{x}})^2$ and $SS_{Err} = \Sigma df_i s_i^2$ values from the preliminary calculations.
(2) Enter $df_{Trt} = k-1$ and $df_{Err} = \Sigma df_i = \Sigma(n_i - 1) = \Sigma n_i - k$.
(3) Add the SS and df columns to find SS_{Tot} and df_{Tot}. [The df_{Tot} must equal $\Sigma n_i - 1$.]
(4) Calculate $MS_{Trt} = SS_{Trt}/df_{Trt}$ and $MS_{Err} = SS_{Err}/df_{Err}$.
(5) Calculate $F = MS_{Trt}/MS_{Err}$.

As a final check, calculate s^2 (i.e., the variance of all the scores in one large group) two different ways as indicated below. If these answers agree, the problem is probably correct.
  * from the "total" column in the table for the preliminary calculations:
    $s^2 = [n\Sigma x^2 - (\Sigma x)^2]/[n(n-1)]$
      $= [48(555.77)-(161.9)^2]/[48(47)] = 465.35/2256 = .206$
  * from the "total" row of the ANOVA table
    $s^2 = SS_{Tot}/df_{Tot}$
      $= 9.694/47 = .206$

. The following preliminary values are identified and/or calculated.

	R	O	Y	Br	Bl	G	total
n	21	8	26	33	5	7	100
Σx	19.104	7.401	23.849	30.123	4.507	6.846	91.470
Σx^2	17.934278	6.862429	21.904543	27.546793	4.075729	6.018734	83.802506
$\bar{x}$	.90971	.92513	.91272	.91282	.90140	.92657	.91470
s^2	.000755	.002226	.001144	.001562	.003280	.001499	

$\bar{\bar{x}} = \Sigma n_i \bar{x}_i/\Sigma n_i$
  $= [21(.90971) + 8(.92513) + 26(.91272) + 33(.91282) + 5(.90140) +7(.92657)]/100$
  $= 91.470/100 = .91470$ [NOTE: $\bar{\bar{x}}$ must always agree with the $\bar{x}$ in the "total" column.]
$\Sigma n_i(\bar{x}_i - \bar{\bar{x}})^2 = 21(.90971-.91470)^2 + 8(.92513-.91470)^2 + 26(.91272-.91470)^2$
        $+ 33(.91282-.91470)^2 + 5(.90140-.91470)^2 + 7(.92657-.91470)^2$
      $= .00355$
$\Sigma df_i s_i^2 = 20(.000755) + 7(.002226) + 25(.001144)$
        $+ 32(.001562) + 4(.003280) + 6(.001499)$
      $= .13135$
$s_B^2 = \Sigma n_i(\bar{x}_i - \bar{\bar{x}})^2/(k-1)$
  $= .00355/5$
  $= .00071$
$s_p^2 = \Sigma df_i s_i^2/\Sigma df_i$
  $= .13135/94$
  $= .001398$

H_o: $\mu_R = \mu_O = \mu_Y = \mu_{Br} = \mu_{Bl} = \mu_G$
H_1: at least one mean is different
$\alpha = .05$
C.R. F $> F^5_{94,.05} = 2.2899$
calculations:

$F = s_B^2/s_p^2$
   $= .00071/.001398$
   $= .5081$

.05

2.2899 $F^5_{9.}$

conclusion:

Do not reject H_o; there is not sufficient evidence to conclude that at least one mean is different.

No. If the intent is to make the different colors have the same mean, there is no evidence that this is not being accomplished. Corrective action is not required.

NOTE: An ANOVA table may be completed as follows:

	source	SS	df	MS	F
$F = MS_{Trt}/MS_{Err}$	Trt	.00355	5	.00071	.51
$= .00071/.00140$	Error	.13135	94	.00140	
$= .51$	Total	.13490	99		

(1) Enter $SS_{Trt} = \Sigma n_i(\bar{x}_i - \bar{\bar{x}})^2$ and $SS_{Err} = \Sigma df_i s_i^2$ values from the preliminary calculations.
(2) Enter $df_{Trt} = k-1$ and $df_{Err} = \Sigma df_i = \Sigma(n_i-1) = \Sigma n_i - k$.
(3) Add the SS and df columns to find SS_{Tot} and df_{Tot}. [The df_{Tot} must equal $\Sigma n_i - 1$.]
(4) Calculate $MS_{Trt} = SS_{Trt}/df_{Trt}$ and $MS_{Err} = SS_{Err}/df_{Err}$.
(5) Calculate $F = MS_{Trt}/MS_{Err}$.

As a final check, calculate s^2 (i.e., the variance of all the scores in one large group) two different ways as indicated below. If these answers agree, the problem is probably correct.

* from the "total" column in the table for the preliminary calculations:
   $s^2 = [n\Sigma x^2 - (\Sigma x)^2]/[n(n-1)]$
      $= [100(83.802506) - (91.470)^2]/[100(99)] = 13.4897/9900 = .001363$
* from the "total" row of the ANOVA table
   $s^2 = SS_{Tot}/df_{Tot}$
      $= .13490/99 = .001363$

13. a. Adding 2° to each temperature in the 18-20 age group will raise $\bar{x}_1$ by 2° to 99.940, but adding a constant to each value will not affect the spread of the values -- i.e., $s_1 = .568$ does not change. Re-working exercise #3 produces the following.
   $\bar{\bar{x}} = \Sigma\bar{x}_i/k$
      $= 296.32/3 = 98.773$
   $s_{\bar{x}}^2 = \Sigma(\bar{x}_i - \bar{\bar{x}})^2/(k-1)$
      $= [(99.94-98.773)^2 + (98.58-98.773)^2 + (97.8-98.773)^2]/2 = 2.346/2 = 1.1729$
   $s_p^2 = \Sigma df_i s_i^2/\Sigma df_i$
      $= [4(.568)^2 + 4(.701)^2 + 4(.752)^2]/12 = .4598$
   NOTE: The manual used the rounded s's provided. Using the original raw data yields a slightly different value.
   $F = ns_{\bar{x}}^2/s_p^2$
      $= 5(1.1729)/.4598 = 5.8647/.4598 = 12.7548$
   C.R. is F $> F^2_{12,.05} = 3.8853$
      since $12.7458 > 3.8853$, reject $H_o:\mu_{18-20} = \mu_{21-29} = \mu_{30+}$ and conclude that at least one of the means is different.

   b. For any statistical procedure to be valid, it should not depend upon the units used in the problem. To reach one conclusion if all measurements are made in inches, for example, and another if they are all made in feet would be inconsistent. Accordingly, changing the temperature readings from the Fahrenheit scale to the Celsius scale will not change the calculated F statistic in exercise #3. Mathematically, $C = (5/9) \cdot F - 160/9$. Since the F test is based on variances, and variances are not affected by additive changes, the -160/9 has no effect. Multiplying each score by a constant, however, multiplies the variance by the square of that constant. Since both the numerator and the denominator of the calculated F statistic are variances, the multiplicative constant of (5/9) will multiply both the numerator and the denominator of the calculated F ratio by $(5/9)^2$ -- and the calculated F value is not changed.

c. No change. Because adding the same constant c to each score increases all the means by c, each $\bar{x}_i\text{-}\bar{\bar{x}}$ difference remains unchanged and s_B^2 is not changed. Because adding the same constant to each score does not affect the spread of the scores, each s_i^2 remains unchanged and s_p^2 is not changed. Since both the numerator and the denominator are unchanged, the calculated F ratio remains the same.

d. No change. Because multiplying each score by the same constant c multiplies all the means by c, each $\bar{x}_i\text{-}\bar{\bar{x}}$ difference changes by a factor of c and s_B^2 is changed by a factor of c^2. Because multiplying each score by the same constant c changes standard deviations by a factor of c and variances by a factor of c^2, each s_i^2 and the s_p^2 are changed by a factor of c^2. Since both the numerator and the denominator are changed by a factor of c^2, the calculated F ratio remains the same.

e. No change. Changing the order of the samples will change the order of the addends in the numerator $\Sigma n_i(\bar{x}_i\text{-}\bar{\bar{x}})^2/(k\text{-}1)$ and the denominator $\Sigma df_i s_i^2/\Sigma df_i$ of the F ratio. Since changing the order of the addends does not affect the sum, both the numerator and the denominator are unchanged and the calculated F ratio remains the same.

5. The following preliminary values are identified and/or calculated.

	1	4
n	10	10
Σx	1493	1445
Σx^2	225,541	236,287
$\bar{x}$	149.3	144.5
s^2	292.900	3053.833

a. $n_1 \le 30$ and $n_4 \le 30$, use t (assume $\sigma_1=\sigma_4$ and use s_p^2)
$\bar{x}_1\text{-}\bar{x}_4 = 149.3 - 144.5 = 4.8$
$s_p^2 = [9(292.000) + 9(2053.833)]/(9 + 9) = 1673$

$H_o: \mu_1\text{-}\mu_4 = 0$
$H_1: \mu_1\text{-}\mu_4 \ne 0$
$\alpha = .05$
C.R. $t < -t_{18,.025} = -2.101$
$\quad t > t_{18,.025} = 2.101$
calculations:
$t_{\bar{x}_1\text{-}\bar{x}_4} = (\bar{x}_1\text{-}\bar{x}_4 - \mu_{\bar{x}_1\text{-}\bar{x}_4})/s_{\bar{x}_1\text{-}\bar{x}_4}$
$\quad = (4.8 - 0)/\sqrt{1673/10 + 1673/10}$
$\quad = 4.8/18.294$
$\quad = .262$

conclusion:
Do not reject H_o; there is not sufficient evidence to conclude that $\mu_1\text{-}\mu_2 \ne 0$.

b. The following preliminary values are identified and/or calculated.
$k = 2$ $\qquad\qquad\qquad\qquad$ $\bar{\bar{x}} = \Sigma\bar{x}_i/k$
$n = 10$ $\qquad\qquad\qquad\qquad\quad$ $= 146.9$
$s_{\bar{x}}^2 = \Sigma(\bar{x}_i\text{-}\bar{\bar{x}})^2/(k\text{-}1)$ $\qquad$ $s_p^2 = \Sigma s_i^2/k$
$\quad = 11.52/1 = 11.52$ $\qquad\qquad$ $= 3346.733/2 = 1673.3365$

H_o: $\mu_1 = \mu_2$
H_1: at least one mean is different
$\alpha = .05$
C.R. $F > F^1_{18,.05} = 4.4139$
calculations:
$$F = ns_{\bar{x}}^2/s_p^2$$
$$= 10(11.52)/1673.3665$$
$$= .0688$$

.05

4.4139 F^1_{18}

conclusion:
   Do not reject H_o; there is not sufficient evidence to conclude that at least one mean is different.

c. critical value: $t^2 = (\pm 2.101)^2 = 4.414 = F$
   calculated statistic: $t^2 = (.262)^2 = .069 = F$

11-3   Two-Way ANOVA

NOTE: The formulas and principles in this section are logical extensions of the previous sections.
   $SS_{Row} = \Sigma n_i(\bar{x}_i - \bar{\bar{x}})^2$ for $i = 1,2,3...$ [for each row]
   $SS_{Col} = \Sigma n_j(\bar{x}_j - \bar{\bar{x}})^2$ for $j = 1,2,3...$ [for each column]
   $SS_{Tot} = \Sigma(x - \bar{\bar{x}})^2$ [for all the x's]
When there is only one observation per cell...
   the unexplained variation is
      $SS_{Err} = SS_{Tot} - SS_{Row} - SS_{Col}$
   and there is not enough data to measure interaction.
When there is more than one observation per cell...
   the unexplained variation (i.e., the failure of items in the same cell to respond the same) is
      $SS_{Err} = \Sigma(x - \bar{x}_{ij})^2 = \Sigma df_{ij}s_{ij}^2$ [for each cell -- i,e., for each i,j (row,col) combination]
   and the interaction sum of squares is
      $SS_{Int} = SS_{Tot} - SS_{Row} - SS_{Col} - SS_{Err}$.

Since the data will be analyzed from statistical software packages, however, the above formulas need not be used by hand.

1. H_o: there is no interaction effect between gender and test
   H_1: there is an interaction effect between gender and test
   $\alpha = .05$
   C.R. $F > F^1_{36,.05} = 4.0847$
   calculations:
      $F = MS_{INT}/MS_E$
      $= 31528/10465$
      $= 3.01$

.05

4.0847 F^1_{36}

   conclusion:
      Do not reject H_o; there is not sufficient evidence to conclude that there is an interaction effect between the variables.

. H_o: $\mu_{verbal} = \mu_{math}$
H_1: the means are different
$\alpha = .05$
C.R. $F > F^1_{36,.05} = 4.0847$
calculations:
   $F = MS_{TEST}/MS_E$
      $= 6027/10465$
      $= .58$

conclusion:
   Do not reject H_o; there is not sufficient evidence to conclude that the means are different.

. H_o: $\mu_M = \mu_F$
H_1: the means are different
$\alpha = .05$
C.R. $F > F^1_{12,.05} = 4.7472$
calculations:
   $F = MS_{GENDER}/MS_E$
      $= 29.4/37.8$
      $= .78$

conclusion:
   Do not reject H_o; there is not sufficient evidence to conclude that the means are different.

. H_o: $\mu_1 = \mu_2 = ... = \mu_{24}$
H_1: at least one μ_i is different
$\alpha = .05$
C.R. $F > F^{23}_{69,.05} = 1.7001$
calculations:
   $F = MS_{SUBJ}/MS_E$
      $= 140.5/36.3$
      $= 3.87$

conclusion:
   Reject H_o; there is sufficient evidence to conclude that at least one μ_i is different --i.e., the choice of subject has an effect on the hearing test score.
This makes practical sense in that we would not expect all 24 human subjects to have the same level of hearing

. H_o: $\mu_1 = \mu_2 = \mu_3 = \mu_4$
H_1: at least one mean is different
$\alpha = .05$
C.R. $F > F^3_{6,.05} = 4.7571$
calculations:
   $F = MS_{Operator}/MS_E$
      $= 19.86/8.03$
      $= 2.47$

conclusion:
   Do not reject H_o; there is not sufficient evidence to conclude that at least one operator mean is different.

11. Passing through the data set (eliminating the pulse rates of 8 and 15 as obvious errors) in order until accumulating $n=9$ persons in each category produces the following values.   While it is assumed the ANOVA table will be obtained using a statistical software package, the actual calculations necessary to construct the table by hand are also given.

		GENDER					SUMMARY STATISTICS		
		male			female				
	63 52 60	73 67 72					$\Sigma x_2 = 561$	$\Sigma x_2 = 687$	
yes	66 60 60	82 90 60					$\Sigma x^2 = 35183$	$\Sigma x^2 = 53195$	
	68 69 63	88 75 80					$\bar{x}_2 = 62.33$	$\bar{x}_2 = 76.33$	$\bar{x} = 69.333$
SMOKE							$s^2 = 26.75$	$s^2 = 94.25$	
	69 67 66	97 88 83					$\Sigma x_2 = 594$	$\Sigma x_2 = 680$	
no	55 47 55	77 60 78					$\Sigma x^2 = 40322$	$\Sigma x^2 = 52644$	
	75 76 84	70 67 60					$\bar{x}_2 = 66.00$	$\bar{x}_2 = 75.56$	$\bar{x} = 70.778$
							$s^2 = 139.75$	$s^2 = 158.28$	

$$\bar{x} = 64.167 \qquad \bar{x} = 75.944 \qquad \bar{\bar{x}}_2 = 70.056$$
$$s^2 = 133.2?$$

$$SS_{Smoke} = \Sigma n_i(\bar{x}_i - \bar{\bar{x}})^2 = 18(60.333 - 70.056)^2 + 18(70.778 - 70.056)^2 = 18.778$$
$$SS_{Gender} = \Sigma n_j(\bar{x}_j - \bar{\bar{x}})^2 = 18(64.167 - 70.056)^2 + 18(75.944 - 70.056)^2 = 1248.444$$
$$SS_{Tot} = \Sigma(x - \bar{\bar{x}})^2 = df \cdot s^2 = 35(133.25) = 4663.889$$
$$SS_{Err} = \Sigma(x - \bar{x}_{ij})^2 = \Sigma df_{ij}s_{ij}^2 \; 8(26.75) + 8(94.25) + 8(139.75) + 8(158.28) = 3352.222$$
$$SS_{Int} = SS_{Tot} - SS_{Smoke} - SS_{Gender} - SS_{Err}$$
$$= 4663.889 - 18.778 - 1248.444 - 3352.222 = 44.444$$

The resulting ANOVA tables is used for the test of hypothesis in exercises #11-13.

Source	df	SS	MS	F
Smoke	1	18.778	18.778	.1793
Gender	1	1248.444	1248.444	11.9175
Interaction	1	44.444	44.444	.4243
Error	32	3352.222	104.757	
Total	35	4663.889		

H_o: there is no smoking/gender interaction
H_1: there is smoking/gender interaction
$\alpha = .05$ [assumed]
C.R. $F > F^1_{32,.05} = 4.1709$
calculations:
$\quad F = MS_{Int}/MS_{Err}$
$\quad = 44.444/104.757$
$\quad = .4243$

conclusion:
Do not reject H_o; there is not sufficient evidence to conclude that there is a smoking/gender interaction.

13. H_o: $\mu_{smoker} = \mu_{non-smoker}$
H_1: the means are different
$\alpha = .05$ [assumed]
C.R. $F > F^1_{32,.05} = 4.1709$
calculations:
$\quad F = MS_{Smoke}/MS_{Err}$
$\quad = 18.778/104.757$
$\quad = .1793$

conclusion:
Do not reject H_o; there is not sufficient evidence to conclude that the pulse rates for smoker and non-smokers are different.

. The group means for the n = 10 persons in each of the 6 cells of Table 11-4 are as follows. The two-way ANOVA indicated no interaction effect, a significant gender effect, and no M&M effect. In each of parts (a)-(c) below, changes are made to this original table in such a way that the variability within each cell (and hence the SS error and the MS error) are unchanged and the denominator of the F statistic remains 40291

		GENDER		
		female	male	
	red	938.6	982.4	960.5
M&M	green	833.4	1126.6	980.0
	blue	1006.1	1085.9	1046.0
		926.0	1065.0	995.5

a. The desired effect can be obtained without changing any of the marginal totals by
   (1) subtracting 100 from each red-female score while adding 100 to each red-male score and
   (2) adding 100 to each blue-female score while subtracting 100 from each blue-male score.
   The following table of group means shows this result.

		GENDER		
		female	male	
	red	838.6	1082.4	960.5
M&M	green	833.4	1126.6	980.0
	blue	1106.1	985.9	1046.0
		926.0	1065.0	995.5

b. The desired effect can be obtained changing only the M&M marginal totals by
   (1) subtracting 100 from each red score (both male and female)
   (2) adding 100 to each blue score (both male and female).
   The following table of group means shows this result.

		GENDER		
		female	male	
	red	838.6	882.4	860.5
M&M	green	833.4	1126.6	980.0
	blue	1106.1	1185.9	1146.0
		926.0	1065.0	995.5

c. The desired effect can be obtained changing only the gender marginal totals by
   (1) subtracting 50 from each male score (all colors)
   (2) adding 50 to each female score (all colors).
   The following table of group means shows this result.

		GENDER		
		female	male	
	red	988.6	932.4	960.5
M&M	green	883.4	1076.6	980.0
	blue	1056.1	1035.9	1046.0
		976.0	1015.0	995.5

Review Exercises

1. H_o: there is no interaction between engine size and transmission type
   H_1: there is an interaction between engine size and transmission type
   $\alpha = .05$
   C.R. $F > F^2_{6,.05} = 5.1433$
   calculations:
   $\quad F = MS_{INT}/MS_E$
   $\quad\quad = .6/11.3$
   $\quad\quad = .05$
   conclusion:

   Do not reject H_o; there is not sufficient evidence to reject the claim that fuel consumption is not affected by an interaction between the variables.

2. H_o: $\mu_{1.5} = \mu_{2.5} = \mu_{2.5}$
   H_1: at least one mean is different
   $\alpha = .05$
   C.R. $F > F^2_{6,.05} = 5.1433$
   calculations:
   $\quad F = MS_{Engine}/MS_E$
   $\quad\quad = 21.6/11.3$
   $\quad\quad = 1.90$

   conclusion:
   Do not reject H_o; there is not sufficient evidence to reject the claim that fuel consumption is not affected by engine size.

3. H_o: $\mu_A = \mu_M$
   H_1: the means are different
   $\alpha = .05$
   C.R. $F > F^1_{6,.05} = 5.9874$
   calculations:
   $\quad F = MS_{Trans}/MS_E$
   $\quad\quad = 40.3/11.3$
   $\quad\quad = 3.56$
   conclusion:
   Reject H_o; there is sufficient evidence to reject the claim that fuel consuption is not affected by type of transmission and to conclude that the means are different.

4. Since each group has equal size n, the simplified forms can be used.
   The following preliminary values are identified and/or calculated.

	oceanside	oceanfront	bayside	bayfront
n	6	6	6	6
Σx	2294	2956	1555	3210
Σx^2	944,422	1,494,108	435,465	1,772,790
$\bar{x}$	382.333	492.667	259.167	535.000
s^2	13469.867	7557.067	6492.167	11088.000

   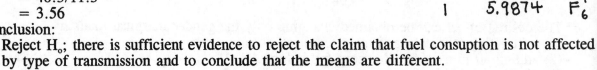

   $k = 4$
   $n = 6$
   $s^2_{\bar{x}} = \Sigma(\bar{x}_i - \bar{\bar{x}})^2/(k-1)$
   $\quad = 45762.243/3$
   $\quad = 15254.081$

   $\bar{\bar{x}} = \Sigma \bar{x}_i/k$
   $\quad = 417.292$
   $s^2_p = \Sigma s^2_i/k$
   $\quad = 38607.100/4$
   $\quad = 9651.775$

H_o: $\mu_{OS} = \mu_{OF} = \mu_{BS} = \mu_{BF}$
H_1: at least one mean is different
$\alpha = .05$
C.R. $F > F^3_{20,.05} = 3.0984$
calculations:
$\quad F = ns_{\bar{x}}^2/s_p^2$
$\quad\quad = 6(15254.081)/9651.775$
$\quad\quad = 9.4827$
conclusion:
$\quad$ Reject H_o; there is sufficient evidence to conclude that at least one mean is different.

5. a. H_o: $\mu_A = \mu_M$
$\quad$ H_1: the means are different
$\quad$ $\alpha = .05$
$\quad$ C.R. $F > F^1_{2,.05} = 18.513$
$\quad$ calculations:
$\quad\quad F = MS_{Trans}/MS_E$
$\quad\quad\quad = .667/.667$
$\quad\quad\quad = 1.00$
$\quad$ conclusion:
$\quad\quad$ Do not reject H_o; there is not sufficient evidence to conclude that the amounts of emitted greenhouse gases are affected by the type of transmission.

b. H_o: $\mu_4 = \mu_6 = \mu_8$
$\quad$ H_1: at least one mean is different
$\quad$ $\alpha = .05$
$\quad$ C.R. $F > F^2_{2,.05} = 19.000$
$\quad$ calculations:
$\quad\quad F = MS_{Cylinder}/MS_E$
$\quad\quad\quad = 4.667/.667$
$\quad\quad\quad = 7.000$
$\quad$ conclusion:
$\quad\quad$ Do not reject H_o; there is not sufficient evidence to conclude that the amounts of emitted greenhouse gases are affected by the number of cylinders.

c. No; we cannot conclude there in no effect, but only that there is not enough evidence to be 95% certain that there is an effect. The very small sample sizes make it unlikely that the data would provide enough evidence to support any claim, no matter how valid it might be.

6. H_o: $\mu_A = \mu_B = \mu_C$
H_1: at least one mean is different
$\alpha = .05$
C.R. $F > F^2_{14,.05} = 3.7389$
calculations:
$\quad F = MS_{Factor}/MS_E$
$\quad\quad = .0038286/.0000816$
$\quad\quad = 46.90$
conclusion:
$\quad$ Reject H_o; there is sufficient evidence to reject the claim that the three groups have the same mean level and to conclude that at least one mean is different.

Cumulative Review Exercises

1. summary statistics: $n = 52$ $\Sigma x = 5.22$ $\Sigma x^2 = 4.0416$
   a. $\bar{x} = (\Sigma x)/n = 5.22/52 = .100$
   b. $s^2 = [n(\Sigma x^2) - (\Sigma x)^2]/[n(n-1)] = [52(4.0416) - (5.22)^2]/[52(51)] = .06897$
      $s = .263$
   c. from the ordered list: $x_1 = 0$
      $Q_1 = P_{25} = (x_{13} + x_{14})/2 = (0 + 0)/2 = 0$
      $\tilde{x} = P_{50} = (x_{26} + x_{27})/2 = (0 + 0)/2 = 0$
      $Q_3 = P_{75} = (x_{39} + x_{40})/2 = (.01 + .01)/2 = .01$
      $x_{52} = 1.41$
   d. $\bar{x} + 2s = .100 + 2(.263) = .626$
      The outliers are the two points above .626: .92 and 1.41
   e. One possible histogram is as follows, with a class width of .25 inches.

   f. No; the distributions are not approximately normal.
   g. let R = rain falls on a selected Monday in Boston
      Since there was rain on 19 of the 52 observed Mondays, we estimate $P(R) = 19/52 = .365$.

2. The ordered values and summary statistics for each agency are as follows.
   BAC: 40 52 55 63 68 72 75 77 $\Sigma x = 502$ $\Sigma x^2 = 32660$ $\bar{x} = 62.75$ $s^2 = 165.6429$
   SFA: 59 66 69 70 73 75 82 84 $\Sigma x = 578$ $\Sigma x^2 = 42232$ $\bar{x} = 72.25$ $s^2 = 67.3571$
   DFA: 42 53 57 61 69 73 73 74 $\Sigma x = 502$ $\Sigma x^2 = 32438$ $\bar{x} = 62.75$ $s^2 = 133.9286$

   a. For each agency, n=8: $P_{25} = (x_2+x_3)/2$ $P_{50} = (x_4+x_5)/2$ $P_{25} = (x_6+x_7)/2$
      The five number summaries x_1, P_{25}, P_{50}, P_{75}, x_n are as follows.
      BAC: 40 53.5 65.5 73.5 77   SFA: 59 67.5 71.5 78.5 84   DFA: 42 55.0 65.0 73.0 74

While there is considerable overlap of scores, the boxplots seem to indicate that the Solomon

& Ford agency scores tend to be higher and more consistent.

b. BAC: $\bar{x} = 62.75$ $s^2 = 165.6429$ $s = 12.87$
   SFA: $\bar{x} = 72.25$ $s^2 = 67.3571$ $s = 8.21$
   DFA: $\bar{x} = 62.75$ $s^2 = 133.9286$ $s = 11.57$

c. Let Barnum agency scores be group 1.
   original claim: $\mu_1 - \mu_2 = 0$ [small samples and σ unknown, first test H_o: $\sigma_1^2 = \sigma_2^2$]

   H_o: $\sigma_1^2 = \sigma_2^2$
   H_1: $\sigma_1^2 \neq \sigma_2^2$
   $\alpha = .05$ [assumed]
   C.R. $F < F_{7,.975}^7 = [.2002]$
   $\quad\quad F > F_{7,.025}^7 = 4.9949$
   calculations:
   $\quad F = s_1^2/s_2^2$
   $\quad\quad = 165.6429/67.3571$
   $\quad\quad = 2.4592$
   conclusion:

   Do not reject H_o; there is not sufficient evidence to conclude that $\sigma_1^2 \neq \sigma_2^2$.
   Now proceed using s_p^2 for both s_1^2 and s_2^2
   $\bar{x}_1 - \bar{x}_2 = 62.75 - 72.25 = -9.50$
   $s_p^2 = (df_1 \cdot s_1^2 + df_2 \cdot s_2^2)/(df_1 + df_2)$
   $\quad\quad = (7 \cdot 165.6429 + 7 \cdot 67.3571)/(7 + 7) = 1631/14 = 116.5$

   H_o: $\mu_1 - \mu_2 = 0$
   H_1: $\mu_1 - \mu_2 \neq 0$
   $\alpha = .05$
   C.R. $t < -t_{14,.025} = -2.145$
   $\quad\quad t > t_{14,.025} = 2.145$
   calculations:
   $\quad t_{\bar{x}_1-\bar{x}_2} = (\bar{x}_1 - \bar{x}_2 - \mu_{\bar{x}_1-\bar{x}_2})/s_{\bar{x}_1-\bar{x}_2}$
   $\quad\quad = (-9.50 - 0)/\sqrt{116.5/8 + 116.5/8}$
   $\quad\quad = -9.50/5.397$
   $\quad\quad = -1.760$
   conclusion:

   Do not reject H_o; there is not sufficient evidence to reject the claim that $\mu_1 - \mu_2 = 0$.

d. For each agency it is true that $n \leq 30$ and σ unknown, use t

BAC	SFA	DFA
$\bar{x} \pm t_{7,.025} \cdot s/\sqrt{n}$	$\bar{x} \pm t_{7,.025} \cdot s/\sqrt{n}$	$\bar{x} \pm t_{7,.025} \cdot s/\sqrt{n}$
$62.75 \pm 2.365 \cdot 12.87/\sqrt{8}$	$72.25 \pm 2.365 \cdot 8.207/\sqrt{8}$	$62.75 \pm 2.365 \cdot 11.57/\sqrt{8}$
62.75 ± 10.76	72.25 ± 6.86	62.75 ± 9.68
$51.99 < \mu < 73.51$	$65.39 < \mu < 79.11$	$53.07 < \mu < 72.43$

   Since the intervals overlap, it is possible that the agencies could have a common mean (say, for example, $\mu = 70$). The interval for Solomon & Ford, however, was slightly higher and narrower than the other two.

e. Since each group has equal size n, the simplified forms can be used.
   $\quad k = 3 \quad\quad\quad\quad\quad\quad \bar{\bar{x}} = \Sigma\bar{x}_i/k$
   $\quad n = 8 \quad\quad\quad\quad\quad\quad\quad\quad = 65.917$
   $\quad s_{\bar{x}}^2 = \Sigma(\bar{x}_i - \bar{\bar{x}})^2/(k-1) \quad s_p^2 = \Sigma s_i^2/k$
   $\quad\quad = 60.167/2 \quad\quad\quad\quad\quad = 366.9286/3$
   $\quad\quad = 30.083 \quad\quad\quad\quad\quad\quad = 122.31$

H_o: $\mu_{BAC} = \mu_{SFA} = \mu_{DFA}$
H_1: at least one mean is different
$\alpha = .05$
C.R. $F > F^2_{21,.05} = 3.4668$
calculations:
$\quad F = ns^2_{\bar{x}}/s^2_p$
$\qquad = 8(30.083)/122.31$
$\qquad = 1.9677$

conclusion:
   Do not reject H_o; there is not sufficient evidence to conclude that at least one mean is different.

On the basis of these results, I cannot be 95% confident that there is any difference in the mean agency scores. Since one of the three must be selected, however, I would choose Solomon & Ford because their scores have both the highest mean and the lowest standard deviation.

3. a. let x = the weight of a newborn baby
   $P(x > 8.00)$
   $\quad = P(z > .42)$
   $\quad = .5000 - .1618$
   $\quad = .3372$

   b. The $\bar{x}$'s are normally distributed because the original distribution is normal.
   $\quad \mu_{\bar{x}} = \mu = 7.54$
   $\quad \sigma_{\bar{x}} = \sigma/\sqrt{n} = 1.09/\sqrt{16} = .2725$
   $\quad P(\bar{x} > 8.00)$
   $\qquad = P(z > 1.69)$
   $\qquad = .5000 - .4545$
   $\qquad = .0455$

   c. let B = a baby weighs more than 7.54 lbs
   $\quad P(B) = .5000$, for each birth
   $\quad P(B_1 \text{ and } B_2 \text{ and } B_3) = P(B_1) \cdot P(B_2) \cdot P(B_3)$
   $\qquad\qquad\qquad\qquad\qquad = (.5000) \cdot (.5000) \cdot (.5000)$
   $\qquad\qquad\qquad\qquad\qquad = .125$

Chapter 12

Statistical Process Control

2-2 Control Charts for Variation and Mean

NOTE: In this section, k = number of sample subgroups
                      n = number of observations per sample subgroup

. There are k·n = (25 years)·(6 observations/year) = 150 observations on the run chart.
$\bar{\bar{x}} = \Sigma\bar{x}/k = 1811.9/25 = 72.48$

There appears to be a slight upward trend accompanied by an increase in variability. Tourists will have to wait longer to see an eruption, and the times for eruptions will be predicted with less accuracy.

NOTE: Although the line $\bar{\bar{x}} = 72.48$ is not specified in the text as a necessary part of the run chart, such a line is included in the Minitab examples and is included in this manual.

3. $\bar{\bar{x}} = \Sigma\bar{x}/k = 1811.9/25 = 72.48$
   $\bar{R} = \Sigma R/k = 564/25 = 22.56$
   $LCL = \bar{\bar{x}} - A_2\bar{R} = 72.476 - (.483)(22.56) = 72.476 - 10.896 = 61.58$
   $UCL = \bar{\bar{x}} + A_2\bar{R} = 72.476 + (.483)(22.56) = 72.476 + 10.896 = 83.37$

The process mean is not within statistical control. To be out of control, the process must meet one of three out-of-control criteria. This process actually meets two of the three out-of-control criteria. (1) There is a pattern to the $\bar{x}$ values -- in this case, an upward drift. (2) There is a point lying beyond the upper or lower control limits -- in this case, there are two points above the UCL. The practical implication for tourists is that the progressively longer intervals indicate that the time between eruptions is not stable -- but that it is growing longer.

5. $\bar{R} = \Sigma R/k = 7.292/20 = .365$ 　　　　 $LCL = D_3\bar{R} = 0.000(.3646) = 0.000$
   　　　　　　　　　　　　　　　　　　　　　　　 $UCL = D_4\bar{R} = 2.114(.3646) = 0.771$

The process variation is not within statistical control. To be out of control, the process must meet one of three out-of-control criteria. This process actually meets all three out-of-control criteria. (1) There is a pattern to the R values -- in this case, an upward drift. (2) There is a point lying beyond the upper or lower control limits -- in this case, there are 3 points above the UCL. (3) The run rule of eight applies -- in this case, the first 13 points are all below the center line.

The following chart summarizes the information necessary for exercises #7, #8 and #9.

sample	consumption			$\bar{x}$	R
1a	4762	3875	2657	3764.7	2105
1b	4358	2201	3187	3248.7	2157
2a	4504	3237	2198	3113.0	2306
2b	2511	3020	2857	2796.0	509
3a	3952	2785	2118	2952.7	1834
3b	2658	2139	3071	2622.7	932
4a	3863	3013	2023	2966.3	1840
4b	2953	3456	2647	3018.7	809

. There are $k \cdot n = $ (8 samples) $\cdot$ (3 observations/sample) $= 24$ observations on the run chart.
$\bar{x} = \Sigma \bar{x}/k = 24682.7/8 = 3085.3$

consumption (kwh)

No; there does not appear to be a pattern suggesting that the process is not within statistical control.

NOTE: Although the line $\bar{x} = 3085.3$ is not specified in the text as a necessary part of the run chart, such a line is included in the Minitab examples and is included in this manual.

. $\bar{\bar{x}} = \Sigma \bar{x}/k = 24682.7/8 = 3085.3$
$\bar{R} = \Sigma R/k = 12492/8 = 1561.5$
$LCL = \bar{\bar{x}} - A_2\bar{R} = 3085.3 - (1.023)(1561.5) = 3085.3 - 1597.4 = 1487.9$
$UCL = \bar{\bar{x}} + A_2\bar{R} = 3085.3 + (1.023)(1561.5) = 3085.3 + 1597.4 = 4682.7$

consumption (kwh)

The process mean is within statistical control. A process is not in statistical control when there is a pattern and/or unusual data point that cannot be attributed to reasonable random effects. For household energy data, any of the following factors could make the process go out of statistical control: unusual weather, family adjustments (increase or decrease in size, adoption of conservation habits), house adjustments (new insulation).

The following chart summarizes the information necessary for exercises #11 and #12.

week	$\bar{x}$	R	week	$\bar{x}$	R	week	$\bar{x}$	R	week	$\bar{x}$	R
1	.019	.05	14	.000	.00	27	.036	.14	40	.067	.2
2	.027	.10	15	.097	.40	28	.027	.11	41	.003	.0
3	.101	.71	16	.237	.87	29	.010	.05	42	.003	.0
4	.186	.64	17	.067	.47	30	.017	.12	43	.181	.6
5	.016	.05	18	.054	.24	31	.082	.44	44	.243	1.4
6	.091	.64	19	.036	.14	32	.044	.18	45	.290	1.2
7	.051	.30	20	.237	.92	33	.164	.64	46	.139	.9
8	.001	.01	21	.049	.27	34	.134	.85	47	.116	.7
9	.039	.16	22	.003	.01	35	.009	.03	48	.109	.4
10	.121	.39	23	.000	.00	36	.017	.12	49	.017	.0
11	.181	.78	24	.101	.71	37	.050	.26	50	.000	.0
12	.029	.17	25	.097	.33	38	.057	.40	51	.106	.7
13	.416	1.41	26	.000	.00	39	.017	.12	52	.146	.4

There are 52 $\bar{x}$'s: $\bar{\bar{x}} = .0835$ and $s_{\bar{x}} = .0875$.   There are 52 R's: $\bar{R} = .3915$ and $s_R = .3828$

11. $\bar{R} = .3915$ $LCL = D_3\bar{R} = 0.076(.3915) = .030$
   $UCL = D_4\bar{R} = 1.924(.3915) = .753$

The process variation is not within statistical control. Since there are points above the upper control limit and points below the lower control limit, we reject the claim that rainfall amounts exhibit stable variation. While only one point either above or below the appropriate control limit is sufficient evidence of unstable variation, there are multiple points beyond each limit.

13. $\bar{s} = \Sigma s/k = 501.2/25 = 20.048$ $LCL = B_3\bar{s} = .118(20.048) = 2.366$
   $UCL = B_4\bar{s} = 1.882(20.048) = 37.730$

load (lbs)

UCL = 37.73
$\bar{s} = 20.048$
LCL = 2.336

This is similar to the R chart given in this section, and both charts indicate the process variation is within statistical control.

-3 Control Charts for Attributes

. This process is within statistical control. Since the first third of the sample means are generally less than the overall mean, the middle third are generally more than the overall mean, and the final third are generally less than the overall mean, however, one may wish to check future analyses to see whether such a patter tends to repeat itself.

. This process is out of statistical control. There is an upward trend, and there is a point above the upper control limit.

. $\bar{p}$ = (Σx)/(Σn) = (30+29+...+23)/(13)(100,000) = 332/1,300,000 = .000255

$\sqrt{\bar{p}\cdot\bar{q}/n}$ = $\sqrt{(.000255)(.999745)/100000}$ = .0000505

LCL = $\bar{p}$ - 3$\sqrt{\bar{p}\cdot\bar{q}/n}$ = .000255 - 3(.0000505) = .000255 - .000152 = .000104

UCL = $\bar{p}$ + 3$\sqrt{\bar{p}\cdot\bar{q}/n}$ = .000255 + 3(.0000505) = .000255 + .000152 = .000407

NOTE: The 13 sample proportions are: .00030 .00029 .00029 .00027 ... .00023

There are three out-of-control criteria, and meeting any one of them means that the process is not within statistical control. This process meets two of those criteria. (1) There is a pattern -- in this case, a downward trend. (3) The run of 8 rule applies -- in this case, the last 9 points are all below the center line. By definition, therefore, process is statistically unstable. In this instance, however, that is good and means that the age 0-4 death rate for infectious diseases is decreasing.

. $\bar{p}$ = (Σx)/(Σn) = (3+3+1...+5)/(52)(7) = 127/364 = .3489

$\sqrt{\bar{p}\cdot\bar{q}/n}$ = $\sqrt{(.3489)(.6511)/7}$ = .1801

LCL = $\bar{p}$ - 3$\sqrt{\bar{p}\cdot\bar{q}/n}$ = .3489 - 3(.1801) = .3489 - .5404 = -.192 [truncated at 0.000]

UCL = $\bar{p}$ + 3$\sqrt{\bar{p}\cdot\bar{q}/n}$ = .3489 + 3(.1801) = .3489 + .5404 = .889

NOTE: The 52 sample proportions are: 3/7 3/7 1/7 ... 5/7

The process is within statistical control. Viewing the proportion of days each week with measurable rainfall as a statistical process, it does not need correction. Whether the proportions are too high or too low for individual preferences or other needs is another issue.

9. $\bar{p} = (\Sigma x)/(\Sigma n) = (25+24+...+31)/(13)(100,000) = 375/1,300,000 = .000288$
    $n\cdot\bar{p} = (100,000)(.000288) = 28.846 \ [= \bar{x}]$
    $\sqrt{n\cdot\bar{p}\cdot\bar{q}} = \sqrt{(100000)(.000288)(.999712)} = 5.370$
    $\text{LCL} = n\cdot\bar{p} - 3\sqrt{n\cdot\bar{p}\cdot\bar{q}} = 28.846 - 3(5.370) = 28.846 - 16.110 = 12.736$
    $\text{UCL} = n\cdot\bar{p} + 3\sqrt{n\cdot\bar{p}\cdot\bar{q}} = 28.846 + 3(5.370) = 28.846 + 16.110 = 44.956$

The charts are identical except for the labels. Since the labels on the np chart are n = 100,00 times the labels on the chart for p, the center line is 100,000(.000288) = 28.486 and the uppe control limit is 100,000(.000450) = 44.956 and the lower control limit is 100,000(.000127) = 12.736. The placement of the points does not change.

Review Exercises

The chart at the right summarizes the information necessary for exercises #1, #2 and #3.

year	$\bar{x}$	R
1980	12.480	9.33
1981	12.106	2.99
1982	10.707	6.26
1983	9.430	5.11
1984	13.372	7.09
1985	10.468	4.32
1986	15.084	6.34
1987	11.684	4.50
1988	10.240	3.99
1989	10.734	4.50
1990	10.706	1.34

$\bar{\bar{x}} = \Sigma\bar{x}/k$
$= 126.922/11$
$= 11.538$

$\bar{R} = \Sigma R/k$
$= 55.86/11$
$= 5.078$

1. sulfate (kg/hectare)

No; there does not appear to be a pattern suggesting that the process is not within statistical control.

2. $\bar{R} = \Sigma R/k = 55.86/11 = 5.078$ $LCL = D_3\bar{R} = 0(5.078) = 0$
                                                  $UCL = D_4\bar{R} = 2.114(5.078) = 10.73$

sulfate (kg/hectare)

The process variation is within statistical control.

3. $\bar{\bar{x}} = \Sigma\bar{x}/k = 126.922/11 = 11.538$
$\bar{R} = \Sigma R/k = 55.86/11 = 5.078$
$LCL = \bar{\bar{x}} - A_2\bar{R} = 11.538 - (.577)(5.078) = 11.538 - 2.930 = 8.608$
$UCL = \bar{\bar{x}} + A_2\bar{R} = 11.538 + (.577)(5.078) = 11.538 + 2.930 = 14.468$

sulfate (kg/hectare)

The process mean is not within statistical control because there is a point above the upper control limit.

4. $\bar{p} = (\Sigma x)/(\Sigma n) = (270+264+...+343)/(13)(100,000) = 4183/1,300,000 = .00322$
$\sqrt{\bar{p}\cdot\bar{q}/n} = \sqrt{(.00322)(.99678)/100000} = .000179$
$LCL = \bar{p} - 3\sqrt{\bar{p}\cdot\bar{q}/n} = .00322 - 3(.000179) = .00322 - .00053 = .00268$
$UCL = \bar{p} + 3\sqrt{\bar{p}\cdot\bar{q}/n} = .00322 + 3(.000179) = .00322 + .00053 = .00375$
NOTE: The 13 sample proportions are: .00270 .00264 .00250 .00278 ... .00343

There are three out-of-control criteria, and meeting any one of them means that the process is not within statistical control. This process meets all three criteria. (1) There is a pattern -- in this case, an upward trend. (2) There is a point outside the lower or upper control limits -- in this case, there are two points below the LCL and one point above the UCL. (3) The run of 8 rule applies -- in this case, the last 8 points are all above the center line. By definition, therefore, process is statistically unstable. In this instance, that means that the age 65+ death rate for infectious diseases is increasing.

5. $\bar{p} = (\Sigma x)/(\Sigma n) = (608+466+...+491)/(15)(1,000) = 7031/15000 = .4687$
$\sqrt{\bar{p} \cdot \bar{q}/n} = \sqrt{(.4687)(.5313)/1000} = .01578$
LCL $= \bar{p} - 3\sqrt{\bar{p} \cdot \bar{q}/n} = .4687 - 3(.01578) = .4687 - .0473 = .4214$
UCL $= \bar{p} + 3\sqrt{\bar{p} \cdot \bar{q}/n} = .4687 + 3(.01578) = .4687 + .0473 = .5161$
NOTE: The 15 sample proportions are: .608 .466 .552 .382 ... .491

There are three out-of-control criteria, and meeting any one of them means that the process is not within statistical control. This process meets the first two criteria. (1) There is a pattern -- in this case, an alternating trend. (2) There is a point outside the lower or upper control limits -- in this case, most of the points are beyond the control limits. By definition, therefore, process is statistically unstable. The pattern is caused by the fact that the national elections occur every two years, every other national election involving a presidential election and bring out the voters in larger numbers.

umulative Review Exercises

1. a. $\bar{p} = (\Sigma x)/(\Sigma n) = (10+8+...+11)/(20)(400) = 150/8000 = .01875$
   $\sqrt{\bar{p}\cdot\bar{q}/n} = \sqrt{(.01875)(.98125)/400} = .006782$
   $LCL = \bar{p} - 3\sqrt{\bar{p}\cdot\bar{q}/n} = .01875 - 3(.006782) = .01875 - .02035 = 0$ [since $LCL \geq 0$]
   $UCL = \bar{p} + 3\sqrt{\bar{p}\cdot\bar{q}/n} = .01875 + 3(.006782) = .01875 + .02035 = .0391$
   NOTE: The 20 sample proportions are: .020 .010 .015 .030 ... .015

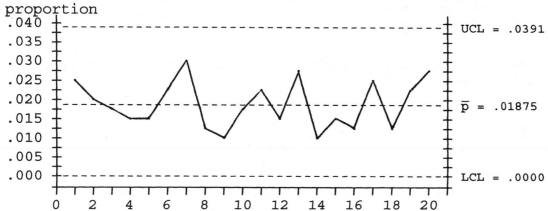

The process is within statistical control, and so the data can be treated as coming from a binomial population with a fixed mean and variance.

   b. Using all the data combined, $\hat{p} = x/n = 150/8000$
   $\hat{p} \pm z_{.025}\cdot\sqrt{\hat{p}\hat{q}/n}$
   $.01875 \pm 1.960\cdot\sqrt{(.01875)(.98125)/8000}$
   $.01875 \pm .00297$
   $.0158 < p < .0217$

   c. original claim $p \leq .01$
   $H_o: p \leq .01$
   $H_1: p > .01$
   $\alpha = .05$
   C.R. $z > 1.645$
   calculations:
   $z_{\hat{p}} = (\hat{p} - \mu_{\hat{p}})/\sigma_{\hat{p}}$
   $= (.01875 - .01)/\sqrt{(.01875)(.98125)/8000}$
   $= .00875/.001517$
   $= 5.770$

   conclusion:
   Reject H_o; there is sufficient evidence to reject the claim that $p \leq .01$ and to conclude that $p > .01$.

2. a. let A_i = the ith point is above the center line
   $P(A_i) = .5$ for each value of i
   $P(A) = P(\text{all 8 above}) = P(A_1)\cdot P(A_2)\cdot...\cdot P(A_8) = (.5)^8 = .00391$
   b. let B_i = the ith point is below the center line
   $P(B_i) = .5$ for each value of i
   $P(B) = P(\text{all 8 below}) = P(B_1)\cdot P(B_2)\cdot...\cdot P(B_8) = (.5)^8 = .00391$
   c. Notice that the events in parts (a) and (b) are mutually exclusive.
   $P(A \text{ or } B) = P(A) + P(B) - P(A \text{ and } B)$
   $= .00391 + .00391 - 0$
   $= .00781$

3. The following chart summarizes the information in the same manner used for the kwh consumption data in section 12-2 exercises #7, #8 and #9. The analysis also follows the format of those exercises.

sample	temperature			$\overline{x}$	R
1a	32	35	59	42.0	27
1b	76	66	42	61.3	34
2a	22	33	56	37.0	34
2b	70	63	42	58.3	28
3a	30	38	55	41.0	25
3b	71	61	38	56.7	33
4a	32	40	57	43.0	25
4b	72	65	45	60.7	27

a. Run Chart: There are $k \cdot n$ = (8 samples)·(3 observations/sample) = 24 observations.
$\overline{\overline{x}} = \Sigma\overline{x}/k = 400/8 = 50.00$

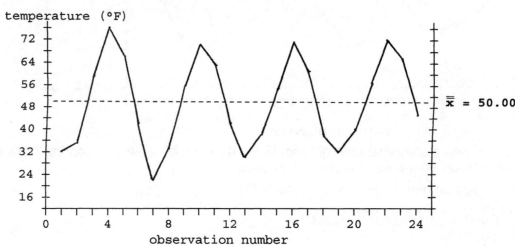

The process is not within statistical control because there is a pattern of cycles indicating something other than random variation is at work.

b. R Chart: $\overline{R} = \Sigma R/k = 233/8 = 29.125$ $LCL = D_3\overline{R} = 0(29.125) = 0$
$UCL = D_4\overline{R} = 2.544(29.125) = 74.97$

The process variation is within statistical control.

c. $\bar{x}$ Chart: $\bar{\bar{x}} = \Sigma\bar{x}/k = 400/8 = 50.00$
$\bar{R} = \Sigma R/k = 233/8 = 29.125$
$LCL = \bar{\bar{x}} - A_2\bar{R} = 50.00 - (1.023)(29.125) = 50.00 - 29.79 = 20.21$
$UCL = \bar{\bar{x}} + A_2\bar{R} = 50.00 + (1.023)(29.125) = 50.00 + 29.79 = 79.79$

The process mean is not within statistical control because there is an alternating pattern of points above and below the overall mean.

4. Letting x represent temperature (°F) and y represent consumption (kwh), the following numerical summary contains all the necessary information for the n=24 paired values.
$\Sigma x = 1200$ $\Sigma y = 74045$
$\Sigma x^2 = 65850$ $\Sigma y^2 = 242624810$ $\Sigma xy = 3562714$
$n(\Sigma x^2)-(\Sigma x)^2 = 140400$ $n(\Sigma y^2)-(\Sigma y)^2 = 340333415$ $n(\Sigma xy)-(\Sigma x)(\Sigma y) = -3348864$

a. $r = -3348864/[\sqrt{140400}\cdot\sqrt{340333415}] = -.484$
   The critical values being $r = \pm.396$ (for n=25 at the $\alpha=.05$ level), we conclude there is a significant negative linear correlation between outside temperature and energy consumption.

b. $b_1 = -3348864/140400 = -23.852$
   $b_0 = (74045/24) - (-23.852)(1200/24) = 4277.824$
   The regression line is $\hat{y} = 4277.824 - 23.852x$

c. $\hat{y}_{60} = 4277.824 - 23.852(60) = 2847$ kwh

Nonparametric Statistics

13-2 Sign Test

NOTE for n ≤ 25: Table A-7 gives only x_L, the <u>lower</u> critical value for the sign test. Accordingly, the text lets x be the <u>smaller</u> of the number of +'s or the number of -'s and warns the user to use common sense to avoid concluding the reverse of what the data indicates.

An alternative approach maintains the natural agreement between the alternative hypothesis and the critical region and is consistent with the logic and notation of parametric tests. Let x <u>always</u> be the number of +'s. The problem's symmetry means that the upper critical value is $x_U = n-x_L$.

Since this alternative approach builds directly on established patterns and provides insight into rationale of the sign test, its C.R. and calculations and picture [notice that if H_o is true then $\mu_x = n/2$] for each exercise are given to the right of the method using only lower critical values.

NOTE for n > 25: The correction for continuity is a conservative adjustment intending to make less likely a false rejection of H_o by shifting the x value .5 units toward the middle. When x is the smaller of the number of +'s or the number of -'s, this always involves replacing x with x+.5. In the alternative approach, x is replaced with either x+.5 or x-.5 according to which one shifts the value toward the middle -- i.e., with x+.5 when $x < \mu_x = (n/2)$,
and with x-.5 when $x > \mu_x = (n/2)$.

1. Let the reported heights be group 1.
   claim: median difference ≠ 0

female	1	2	3	4	5	6	7	8	9	10
R-M	+	-	+	+	+	-	+	-	+	+

   n = 10 +'s and -'s

   H_o: median difference = 0
   H_1: median difference ≠ 0
   α = .05
   C.R. x ≤ $x_{L,10,.025}$ = 1   <u>OR</u>   C.R. x ≤ $x_{L,10,.025}$ = 1
   　　　　　　　　　　　　　　　　　　　　　　x ≥ $x_{U,10,.025}$ = 10-1 = 9
   calculations: calculations:
   　x = 3   　x = 7
   conclusion:
   　Do not reject H_o; there is not sufficient evidence to conclude that the median difference is different from 0.

NOTE: Compare the above results to those obtained in exercise #5 of section 8-3, where a parametric test was used on the same data. Although the text does not so indicate, several exercises in this chapter are re-tests of data that was previously analyzed parametrically; comparisons of the two results are usually informative and provide statistical insights.

3. Let x = the number who did <u>not</u> vote.
   claim: p < .5
      701 +'s (voted)   301 -'s (did not vote)
      n = 1002 +'s or -'s
   Since n > 25, use z with
     μ_x = n/2 = 1002/2 = 501
     σ_x = $\sqrt{n}$/2 = $\sqrt{1002}$/2 = 15.827
   H_o: p ≥ .5
   H_1: p < .5
   α = .05 [assumed]
   C.R. z < $-z_{.05}$ = -1.645

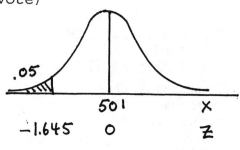

   calculations:
     x = 301
     z_x = $[(x+.5)-\mu_x]/\sigma_x$
        = [301.5 - 501]/15.827
        = -199.5/15.827
        = -12.605
   conclusion:
     Reject H_o; there is sufficient evidence to conclude that p < .5 (i.e., that the majority of the people say they <u>did</u> vote in the election).
   NOTE: This problem may be worked more naturally in the manner in which it is stated as follows. See the second NOTE at the beginning of the chapter.
   Let x = the number who voted.
   claim: p > .5
      701 +'s (voted)   301 -'s (did not vote)
      n = 1002 +'s or -'s
   Since n > 25, use z with
     μ_x = n/2 = 1002/2 = 501
     σ_x = $\sqrt{n}$/2 = $\sqrt{1002}$/2 = 15.827
   H_o: p ≤ .5
   H_1: p > .5
   α = .05 [assumed]
   C.R. z > $z_{.05}$ = 1.645
   calculations:
     x = 701
     z_x = $[(x-.5)-\mu_x]/\sigma_x$
        = [700.5 - 501]/15.827
        = 199.5/15.827
        = 12.605
   conclusion:
     Reject H_o; there is sufficient evidence to conclude that p > .5 (i.e., that the majority of the people say they voted in the election).

5. Let weights above 12 be +'s.
   claim: median > 12
   ```
   33 +'s        1 -           2 0's
   n = 34 +'s or -'s
   ```
   Since n > 25, use z with
   $\mu_x = n/2 = 34/2 = 17$
   $\sigma_x = \sqrt{n}/2 = \sqrt{34}/2 = 2.915$

   H_o: median ≤ 12
   H_1: median > 12
   $\alpha = .05$ [assumed]

   C.R. $z < -z_{.05} = -1.645$ OR C.R. $z > z_{.05} = 1.645$
   calculations: calculations:
   x = 1   x = 33
   $z_x = [(x+.5)-\mu_x]/\sigma_x$ $z_x = [(x-.5)-\mu_x]/\sigma_x$
   $= [1.5 - 17]/2.915$ $= [32.5 - 17]/2.915$
   $= -15.5/2.915$ $= 15.5/2.915$
   $= -5.316$ $= 5.316$

   conclusion:
   Reject H_o; there is sufficient evidence to conclude that median > 12.

   Yes; assuming that the intent of the company is to produce a product with a median weight slightly larger than the stated weight of 12 ounces, it appears that the cans are being filled correctly.

7. Let weights above .9085 be +'s.
   claim: median > .9085
   ```
   18 +'s        15 -'s        no 0's
   n = 33 +'s or -'s
   ```
   Since n > 25, use z with
   $\mu_x = n/2 = 33/2 = 16.5$
   $\sigma_x = \sqrt{n}/2 = \sqrt{33}/2 = 2.872$

   H_o: median ≤ .9085
   H_1: median > .9085
   $\alpha = .05$ [assumed]

   C.R. $z < -z_{.05} = -1.645$ OR C.R. $z > z_{.05} = 1.645$
   calculations: calculations:
   x = 15   x = 18
   $z_x = [(x+.5)-\mu_x]/\sigma_x$ $z_x = [(x-.5)-\mu_x]/\sigma_x$
   $= [15.5 - 16.5]/2.872$ $= [17.5 - 16.5]/2.872$
   $= -1/2.872$ $= 1/2.872$
   $= -.348$ $= .348$

   conclusion:
   Do not reject H_o; there is not sufficient evidence to conclude that median > .9085.

   Yes; based on this results (including the fact that there were more +'s than -'s), the package appears to be labeled correctly.

9. Let the right hands be group 1.

claim: $\tilde{\mu}_R - \tilde{\mu}_L = 0$

```
    1 +        13 -'s
    n = 14 +'s or -'s
```

H_o: $\tilde{\mu}_R - \tilde{\mu}_L = 0$

H_1: $\tilde{\mu}_R - \tilde{\mu}_L \neq 0$

$\alpha = .05$

C.R. $x \leq x_{L,14,.025} = 2$

$\quad x \geq x_{U,14,.025} = 14\text{-}2 = 12$

calculations:

$\quad x = 1$

conclusion:

Reject H_o; there is sufficient evidence to reject the claim that $\tilde{\mu}_R - \tilde{\mu}_L = 0$ and to conclude that $\tilde{\mu}_R - \tilde{\mu}_L \neq 0$ (in fact, that $\tilde{\mu}_R - \tilde{\mu}_L < 0$).

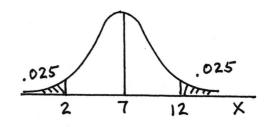

11. Let the before times be group 1.

claim: median difference = 0

```
    6 +'s        22 -'s        2 0's
    n = 28 +'s or -'s
```

Since n > 25, use z with

$\quad \mu_x = n/2 = 28/2 = 14$

$\quad \sigma_x = \sqrt{n}/2 = \sqrt{14}/2 = 2.646$

H_o: median difference = 0

H_1: median difference $\neq$ 0

$\alpha = .01$

C.R. $z < -z_{.005} = -2.575$

$\quad z > z_{.005} = 2.575$

calculations:

$\quad x = 6$

$\quad z_x = [(x+.5)-\mu_x]/\sigma_x$

$\quad\quad = [6.5 - 14]/2.646$

$\quad\quad = -7.5/2.646$

$\quad\quad = -2.835$

conclusion:

Reject H_o; there is sufficient evidence reject the claim that median difference = 0 and to conclude that median difference $\neq$ 0 (in fact that median difference < 0 -- i.e., the reaction times are less in the before group).

Yes; based on these results it appears that the insurance company is justified in charging higher rates for those who drink and drive.

13. Let scores above 100 be +'s.

claim: median $\geq$ 100

```
    40 +'s        60 -'s        21 0's
```

a. Using the usual method: discard the 0's.

```
    n = 100 +'s or -'s
```

Since n > 25, use z with

$\quad \mu_x = n/2 = 100/2 = 50$

$\quad \sigma_x = \sqrt{n}/2 = \sqrt{100}/2 = 5.000$

H_o: median $\geq$ 0

H_1: median < 0

$\alpha = .05$

C.R. $z < -z_{.05} = -1.645$

calculations:

x = 40

$z_x = [(x+.5)-\mu_x]/\sigma_x$

    = [40.5 - 50]/5.000

    = -9.5/5.000

    = -1.900

conclusion:

Reject H_o; there is sufficient evidence to conclude that median < 100.

b. Using the second method: count half the zeros each way (and discarding the odd zero).

    50 +'s   70 -'s

     n = 120 +'s or -'s

Since n > 25, use z with

$\mu_x = n/2 = 120/2 = 60$

$\sigma_x = \sqrt{n}/2 = \sqrt{120}/2 = 5.477$

H_o: median ≥ 0

H_1: median < 0

$\alpha = .05$

C.R. z < $-z_{.05}$ = -1.645

calculations:

x = 50

$z_x = [(x+.5)-\mu_x]/\sigma_x$

    = [50.5 - 60]/5.477

    = -9.5/5.477

    = -1.734

conclusion:

Reject H_o; there is sufficient evidence to conclude that median < 100.

c. Using the third method: count the zeros in a one-tailed test to favor H_o.

    61 +'s   60 -'s

     n = 161 +'s or -'s

Since n > 25, use z with

$\mu_x = n/2 = 121/2 = 60.5$

$\sigma_x = \sqrt{n}/2 = \sqrt{121}/2 = 5.500$

H_o: median ≥ 0

H_1: median < 0

$\alpha = .05$

C.R. z < $-z_{.05}$ = -1.645

calculations:

x = 60

$z_x = [(x+.5)-\mu_x]/\sigma_x$

    = [60.5 - 60.5]/5.500

    = 0/5.500 = 0

conclusion:

Do not reject H_o; there is not sufficient evidence to conclude that median < 100.

15. Use the binomial table for p=.5 and reject for lower-tailed x values summing to less than .03.

For $1 \leq n \leq 5$, P(x=0) > .03

For $6 \leq n \leq 8$, P(x$\leq$0) $\leq$ .03 but P(x$\leq$1) > .03

For $9 \leq n \leq 11$, P(x$\leq$1) $\leq$ .03 but P(x$\leq$2) > .03

For $12 \leq n \leq 13$, P(x$\leq$2) $\leq$ .03 but P(x$\leq$3) > .03

For $14 \leq n \leq 15$, P(x$\leq$3) $\leq$ .03 but P(x$\leq$4) > .03

In summary, the desired critical values are as follows.

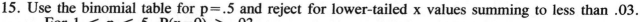

```
n:  1  2  3  4  5  6  7  8  9  10  11  12  13  14  15
x:  *  *  *  *  *  0  0  0  1   1   1   2   2   3   3
```

3-3 Wilcoxon Signed-Ranks Test for Matched Pairs

NOTE: Table A-8 gives only T_L, the <u>lower</u> critical value for the signed-ranks test. Accordingly, the text lets T be the <u>smaller</u> of the sum of positive ranks or the sum of the negative ranks and warns the user to use common sense to avoid concluding the reverse of what the data indicates.

An alternative approach maintains the natural agreement between the alternative hypothesis and the critical region and is consistent with the logic and notation of parametric tests. Let T <u>always</u> be the sum of the positive ranks. By symmetry, the upper critical value is $T_U = \Sigma R - T_L$.

Since this alternative approach builds directly on established patterns and provides insight into rationale of the signed-ranks test, its C.R. and calculations and picture [notice that if H_o is true then $_T = \Sigma R / 2$] for each exercise are given to the right of the method using only lower critical values.

NOTE 2: This manual follows the text and the directions to the exercises of this section by using "the populations have the same distribution" as the null hypothesis. To be more precise, the signed-rank test doesn't test "distributions" but tests the "location" (i.e., central tendency -- as opposed to variation) of distributions. The test discerns whether one group taken as a whole tends to have higher or lower scores than another group taken as a whole. The test does not discern whether one group is more variable than another. This distinction is reflected in the wording of the conclusion when rejecting H_o. Notice also that the signed-rank test measures overall differences between the groups and <u>not</u> whether the two groups give the same results for individual subjects. If the group 1 scores, for example, were higher for half the subjects and lower for the other half of the subjects by the same amounts, then $\Sigma R-$ would equal $\Sigma R+$ (so we could not reject H_o), but the distributions would be very different.

NOTE 3: This manual uses a minus sign preceding ranks associated with negative differences. While the ranks themselves are not negative, the use of the minus sign helps to organize the information.

1. Let the reported heights be group 1.
    claim: the populations have the same distribution

female	1	2	3	4	5	6	7	8	9	10
R-M	-.5	-.1	.2	1.6	1.9	-.4	.3	-.3	.6	1.8
R	6	-1	2	8	10	-5	3.5	-3.5	7	9

    n = 10 non-zero ranks
    $\Sigma R- = 9.5$
    $\Sigma R+ = 45.5$
    $\Sigma R = 55.0$
    check: $\Sigma R = n(n+1)/2 = 10(11)/2 = 55$
    H_o: the populations have the same distribution
    H_1: the populations have different distributions
    $\alpha = .01$
    C.R. $T \le T_{L,10,.005} = 3$ <u>OR</u>   C.R. $T \le T_{L,10,.005} = 3$
                                                    $T \ge T_{U,10,.005} = 55-3 = 52$
    calculations: calculations:
      $T = 9.5$ $T = 45.5$
    conclusion:
        Do not reject H_o; there is not sufficient evidence to reject the claim that the populations have the same distribution.

3. Let the right hands be group 1.
   claim: the populations have the same distribution

```
R-L |  -33  -74  -75  -42  -80  -36   -6  -10  -28  -86   33  -61  -56  -41
R   | -4.5  -11  -12   -8  -13   -6   -1   -2   -3  -14  4.5  -10   -9   -7
n = 14 non-zero ranks
ΣR- = 100.5
ΣR+ =   4.5
ΣR  = 105.0
check: ΣR = n(n+1)/2 = 14(15)/2 = 105
```

H_o: the populations have the same distribution
H_1: the populations have different distributions
$\alpha = .05$
C.R. $T \leq T_{L,14,.025} = 21$
     $T \geq T_{U,14,.025} = 105\text{-}21 = 84$
calculations:
   $T = 4.5$
conclusion:
   Reject H_o; there is sufficient evidence to reject the claim that the populations have the same distribution and to conclude that they have different distributions (in fact, right hand times are lower).

5. Let the first trial be group 1.
   claim: the populations have different distributions

```
child| A    B    C    D    E    F    G    H    I    J    K    L    M    N    (
F-S  | 0   13    5   15   15  126   28   -2   -5   31    3   51    3   14   3
R    | -    6  4.5  8.5  8.5   14   10   -1 -4.5   11  2.5   13  2.5    7   1
n = 14 non-zero ranks
ΣR- =   5.5
ΣR+ =  99.5
ΣR  = 105.0
check: ΣR = n(n+1)/2 = 14(15)/2 = 105
```

H_o: the populations have the same distribution
H_1: the populations have different distributions
$\alpha = .01$
C.R. $T \leq T_{L,14,.005} = 13$ **OR**   C.R. $T \leq T_{L,14,.005} = 13$
                                              $T \geq T_{U,14,.005} = 105\text{-}13 = 92$
calculations: calculations:
   $T = 5.5$ $T = 99.5$
conclusion:
   Reject H_o; there is sufficient evidence conclude that the populations have different distributions (in fact, that the times for the first trial are larger).

7. Let Before be group 1.
   claim: the populations have different distributions (i.e., the drug has an effect)

pair	A	B	C	D	E	F	G	H	I	J	K	L
B-A	9	4	21	3	20	31	17	26	19	10	23	33
R	3	2	8	1	7	11	5	10	6	4	9	12

   n = 12 non-zero ranks
   ΣR- = 0
   ΣR+ = 78
   ΣR = 78
   check: ΣR = n(n+1)/2 = 12(13)/2 = 78

   H_o: the populations have the same distribution
   H_1: the populations have different distributions
   α = .05 [assumed]

   C.R. T ≤ $T_{L,12,.025}$ = 14   OR   C.R. T ≤ $T_{L,12,.025}$ = 14
                                                T ≥ $T_{U,12,.025}$ = 78-14 = 64
   calculations: calculations:
     T = 0   T = 78
   conclusion:
     Reject H_o; there is sufficient evidence to reject the claim that the populations have the same distribution and to conclude that they have different distributions (in fact, the Before group has higher scores).

9. Since T is defined as the lower of ΣR- and ΣR+, its smallest possible value is 0. Its largest value occurs when ΣR- = ΣR+ = (1/2)(ΣR) = (1/2)[n(n+1)/2] = n(n+1)/4.
   a. assuming no zeros, 0 and 10(11)/4 = 27.5
   b. assuming no zeros, 0 and 50(51)/4 = 637.5
   c. for n = 100 non-zero ranks, use the z approximation with
   μ_T = n(n+1)/4
        = 100(101)/4 = 2525
   σ_T = $\sqrt{[n(n+1)(2n+1)/24]}$
        = $\sqrt{[100(101)(201)/24]}$ = 290.84
   For α = .05 in a two-tailed test, the critical z is 1.960.
   The critical T (i.e., the smaller of ΣR- and ΣR+) is found by solving as follows.
   z = (T - μ_T)/σ_T < -1.96
   (T - 2525)/290.84 < -1.96
        T - 2525 < -570.05
             T < 1954.95

   With the assumption of no ties in ranks, the C.R. is T ≤ 1954.
   NOTE: In general, however, assigning the average of the tied ranks would introduce (1/2)'s -- but not (1/3)'s, (1/4)'s, etc., since the average of n consecutive integers is always a either a whole number (if n is odd) or a half number (if n is even). The proper critical value for T would then be 1954.5.

13-4 Wilcoxon Rank-Sum Test for Two Independent Samples

NOTE: As in the previous section, the manual foilows the wording in the text and tests the hypothesis that "the populations have the same distribution" with the understanding that the test detects only differences in location and not differences in variability. In addition, always letting $R = \Sigma R_1$ guarantees agreement and consistency between the directions of H_1 and the C.R. as in the previous chapters.

1. Below are the scores (in order) for each group. The group listed first is considered group 1.
   claim: the populations have the same distribution

Red	R	Bro	R
.870	6	.856	1
.872	8	.858	2
.874	9	.860	3
.882	12	.866	4
.888	13	.867	5
.891	15	.871	7
.897	16.5	.875	10
.898	18.5	.876	11
.908	25	.889	14
.908	25	.897	16.5
.908	25	.898	18.5
.911	29	.900	20
.912	30	.902	21.5
.913	31	.902	21.5
.920	35.5	.904	23
.924	39.5	.909	27.5
.924	39.5	.909	27.5
.933	45	.914	32.5
.936	46.5	.914	32.5
.952	48	.919	34
.983	52	.920	35.5
	569.0	.921	37
		.923	38
		.928	41
		.930	42.5
		.930	42.5
		.932	44
		.936	46.5
		.955	49
		.965	50
		.976	51
		.988	53
		1.033	54
			916.0

$$n_1 = 21 \qquad \Sigma R_1 = 569$$
$$n_2 = 33 \qquad \Sigma R_2 = 916$$

$$n = \Sigma n = 54 \qquad \Sigma R = 1485$$

check: $\Sigma R = n(n+1)/2$
$$= 54(55)/2$$
$$= 1485$$

$$R = \Sigma R_1 = 569$$

$$\mu_R = n_1(n+1)/2$$
$$= 21(55)/2$$
$$= 577.5$$

$$\sigma_R^2 = n_1 n_2 (n+1)/12$$
$$= (21)(33)(55)/12$$
$$= 3176.25$$

H_o: the populations have the same distribution
H_1: the populations have different distributions
$\alpha = .05$
C.R. $z < -z_{.025} = -1.96$
    $z > z_{.025} = 1.96$
calculations:
$$z_R = (R - \mu_R)/\sigma_R$$
$$= (569 - 577.5)/\sqrt{3176.25}$$
$$= -8.5/56.358$$
$$= -.151$$

conclusion:
   Do not reject H_o; there is not sufficient evidence to reject the claim that the populations have the same distribution.

3. Below are the scores (in order) for each group. The group listed first is considered group 1.
   claim: the populations have the same distribution

O-C	R	Con	R
210	1	334	7.5
287	2	349	11
288	3	402	12
304	4	413	14
305	5	429	15
308	6	445	16
334	7.5	460	18.5
340	9	476	20.5
344	10	483	21
407	13	501	22
455	17	519	23
463	19	594	24
	96.5		203.5

$n_1 = 12 \qquad \Sigma R_1 = 96.5$

$n_2 = 12 \qquad \Sigma R_2 = 203.5$

$n = \Sigma n = 24 \qquad \Sigma R = 300.0$

check: $\Sigma R = n(n+1)/2$
$\qquad\qquad = 24(25)/2$
$\qquad\qquad = 300$

$R = \Sigma R_1 = 96.5$

$\mu_R = n_1(n+1)/2$
$\qquad = 12(25)/2$
$\qquad = 150$

$\sigma_R^2 = n_1 n_2 (n+1)/12$
$\qquad = (12)(12)(25)/12$
$\qquad = 300$

H_o: the populations have the same distribution
H_1: the populations have different distributions
$\alpha = .01$
C.R. $z < -z_{.005} = -2.575$
$\qquad z > z_{.005} = 2.575$
calculations:
$z_R = (R - \mu_R)/\sigma_R$
$\qquad = (96.5 - 150)/\sqrt{300}$
$\qquad = -53.5/17.321 = -3.089$

conclusion:
   Reject H_o; there is sufficient evidence to reject the claim that the populations have the same distribution and to conclude that they have different distributions (in fact, population 1 has smaller volumes).

Based on this result, we can be 99% confident that there are biological factors related to obsessive-compulsive disorders.

5. See the data summary on the data page.
   claim: the populations have different distributions
   preliminary calculations:

$n_1 = 36 \qquad \Sigma R_1 = 1958 \qquad$ check: $\Sigma R = n(n+1)/2$

$n_2 = 36 \qquad \Sigma R_2 = 670 \qquad\qquad\qquad = 72(73)/2$

$\qquad\qquad\qquad\qquad\qquad\qquad\qquad\qquad = 2628$

$n = \Sigma n = 72 \qquad \Sigma R = 2628$

$R = \Sigma R_1 = 1958$

$\mu_R = n_1(n+1)/2 = 36(73)/2 = 1314$

$\sigma_R^2 = n_1 n_2 (n+1)/12 = (36)(36)(73)/12 = 7884$

H_o: the populations have the same distribution
H_1: the populations have different distributions
$\alpha = .05$ [assumed]
C.R. $z < -z_{.025} = -1.96$
$\qquad z > z_{.025} = 1.96$

calculations:
$$z_R = (R - \mu_R)/\sigma_R$$
$$= (1958 - 1314)/\sqrt{7884}$$
$$= 644/88.792 = 7.253$$

conclusion:

Reject H_o; there is sufficient evidence to conclude that the populations have different distributions (in fact, that the regular Coke weighs are larger).

Such a difference might be explained by the absence of sugar in Diet Coke.

Below are the scores (in order) for exercises #5, #6 and #7. The group listed first is considered group 1.

	#5				#6				#7			
	CokeR	R	CokeD	R	CokeR	R	PepsR	R	UMA	R	DCC	R
01	.7901	33	.7758	1	.7901	1	.8139	9	18.95	1	28.35	6
02	.8044	38	.7760	2	.8044	2	.8156	15	24.00	2	33.35	7
03	.8062	39	.7771	3	.8062	3	.8170	20.5	27.60	3	35.00	9
04	.8073	40	.7773	4	.8073	4	.8172	23.0	27.95	4	38.45	10
05	.8079	41	.7802	5	.8079	5	.8172	23.0	28.00	5	39.15	12
06	.8110	42	.7806	6	.8110	6	.8192	29.5	34.95	8	41.95	14.5
07	.8126	43	.7811	7	.8126	7	.8192	29.5	38.70	11	42.65	16
08	.8128	44	.7813	8	.8128	8	.8200	34	41.60	13	44.00	17
09	.8143	45	.7822	10.0	.8143	10	.8211	37.5	41.95	14.5	44.65	18
10	.8150	46.5	.7822	10.0	.8150	11.5	.8211	37.5	45.60	19	50.15	21
11	.8150	46.5	.7822	10.0	.8150	11.5	.8211	37.5	48.70	20	50.65	22
12	.8152	48.5	.7826	12	.8152	13.5	.8214	40	50.70	23	51.35	24
13	.8152	48.5	.7830	13	.8152	13.5	.8216	41	52.00	25	53.35	26.5
14	.8161	50.5	.7833	14	.8161	16.5	.8227	43.5	54.70	29	53.35	26.5
15	.8161	50.5	.7837	15	.8161	16.5	.8227	43.5	57.35	30	54.35	28
16	.8163	52	.7839	16	.8163	18	.8227	43.5	60.00	35.5	57.55	31
17	.8165	53	.7844	17	.8165	19	.8227	43.5	60.00	35.5	58.65	32
18	.8170	54	.7852	18.5	.8170	20.5	.8233	47.5	67.00	45	59.35	33
19	.8172	55	.7852	18.5	.8172	23.0	.8233	47.5	67.70	46	60.00	35.5
20	.8176	56	.7859	20	.8176	25	.8244	50.5	70.00	48	60.00	35.5
21	.8181	57	.7861	21	.8181	26	.8244	50.5	72.70	49	61.00	39.0
22	.8189	58	.7868	22	.8189	27	.8247	54.0	73.00	50	61.00	39.0
23	.8192	59.5	.7870	23	.8192	29.5	.8247	54.0	75.35	52	61.00	39.0
24	.8192	59.5	.7872	24	.8192	29.5	.8251	56.5	76.00	55	62.00	41
25	.8194	61.5	.7874	25.5	.8194	32.5	.8256	58	77.00	56	63.60	42
26	.8194	61.5	.7874	25.5	.8194	32.5	.8258	59	78.00	58	64.55	43
27	.8207	63	.7879	27.5	.8207	35	.8260	60.5	79.20	59	65.65	44
28	.8211	64	.7879	27.5	.8211	37.5	.8260	60.5	80.55	62	69.35	47
29	.8229	65	.7881	29	.8229	46	.8271	63	82.00	63	75.00	51
30	.8244	66.5	.7885	30	.8244	50.5	.8291	65.5	83.35	64	75.65	53
31	.8244	66.5	.7892	31	.8244	50.5	.8291	65.5	84.30	65	75.95	54
32	.8247	68	.7896	32	.8247	54.0	.8302	68	86.00	66	77.80	57
33	.8451	69	.7907	34	.8451	56.5	.8317	69	86.70	67	80.00	60.5
34	.8264	70	.7910	35	.8264	62	.8319	70	88.00	68.5	80.00	60.5
35	.8284	71	.7923	36.5	.8284	64	.8388	71	89.35	70	80.00	68.5
36	.8295	72	.7923	36.5	.8295	67	.8401	72	92.00	71		1163.0
37		1958.0		670.0		934.5		1693.5	92.70	72		
38									94.00	73		
39									95.00	74		
40									102.95	75		
										1687.0		

7. See the data summary on the data page.
   claim: the populations have different distributions
   preliminary calculations:

$n_1 = 40$	$\Sigma R_1 = 1687$	check: $\Sigma R = n(n+1)/2$
$n_2 = 35$	$\Sigma R_2 = 1163$	$= 75(76)/2$
		$= 2850$

   $n = \Sigma n = 75$ $\Sigma R = 2850$

   $R = \Sigma R_1 = 1687$
   $\mu_R = n_1(n+1)/2 = 40(76)/2 = 1520$
   $\sigma_R^2 = n_1 n_2(n+1)/12 = (40)(35)(76)/12 = 8866.67$

   H_o: the populations have the same distribution
   H_1: the populations have different distributions
   $\alpha = .05$ [assumed]
   C.R. $z < -z_{.025} = -1.96$
   $z > z_{.025} = 1.96$
   calculations:

   $z_R = (R - \mu_R)/\sigma_R$
   $= (1687 - 1520)/\sqrt{8866.67}$
   $= 167/94.163$
   $= 1.774$
   conclusion:
   Do not reject H_o; there is not sufficient evidence to conclude that the new textbook prices are different at the two colleges.

9. The rank-sum test is based entirely on the ranks of the observations. Changing the units of measurement will not affect the ranks -- e.g., the largest observation will still have the largest value no matter what units are used. Changing the units of measurement, including weighing the M&M's in ounces instead of grams, will have no affect on the Wilcoxon rank-sum test.

11. The denominators of the Mann-Whitney and Wilcoxon statistics are the same.
    The numerator of the Mann-Whitney statistic, assuming $R = \Sigma R_1$, is
    $U - n_1 n_2/2 = [n_1 n_2 + n_1(n_1+1)/2 - R] - n_1 n_2/2$
    $= n_1 n_2/2 + n_1(n_1+1)/2 - R$
    $= (n_1/2)(n_2+n_1+1) - R$
    $= -[R - n_1(n_1+n_2+1)/2]$
    $= -[\text{numerator of Wilcoxon statistic}]$

    For the M&M example in this section,
    $U = n_1 n_2 + n_1(n_1+1)/2 - R$
    $= (21)(26) + (21)(22)/2 - 469.5$
    $= 307.5$
    $\mu_U = n_1 n_2/2 = (21)(26)/2 = 273$
    $\sigma_U^2 = \sigma_R^2 = n_1 n_2(n+1)/12 = (21)(26)(48)/12 = 2184$
    $z_U = (U - \mu_U)/\sigma_U$
    $= (307.5 - 273)/\sqrt{2184}$
    $= 34.5/46.733$
    $= .738$

    As predicted, z_U has the same absolute value but the opposite sign of z_R.

13-5 Kruskal-Wallis Test

NOTE: As in the previous sections, the manual follows the wording in the text and tests the hypothesis that "the populations have the same distribution" with the understanding that the test detects only differences in location and not differences in variability.

1. summary statistics:

$$n_1 = 12 \quad R_1 = 157$$
$$n_2 = 12 \quad R_2 = 265.5$$
$$n_3 = 12 \quad R_3 = 358.5$$
$$n_4 = 12 \quad R_4 = 395$$

$$n = \Sigma n = 48 \quad \Sigma R = 1176.0$$

check:
$$\Sigma R = n(n+1)/2$$
$$= 18(49)/2$$
$$= 1176$$

H_o: the populations have the same distribution
H_1: the populations have different distributions
$\alpha = .05$
C.R. $H > \chi^2_{3,.05} = 7.815$
calculations:
$$H = [12/n(n+1)] \cdot [\Sigma(R_i^2/n_i)] - 3(n+1)$$
$$= [12/48(49)] \cdot [(157)^2/12 + (265.5)^2/12$$
$$+ (358.5)^2/12 + (395)^2/12] - 3(49)$$
$$= [.005102] \cdot [31640.542] - 147$$
$$= 14.431$$

conclusion:
Reject H_o; there is sufficient evidence to reject the claim that the different years have time intervals with identical populations and to conclude that the populations have different distributions.

3. Below are the scores for each group. The group listed first is group 1, etc.

subco	R	compa	R	midsi	R	fulls	R
595	5	1051	15	629	6	1085	17
1063	16	1193	18	1686	20	971	12
885	10	946	11	880	9	996	14
519	3	984	13	181	1	804	8
422	2	584	4	645	7	1376	19
	36		61		43		70

$$n_1 = 5 \quad R_1 = 36$$
$$n_2 = 5 \quad R_2 = 61$$
$$n_3 = 5 \quad R_3 = 43$$
$$n_4 = 5 \quad R_4 = 70$$

$$n = \Sigma n = 20 \quad \Sigma R = 210$$

check:
$$\Sigma R = n(n+1)/2$$
$$= 20(21)/2$$
$$= 210$$

H_o: the populations have the same distribution
H_1: the populations have different distributions
$\alpha = .05$ [assumed]
C.R. $H > \chi^2_{3,.05} = 7.815$
calculations:
$$H = [12/n(n+1)] \cdot [\Sigma(R_i^2/n_i)] - 3(n+1)$$
$$= [12/20(21)] \cdot [(36)^2/5 + (61)^2/5$$
$$+ (43)^2/5 + (70)^2/5] - 3(21)$$
$$= [.0286] \cdot [2353.2] - 63$$
$$= 4.234$$

conclusion:
Do not reject H_o; there is not sufficient evidence to conclude that left femur loads for the four weight categories are not all the same.

No; the data do not show that the weight of the car has no effect, but the data do not provide sufficient evidence to conclude that the weight of the car does have an effect.

5. Below are the scores (in order) for each group. The group listed first is group 1, etc.

4000	R	1850	R	150	R
125	1	129	4	128	2
129	4	129	4	136	14
131	6	134	10	137	17.5
132	7.5	134	10	137	17.5
132	7.5	136	14	138	21.5
134	10	136	14	139	24
135	12	137	17.5	141	25
138	21.5	137	17.5	142	26
138	21.5	138	21.5	145	27
	91.0		112.5		174.5

$n_1 = 9$ $R_1 = 91$
$n_2 = 9$ $R_2 = 112.5$
$n_3 = 9$ $R_3 = 174.5$

$n = \Sigma n = 27$ $\Sigma R = 378.0$

check:
$\Sigma R = n(n+1)/2$
$= 27(28)/2$
$= 378$

H_o: the populations have the same distribution
H_1: the populations have different distributions
$\alpha = .05$
C.R. $H > \chi^2_{2,.05} = 5.991$
calculations:

$H = [12/n(n+1)] \cdot [\Sigma(R_i^2/n_i)] - 3(n+1)$
$= [12/27(28)] \cdot [(91)^2/9 + (112.5)^2/9 + (174.5)^2/9] - 3(28)$
$= [.0159] \cdot [5709.7] - 84$
$= 6.631$

conclusion:
   Reject H_o; there is sufficient evidence to reject the claim that the populations have the same distribution and to conclude that they populations have different distributions.
NOTE: Compare these results to those of the parametric test in section 11-2, exercise #7.

7. Below are the weights in thousandths of a gram (in order) for each group.
The group listed first is group 1, etc.

red	R	ora	R	yel	R	bro	R	blu	R	gre	R
870	9.5	861	5	868	8	856	2	838	1	890	26
872	12	897	31	876	16.5	858	3	870	9.5	902	39.5
874	13	898	34	877	18	860	4	875	14.5	902	39.5
882	21	903	42	879	19.5	866	6	956	87	911	52
888	24	920	61	879	19.5	867	7	968	90.5	930	72
891	27	942	82	886	22.5	871	11		202.5	949	83.5
897	31	971	92	886	22.5	875	14.5			1002	98
898	34	1009	99	892	28	876	16.5				410.5
908	46		446	893	29	889	25				
908	46			900	36.5	897	31				
908	46			906	44	898	34				
911	52			910	50	900	36.5				
912	54			911	52	902	39.5				
913	55			917	58	902	39.5				
920	61			921	63.5	904	43				
924	67			924	67	909	48.5				
924	67			926	69	909	48.5				
933	75			934	76	914	56.5				
936	77.5			939	79	914	56.5				
952	85			940	80	919	59				
983	95			941	81	920	61				
	998.0			949	83.5	921	63.5				

yel	R	bro	R
960	88	923	65
968	90.5	928	70
978	94	930	72
989	97	930	72
	1392.5	932	74
		936	77.5
		955	86
		965	89
		976	93
		988	96
		1033	100
			1600.5

[continued at the right]

$$n_1 = 21 \quad R_1 = 998.0$$
$$n_2 = 8 \quad R_2 = 446.0$$
$$n_3 = 26 \quad R_3 = 1392.5$$
$$n_4 = 33 \quad R_4 = 1600.5$$
$$n_5 = 5 \quad R_5 = 202.5$$
$$n_6 = 7 \quad R_6 = 410.5$$

$$n = \Sigma n = 100 \quad \Sigma R = 5050.0$$

check:
$$\Sigma R = n(n+1)/2$$
$$= 100(101)/2$$
$$= 5050$$

H_o: the populations have the same distribution
H_1: the populations have different distributions
$\alpha = .05$ [assumed]
C.R. $H > \chi^2_{5,.05} = 11.071$
calculations:

$$H = [12/n(n+1)] \cdot [\Sigma(R_i^2/n_i)] - 3(n+1)$$
$$= [12/100(101)] \cdot [(998)^2/21 + (446)^2/8 + (1392.5)^2/26$$
$$+ (1600.5)^2/33 + (202.5)^2/5 + (410.5)^2/7] - 3(101)$$
$$= [.00119] \cdot [256770.7] - 303$$
$$= 2.074$$

conclusion:

Do not reject H_o; there is not sufficient evidence to reject the claim that the populations have the same distribution.

No; the results do not indicate a problem that requires corrective action.

NOTE: Compare these results to those of the parametric test in section 11-2, exercise #11.

9. $H = [12/n(n+1)] \cdot [\Sigma(R_i^2/n_i)] - 3(n+1)$, which depends only on the rank of each score and the number of scores within each group

a. Adding or subtracting a constant to each score does not change the number of scores within each group. Since adding or subtracting a constant to each score does not affect the order of the scores, their ranks and the calculated H statistic are not affected.

b. Multiplying or dividing each score by a positive constant does not change the number of scores within each group. Since multiplying or dividing each score by a positive constant does not affect the order of the scores, their ranks and the calculated H statistic are not affected.

c. Changing a single sample value so that it becomes an outlier will not change the overall values of the ranks (i.e., they will still vary from 1 to n), but it may change individual ranks by ± 1 and affect how the ranks are distributed among the groups. In general, the effect on the value of the test statistic H will be minimal unless the change places the largest (or smallest) rank in the group which previously had the smallest (or largest) ranks.

11.

rank	t	$T = t^3 - t$
8	3	24
12	5	120
15.5	2	6
18.5	2	6
21	3	24
24	3	24
30.5	4	60
34.5	4	60
37.5	2	6
40	3	24
43.5	2	6
45.5	2	6
		366

correction factor:
$$1 - \Sigma T/(n^3-n) = 1 - 366/(48^3-48)$$
$$= 1 - 366/110544$$
$$= .9966891$$

The original calculated test statistic is H = 14.431.
The corrected calculated test statistic is H = 14.431/.9966891 = 14.479.
No; the corrected value of H does not differ substantially from the original one.
NOTE: Be careful when counting the number of tied ranks; in addition to the easily recognized
".5's" that result from an even number of ties, there may be multiple whole number ranks
resulting from an odd number of ties.

3-6 Rank Correlation

NOTE: This manual calculates $d = R_x - R_y$, thus preserving the sign of d. This convention means
Σd must equal 0 and provides a check for the assigning and differencing of the ranks. In addition,
it must always be true that $\Sigma R_x = \Sigma R_y = n(n+1)/2$.

1. In each case the n = 5 pairs are pairs of ranks, called R_x and R_y below to emphasize that fact.

a.

R_x	R_y	d	d^2
1	1	0	0
3	3	0	0
5	5	0	0
4	4	0	0
2	2	0	0
15	15	0	0

$r_s = 1 - [6(\Sigma d^2)]/[n(n^2-1)]$
$= 1 - [6(0)]/[5(24)]$
$= 1 - 0$
$= 1$

Yes; there appears to be a perfect
positive correlation between
R_x and R_y.

b.

R_x	R_y	d	d^2
1	5	-4	16
2	4	-2	4
3	3	0	0
4	2	2	4
5	1	4	16
15	15	0	40

$r_s = 1 - [6(\Sigma d^2)]/[n(n^2-1)]$
$= 1 - [6(40)]/[5(24)]$
$= 1 - 2$
$= -1$

Yes; there appears to be a perfect
negative correlation between
R_x and R_y.

c.

R_x	R_y	d	d^2
1	2	-1	1
2	5	-3	9
3	3	0	0
4	1	3	9
5	4	1	1
15	15	0	20

$r_s = 1 - [6(\Sigma d^2)]/[n(n^2-1)]$
$= 1 - [6(20)]/[5(24)]$
$= 1 - 1$
$= 0$

No; there does not appear to be
any correlation between
R_x and R_y.

IMPORTANT NOTE: The rank correlation is correctly calculated using the ranks in the Pearson
product moment correlation formula of chapter 9 to produce
$$r_s = [\Sigma R_x R_y - (\Sigma R_x)(\Sigma R_y)]/[\sqrt{\Sigma R_x^2 - (\Sigma R_x)^2} \cdot \sqrt{\Sigma R_y^2 - (\Sigma R_y)^2}]$$
Since $\Sigma R_x = \Sigma R_y = 1+2+...+n = n(n+1)/2$ [always]
and $\Sigma R_x^2 = \Sigma R_y^2 = 1^2+2^2+...+n^2 = n(n+1)(2n+1)/6$ [when there are ties in the ranks],
it can be shown by algebra that the above formula can be shortened to
$$r_s = 1 - [6(\Sigma d^2)]/[n(n^2-1)]$$ when there are no ties in the ranks.

As the presence of ties typically does not make a difference in the first 3 decimals of r_s, this
manual uses the shortened formula exclusively and notes when use of the longer formula gives a
slightly different result.

3. The following table summarizes the calculations.

R_x	R_y	d	d^2
2	2	0	0
6	7	-1	1
3	7	-3	9
5	4	1	1
7	5	2	4
10	8	2	4
9	9	0	0
8	10	-2	4
4	3	1	1
1	1	0	0
55	55	0	24

$$r_s = 1 - [6(\Sigma d^2)]/[n(n^2-1)]$$
$$= 1 - [6(24)]/[10(99)]$$
$$= 1 - .145$$
$$= .855$$

H_0: $\rho_s = 0$
H_1: $\rho_s \neq 0$
$\alpha = .05$
C.R. $r_s < -.648$
$\qquad r_s > .648$
calculations:
$\quad r_s = .855$
conclusion:

Reject H_0; there is sufficient evidence to reject the claim that $\rho_s = 0$ and to conclude that $\rho_s \neq 0$ (in fact, that $\rho > 0$).

Yes; it does appear that salary increases as stress increases.

5. The following table summarizes the calculations.

R_x	R_y	d	d^2
2	5	-3	9
7	2	5	25
6	3	3	9
4	8	-4	16
5	10	-5	25
8	9	-1	1
9	1	8	64
10	7	3	9
3	6	-3	9
1	4	-3	9
55	55	0	176

$$r_s = 1 - [6(\Sigma d^2)]/[n(n^2-1)]$$
$$= 1 - [6(176)]/[10(99)]$$
$$= 1 - 1.067$$
$$= -.067$$

H_0: $\rho_s = 0$
H_1: $\rho_s \neq 0$
$\alpha = .05$
C.R. $r_s < -.648$
$\qquad r_s > .648$
calculations:
$\quad r_s = -.067$
conclusion:

Do not reject H_0; there is not sufficient evidence to conclude that $\rho_s \neq 0$.

No; there does not appear to be a relationship between the stress level of jobs and their physical demands.

7. The following table summarizes the calculations.

x	R_x	y	R_y	d	d^2
33.46	1	5.50	2	-1	1
50.68	2	5.00	2	1	1
87.92	4	8.08	3	1	1
98.84	5	17.00	6	-1	1
63.60	3	12.00	4	-1	1
107.34	6	16.00	5	1	1
	21		21	0	6

$$r_s = 1 - [6(\Sigma d^2)]/[n(n^2-1)]$$
$$= 1 - [6(6)]/[6(35)]$$
$$= 1 - .171$$
$$= .829$$

H_0: $\rho_s = 0$
H_1: $\rho_s \neq 0$
$\alpha = .05$
C.R. $r_s < -.886$
    $r_s > .886$
calculations:
    $r_s = .715$
conclusion:
    Do not reject H_0; there is not sufficient evidence to conclude that $\rho_s \neq 0$.

9. The following table summarizes the calculations.

x	R_x	y	R_y	d	d^2
17.2	1	.19	3	-2	4
43.5	6	.20	4.5	1.5	2.25
30.7	4	.26	8	-4	16
53.1	8	.16	1	7	49
37.2	5	.24	7	-2	4
21.0	2	.20	4.5	-2.5	6.25
27.6	3	.18	2	1	1
46.3	7	.23	6	1	1
	36.0		36.0	0.0	83.50

$$r_s = 1 - [6(\Sigma d^2)]/[n(n^2-1)]$$
$$= 1 - [6(83.5)]/[8(63)]$$
$$= 1 - .994$$
$$= .006$$

H_0: $\rho_s = 0$
H_1: $\rho_s \neq 0$
$\alpha = .05$
C.R. $r_s < -.738$
    $r_s > .738$
calculations:
    $r_s = .006$
conclusion:
    Do not reject H_0; there is not sufficient evidence to conclude that BAC and age are related.
NOTE: Using formula 9-1 (since there are ties) yields $r_s = 0.000$. Compare this to the parametric hypothesis test using Pearson's correlation in section 9-2 exercise #9.

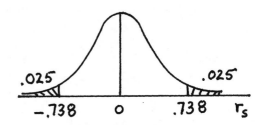

11. Refer to the data page for exercises #11 and #12. Since there are many ties, use formula 9-1 applied to the ranks. The critical values are $\pm z/\sqrt{n-1} = \pm 1.96/\sqrt{49} = \pm.280$.
    $n(\Sigma R_i^2) - (\Sigma R_i)^2 = 50(42894.50) - (1275)^2 = 519100$
    $n(\Sigma R_d^2) - (\Sigma R_d)^2 = 50(42917.00) - (1275)^2 = 520225$
    $n(\Sigma R_h^2) - (\Sigma R_h)^2 = 50(42747.00) - (1275)^2 = 511725$
    $n(\Sigma R_i R_d) - (\Sigma R_i)(\Sigma R_d) = 50(40679.25) - (1275)(1275) = 408337.50$
    $n(\Sigma R_i R_h) - (\Sigma R_i)(\Sigma R_h) = 50(32144.75) - (1275)(1275) = -18387.50$

Below are the relevant values for exercises #11 and #12.

	exercise #11						exercise #12				
int	R_i	dur	R_d	hgt	R_h	nic	R_n	tar	R_t	car	R_c
86	34.5	240	28	140	30.5	1.2	24.0	16	23.5	15	19.0
86	34.5	237	23.0	154	45	1.2	24.0	16	23.5	15	19.0
62	6.5	122	8	140	30.5	1.0	15.0	16	23.5	17	26.0
104	50	267	40.5	140	30.5	0.8	8.5	9	8.0	6	3
62	6.5	113	4	160	49.5	0.1	1	1	1	1	1
95	45	258	37	140	30.5	0.8	8.5	8	5.5	8	6
79	18.0	232	19	150	40.5	0.8	8.5	10	10	10	8.0
62	6.5	105	3	150	40.5	1.0	15.0	16	23.5	17	26.0
94	44	276	50	160	49.5	1.0	15.0	14	16.5	13	14.0
79	18.0	248	34	155	47.0	1.0	15.0	13	14.0	13	14.0
86	34.5	243	32	125	6.5	1.1	20.0	13	14.0	13	14.0
85	32	241	30.0	136	19.5	1.2	24.0	15	19.0	15	19.0
86	34.5	214	14	140	30.5	1.2	24.0	16	23.5	15	19.0
58	3	114	5	155	47.0	0.7	5.5	9	8.0	11	10.5
89	40.0	272	47	130	10.0	0.9	11	11	11	15	19.0
79	18.0	227	17	125	6.5	0.2	2	2	2	3	2
83	28.0	237	23.0	125	6.5	1.4	28.5	18	28.5	18	28.5
82	24.5	238	25.5	139	23.0	1.2	24.0	15	19.0	15	19.0
84	30.5	203	13	125	6.5	1.1	20.0	13	14.0	12	12
82	24.5	270	44.0	140	30.5	1.0	15.0	15	19.0	16	23.5
78	14.5	218	15	140	30.5	1.3	27	17	27	16	23.5
91	43	226	16	135	15.5	0.8	8.5	9	8.0	10	8.0
89	40.0	250	36	141	37	1.0	15.0	12	12	10	8.0
79	18.0	245	33	140	30.5	1.0	15.0	14	16.5	17	26.0
57	2	120	6.5	139	23.0	0.5	3	5	3	7	4.5
100	48	267	40.5	110	3	0.6	4	6	4	7	4.5
62	6.5	103	2	140	30.5	0.7	5.5	8	5.5	11	10.5
87	37	270	44.0	135	15.5	1.4	28.5	18	28.5	15	19.0
70	11	241	30.0	140	30.5	1.1	20.0	16	23.5	18	28.5
88	38	239	27	135	15.5		435.0		435.0		435.0
82	24.5	233	20	140	30.5						
83	28.0	238	25.5	139	23.0						
56	1	102	1	100	1						
81	21.5	271	46	105	2						
74	13	127	10	130	10.0						
102	49	275	48.5	135	15.5						
61	4	140	12	131	12						
83	28.0	264	38	135	15.5						
73	12	134	11	153	43.5						
97	46	268	42	155	47.0						
67	9	124	9	140	30.5						
90	42	270	44.0	150	40.5						
84	30.5	249	35	153	43.5						
82	24.5	237	23.0	120	4						
81	21.5	235	21	138	21						
78	14.5	228	18	135	15.5						
89	40.0	265	39	145	38						
69	10	120	6.5	130	10.0						
98	47	275	48.5	136	19.5						
79	18.0	241	30.0	150	40.5						
	1275.0		1275.0		1275.0						

for exercise #11

$$\Sigma R_i^2 = 42894.50$$

$$\Sigma R_d^2 = 42917.00$$

$$\Sigma R_h^2 = 42747.00$$

$$\Sigma R_i R_d = 40679.25$$

$$\Sigma R_i R_h = 32144.75$$

for exercise #12

$$\Sigma R_n^2 = 8509.00$$

$$\Sigma R_t^2 = 8530.00$$

$$\Sigma R_c^2 = 8519.00$$

$$\Sigma R_n R_t = 8355.50$$

$$\Sigma R_n R_c = 7994.25$$

a. Use the interval (i) and duration (d) values.

H_o: $\rho_s = 0$

H_1: $\rho_s \neq 0$

$\alpha = .05$

C.R. $r_s < -.280$

   $r_s > .280$

calculations:

   $r_s = 408337.50/[\sqrt{519100} \sqrt{520225}]$

   $= .7858$

.025   .025

−.280   O   .280 r_s

conclusion:

   Reject H_o; there is sufficient evidence to conclude that the interval between eruptions is (positively) correlated with the duration of the last eruption.

b. Use the interval (i) and height (h) values.

H_o: $\rho_s = 0$

H_1: $\rho_s \neq 0$

$\alpha = .05$

C.R. $r_s < -.280$

   $r_s > .280$

calculations:

   $r_s = -18387.50/[\sqrt{519100} \sqrt{511725}]$

   $= -.0356$

.025   .025

−.280   O   .280 r_s

conclusion:

   Do not reject H_o; there is not sufficient evidence to conclude that the interval between eruptions is correlated with the height of the last eruption.

c. Based on the results from parts (a) and (b), an eruption's duration is a better predictor of the time interval until the next eruption than is its height. Duration is significantly correlated with interval, but height is not.

3. Refer to the data page for exercises #13 and #14. Since there are many ties, use formula 9-1 applied to the ranks. The critical values for n=30 are ±.364.

   $n(\Sigma R_p^2) - (\Sigma R_p)^2 = 30(9455) - (465)^2 = 67425$

   $n(\Sigma R_w^2) - (\Sigma R_w)^2 = 30(9452.00) - (465)^2 = 67335$

   $n(\Sigma R_c^2) - (\Sigma R_c)^2 = 30(9406.50) - (465)^2 = 65970$

   $n(\Sigma R_p R_w) - (\Sigma R_p)(\Sigma R_w) = 30(9079.0) - (465)(465) = 56145$

   $n(\Sigma R_p R_c) - (\Sigma R_p)(\Sigma R_c) = 30(6475.5) - (465)(465) = -21960$

a. Use the price (p) and weight (w) values.

H_o: $\rho_s = 0$

H_1: $\rho_s \neq 0$

$\alpha = .05$

C.R. $r_s < -.364$

   $r_s > .364$

calculations:

   $r_s = 56145/[\sqrt{67425} \sqrt{67335}]$

   $= .8333$

.025   .025

−.364   O   .364 r_s

conclusion:

   Reject H_o; there is sufficient evidence to conclude that the price of a diamond is (positively) correlated with its weight.

Below are the relevant values for exercises #13 and #14.

pri	R_p	wgt	R_w	col	R_c
6958	11	1.00	1.5	3	8
5885	7	1.00	1.5	5	16.5
6333	8	1.01	3.5	4	11.5
4299	2	1.01	3.5	5	16.5
9589	17	1.02	5	2	4.5
6921	10	1.04	6.5	4	11.5
4426	3	1.04	6.5	5	16.5
6885	9	1.07	8.5	4	11.5
5826	6	1.07	8.5	5	16.5
3670	1	1.11	10	9	30
7176	12	1.12	11	2	4.5
7497	14	1.16	12	5	16.5
5170	4	1.20	13	6	22.5
5547	5	1.23	14	7	26.5
18596	24	1.25	15	1	1.5
7521	15	1.29	16	6	22.5
7260	13	1.50	17	6	22.5
8139	16	1.51	18	6	22.5
12196	21	1.67	19	3	8
14998	23	1.72	20	4	11.5
9736	18	1.76	21	8	28.5
9859	19	1.80	22	5	16.5
12398	22	1.88	23	6	22.5
25322	25	2.03	24.5	2	4.5
11008	20	2.03	24.5	8	28.5
38794	28	2.06	26	2	4.5
66780	30	3.00	27	1	1.5
46769	29	4.01	28.5	3	8
28800	26	4.01	28.5	6	22.5
28868	27	4.05	30	7	26.5
	465		465.0		465.0

exercise #14

gro	R_g	bud	R_b	vie	R_v
81.843	16	18.500	11	8.2	28
194.125	27	140.000	35	6.7	8.5
147.540	24	50.000	18.5	8.1	27
75.600	15	72.000	25	8.3	30.5
12.006	1	.250	1	7.9	25
100.853	18	90.000	29.5	8.3	30.5
67.155	12	104.000	33	6.7	8.5
140.424	22	75.000	26.5	6.4	4
68.759	13	55.000	20.5	7.3	13.5
329.691	35	55.000	20.5	7.7	19.5
217.631	30	22.000	13	7.1	11
198.571	29	3.900	3	8.0	26
138.339	21	10.000	7	8.5	34
181.280	26	6.000	4	7.3	13.5
47.000	5	.325	2	7.7	19.5
19.819	2	70.000	23.5	5.2	2
72.219	14	17.000	10	6.5	5.5
306.124	34	75.000	26.5	6.6	7
197.171	28	39.000	16	7.8	23.0
260.000	33	12.000	8	7.8	23.0
250.147	32	90.000	29.5	7.4	15
20.100	3	45.000	17	6.8	10
107.930	20	8.000	6	8.3	30.5
242.374	31	20.000	12	8.3	30.5
178.091	25	70.000	23.5	9.1	36
96.067	17	25.000	14	8.6	35
103.001	19	15.000	9	7.7	19.5
48.068	7	110.000	34	4.3	1
36.900	4	6.400	5	7.7	19.5
65.000	11	62.000	22	7.6	16.5
63.540	10	90.000	29.5	7.8	23.0
48.265	8	50.000	18.5	7.6	16.5
56.876	9	35.000	15	6.5	5.5
600.743	36	200.000	36	8.4	33
146.261	23	100.000	32	7.2	12
47.474	6	90.000	29.5	5.8	3
	666		666.0		666.0

for exercise #13

$$\Sigma R_p^2 = 9455$$

$$\Sigma R_w^2 = 9452.00$$

$$\Sigma R_c^2 = 9406.50$$

$$\Sigma R_p R_w = 9079.00$$

$$\Sigma R_p R_c = 6475.00$$

for exercise #14

$$\Sigma R_g^2 = 16206$$

$$\Sigma R_b^2 = 16199.00$$

$$\Sigma R_v^2 = 16192.00$$

$$\Sigma R_g R_b = 12743.50$$

$$\Sigma R_g R_v = 13464.50$$

b. Use the price (p) and color (c) values.
H_o: $\rho_s = 0$
H_1: $\rho_s \neq 0$
$\alpha = .05$
C.R. $r_s < -.364$
$\quad r_s > .364$
calculations:
$\quad r_s = -21960/[\sqrt{67425}\,\sqrt{65970}\,]$
$\quad\quad = -.3293$
conclusion:

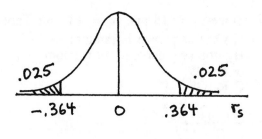

Do not reject H_o; there is not sufficient evidence to conclude that the price of a diamond is correlated with its color rating.

c. A diamond's value is determined more by its weight than by its color. The weight of a diamond is more strongly correlated with its price than is its color. In fact, since the correlation between price and color is not significant, color should ordinarily not be used (at least by itself) when predicting or otherwise considering price.

15. a. $t_{6,.025} = 2.365$; $r_s^2 = (2.447)^2/[(2.447)^2 + 6] = .499$, $r_s = \pm.707$
b. $t_{13,.025} = 2.160$; $r_s^2 = (2.160)^2/[(2.160)^2 + 13] = .264$, $r_s = \pm.514$
c. $t_{28,.025} = 2.048$; $r_s^2 = (2.048)^2/[(2.048)^2 + 28] = .130$, $r_s = \pm.361$
d. $t_{28,.005} = 2.763$; $r_s^2 = (2.763)^2/[(2.763)^2 + 28] = .214$, $r_s = \pm.463$
e. $t_{6,.005} = 3.707$; $r_s^2 = (3.707)^2/[(3.707)^2 + 6] = .696$, $r_s = \pm.834$

3-7 Runs Test for Randomness

NOTE: In each exercise, the item that appears first in the sequence is considered to be of the first type and its count is designated by n_1.

1. $n_1 = 10$ (# of D's) $G = 2$ (# of runs)
$n_2 = 5$ (# of R's)   CV: 3,12 (from Table A-10)
No; the sequence does not appear to be random.

3. $n_1 = 8$ (# of A's) $G = 6$ (# of runs)
$n_2 = 12$ (# of B's)   CV: 6,16 (from Table A-10)
No; the sequence does not appear to be random.

5. Since $n_1 = 12$ and $n_2 = 8$, use Table A-10.
H_o: the sequence is random
H_1: the sequence is not random
$\alpha = .05$
C.R. $G \leq 6$
$\quad G \geq 16$
calculations:
$\quad G = 10$
conclusion:

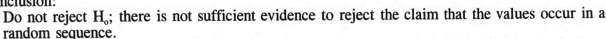

Do not reject H_o; there is not sufficient evidence to reject the claim that the values occur in a random sequence.

A lack of randomness would mean there was a pattern. Recognition of that pattern would give the bettor an advantage and put the casino at a disadvantage.

7. Since $n_1 = 15$ and $n_2 = 13$, use Table A-10.
H_0: the sequence is random
H_1: the sequence is not random
$\alpha = .05$
C.R. $G \leq 9$
$G \geq 21$
calculations:
$G = 12$
conclusion:

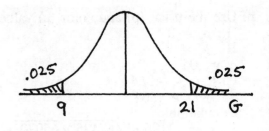

Do not reject H_0; there is not sufficient evidence to reject the claim that the values occur in a random sequence.

Given the listed results, Fred should not conclude that there is anything but randomness determining the marital status of the women he selects.

9. Since $n_1 = 17$ and $n_2 = 14$, use Table A-10.
H_0: the sequence is random
H_1: the sequence is not random
$\alpha = .05$
C.R. $G \leq 10$
$G \geq 23$
calculations:
$G = 14$
conclusion:

Do not reject H_0; there is not sufficient evidence to reject the claim that the values occur in a random sequence.

Yes; it appears that we elect Democrat and Republican candidates in a sequence that is random.

11. Define July 1 to be the middle of the year.

A A A A A A B B B B B B
Since $n_1 = 6$ and $n_2 = 6$, use Table A-10.
H_0: dates after and before the middle occur in a random sequence
H_1: dates after and before the middle do not occur in a random sequence
$\alpha = .05$
C.R. $G \leq 3$
$G \geq 11$
calculations:
$G = 2$
conclusion:

Reject H_0; there is sufficient evidence to reject the claim that dates after and before the middle of the year occur in a random sequence and to conclude that the sequence is not random (in fact, dates on the same side of the middle tend to occur in groups).

13. The sequence in O's and E's is as follows.
O E O O O E E O O O E O O O O E O E E E
E E E O O E O E O O O E E E E E O O O O
E O O O O O O O O E O E E E O O E O E E
O O E O E O E O E E E E E E E E E E O O
E E E E E O E E E O O E E O O O E E O O
Since $n_1 = 49$ and $n_2 = 51$, use the normal approximation.
$$\mu_G = 2n_1 n_2/(n_1 + n_2) + 1$$
$$ = 2(49)(51)/100 + 1 = 50.98$$
$$\sigma_G^2 = [2n_1 n_2(2n_1 n_2 - n_1 - n_2)]/[(n_1 + n_2)^2(n_1 + n_2 - 1)]$$
$$ = [2(49)(51)(4898)]/[(100)^2(99)]$$
$$ = 24.727$$

H_o: the sequence is random
H_1: the sequence is not random
$\alpha = .05$ [assumed]
C.R. $z < -z_{.025} = -1.960$
$\quad\quad z > z_{.025} = 1.960$
calculations:
$\quad G = 43$
$\quad z_G = (G - \mu_G)/\sigma_G$
$\quad\quad = (43 - 50.98)/\sqrt{24.73}$
$\quad\quad = -7.98/4.973$
$\quad\quad = -1.605$

conclusion:
$\quad$ Do not reject H_o; there is not sufficient evidence to conclude that the sequence is not random.

5. Ignoring the 5 values that are equal to 420, there are $n_1 = 34$ B's and $n_2 = 31$ A's.
Since $n_1 = 34$ and $n_2 = 31$, use the normal approximation.
$\quad \mu_G = 2n_1n_2/(n_1+n_2) + 1$
$\quad\quad = 2(34)(31)/65 + 1 = 33.431$
$\quad \sigma_G^2 = [2n_1n_2(2n_1n_2-n_1-n_2)]/[(n_1+n_2)^2(n_1+n_2-1)]$
$\quad\quad = [2(34)(31)(2043)]/[(65)^2(64)] = 15.927$
H_o: the sequence is random
H_1: the sequence is not random
$\alpha = .05$ [assumed]
C.R. $z < -z_{.025} = -1.960$
$\quad\quad z > z_{.025} = 1.960$
calculations:
$\quad G = 35$
$\quad z_G = (G - \mu_G)/\sigma_G$
$\quad\quad = (35 - 33.431)/\sqrt{15.927}$
$\quad\quad = 1.569/3.991$
$\quad\quad = .393$

conclusion:
$\quad$ Do not reject H_o; there is not sufficient evidence to conclude that the sequence is not random.

While the distances were probably estimated rather than measured, there is no evidence that the estimations utilized a pattern related to distances being above and below the median.

7. The minimum possible number of runs is $G = 2$ and occurs when all the A's are together and all the B's are together (e.g., A A B B).
The maximum possible number of runs is $G = 4$ and occurs when the A's and B's alternate (e.g., A B A B).
Because the critical region for $n_1 = n_2 = 2$ is
$\quad\quad$ C.R. $G \le 1$
$\quad\quad\quad G \ge 6$
the null hypothesis of the sequence being random can never be rejected at the .05 level.
Very simply, this means that it is not possible for such a small sample to provide 95% certainty that a non-random phenomenon is occurring.

Review Exercises

1. Let the before readings be group 1.
   claim: median difference = 0

   ```
   B-A  | +  +  +  +  +  +  +  +  +  +  +  +
   n = 12 +'s and -'s
   ```

   H_0: median difference = 0
   H_1: median difference $\neq$ 0
   α = .05
   C.R. $x \leq x_{L,12,.025} = 2$ OR   C.R. $x \leq x_{L,12,.025} = 2$
   $\qquad\qquad\qquad\qquad\qquad\qquad\qquad\quad x \geq x_{U,12,.025} = 12\text{-}2 = 10$

   calculations: calculations:
   $\quad x = 0$ $\quad x = 12$

   conclusion:
      Reject H_0; there is sufficient evidence to reject the claim that median difference = 0 and to conclude that median difference $\neq$ 0 (in fact, median difference > 0 -- i.e., the readings are higher before the medication).

   NOTE: Compare these results to those of the parametric test in section 8-3, exercise #8.

2. Let the before readings be group 1.
   claim: the populations have the same distribution

   ```
   B-A |  9   4   21   3   20   31   17   26   26   10   23   33
   R   |  3   2    7   1    6   11    5  9.5  9.5    4    8   12
   n = 12 non-zero ranks
   ```

   $\Sigma R- = 0$
   $\Sigma R+ = 78.0$
   $\Sigma R = 78.0$
   check: $\Sigma R = n(n+1)/2 = 12(13)/2 = 78$

   H_0: the populations have the same distribution
   H_1: the populations have different distributions
   α = .05
   C.R. $T \leq T_{L,12,.025} = 14$ OR   C.R. $T \leq T_{L,12,.025} = 14$
   $\qquad\qquad\qquad\qquad\qquad\qquad\qquad\quad T \geq T_{U,12,.025} = 78\text{-}14 = 64$

   calculations: calculations:
   $\quad T = 0$ $\quad T = 78.0$

   conclusion:
      Reject H_0; there is sufficient evidence to reject the claim that the populations have the same distribution and to conclude that the populations have different distributions (in fact, that the before readings are higher).

3. If there were no drug effect and each after score was equal to its corresponding before score, there would be a perfect correlation and $r_s = 1.00$. If there were a consistent drug effect and each after score was 50 less than its corresponding before score, there would be a perfect correlation (all the ranks remaining the same) and $r_s = 1.00$. Since $r_s = 1.00$ for both no drug effect and a consistent drug effect, the use of rank correlation would not be appropriate for this problem.

4. Below are the scores (in order) for each group. The group listed first is considered group 1.

placebo	R	calcium	R
96.5	1	102.7	2
104.8	4	104.4	3
106.1	5	106.6	6
107.2	7	108.0	8
108.5	9	109.6	10
113.6	12	113.2	11
116.3	14.5	114.7	13
118.1	16.5	116.3	14.5
120.4	18	118.1	16.5
122.5	21	120.9	19
123.1	22	121.4	20
124.6	25	123.4	23
128.8	27	124.3	24
	182	127.7	26
		129.1	28
			224

$n_1 = 13 \qquad \Sigma R_1 = 182$
$n_2 = 15 \qquad \Sigma R_2 = 224$

$n = \Sigma n = 28 \qquad \Sigma R = 406$

check: $\Sigma R = n(n+1)/2$
$\qquad = 28(29)/2 = 406$

$R = \Sigma R_1 = 182$

$\mu_R = n_1(n+1)/2$
$\qquad = 13(29)/2 = 188.5$

$\sigma_R^2 = n_1 n_2 (n+1)/12$
$\qquad = (13)(15)(29)/12$
$\qquad = 471.25$

H_o: the populations have the same distribution
H_1: the populations have different distributions
$\alpha = .05$
C.R. $z < -z_{.025} = -1.96$
$\qquad z > z_{.025} = 1.96$
calculations:
$z_R = (R - \mu_R)/\sigma_R$
$\qquad = (182 - 188.5)/\sqrt{471.25}$
$\qquad = -6.5/21.708 = -.299$

conclusion:
Do not reject H_o; there is not sufficient evidence to reject the claim that the populations have the same distribution.

5. Below are the scores (in order) for each group.
claim: the populations have different distributions

Pres	R	Pope	R	Mona	R
0	1	3	4	7	6
1	2	6	5	10	7
2	3	11	8	12	9.5
12	9.5	15	12.5	13	11
15	12.5	18	17.5	17	15.5
16	14	19	19	25	22
17	15.5	25	22	33	25
18	17.5		88.0		96.0
21	20				
25	22				
26	24				
36	26				
	167.0				

$n_1 = 12 \qquad R_1 = 167$
$n_2 = 7 \qquad R_2 = 88$
$n_3 = 7 \qquad R_3 = 96$

$n = \Sigma n = 26 \qquad \Sigma R = 351$

check:
$\Sigma R = n(n+1)/2$
$\qquad = 26(27)/2$
$\qquad = 351$

H_o: the populations have the same distribution
H_1: the populations have different distributions
$\alpha = .05$
C.R. $H > \chi^2_{2,.05} = 5.991$
calculations:

$$H = [12/n(n+1)] \cdot [\Sigma(R_i^2/n_i)] - 3(n+1)$$
$$= [12/26(27)] \cdot [(167)^2/12 + (88)^2/7 + (96)^2/7] - 3(27)$$
$$= [.0171] \cdot [4746.9] - 81$$
$$= .144$$

conclusion:

Do not reject H_o; there is not sufficient evidence to conclude that the populations have different distributions.

6. The following table summarizes the calculations.

x	R_x	y	R_y	d	d^2
107	10	111	10.5	-0.5	0.25
96	5	97	3	2	4
103	8	116	12	-4	16
90	3	107	8	-5	25
96	5	99	4	1	1
113	12	111	10.5	1.5	2.25
86	1	85	1	0	0
99	7	108	9	-2	4
109	11	102	6	5	25
105	9	105	7	2	4
96	5	100	5	0	0
89	2	93	2	0	0
	78		78.0	0.0	81.50

$$r_s = 1 - [6(\Sigma d^2)]/[n(n^2-1)]$$
$$= 1 - [6(81.5)]/[12(143)]$$
$$= 1 - .285$$
$$= .715$$

H_o: $\rho_s = 0$
H_1: $\rho_s \neq 0$
$\alpha = .05$
C.R. $r_s < -.591$
  $r_s > .591$
calculations:
  $r_s = .715$

conclusion:

Reject H_o; there is sufficient evidence to conclude that $\rho_s \neq 0$ (in fact, $\rho_s > 0$).

7. a. While the sequence begins and ends in the middle of a cycle, there is a repeated pattern of 4 T answers followed by 1 F answer.

  b. Since $n_1 = 20$ and $n_2 = 5$, use Table A-10.

  H_o: the sequence is random
  H_1: the sequence is not random
  $\alpha = .05$
  C.R. $G \leq 5$
    $G \geq 12$
  calculations:
    $G = 11$

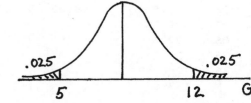

  conclusion:

  Do not reject H_o; there is not sufficient evidence to reject the claim that the T's and F's occur in a random sequence.

c. There are two explanations for the apparent discrepancy between parts (a) and (b). First, the runs test concerns itself only with the number of runs (which in this case is not unusual) and not the number of similar terms within each run (which in this case form a pattern). Secondly, the upper critical value G=12 is unattainable, and the observed G=11 is the maximum number of runs possible for $n_1=20$ and $n_2=5$ -- i.e., the sample size (in particular n_2) is too small to identify significant departure from randomness in the direction of having more runs than would be expected by chance alone.

8. Let the post-course scores be group 1 (i.e., let those with improved grades be $+$'s).
   claim: median improvement > 0

   ```
   62 +'s        35 -'s        3 0's
   n = 97 +'s or -'s
   ```

   Since n > 25, use z with
   $\mu_x = n/2 = 97/2 = 48.5$
   $\sigma_x = \sqrt{n}/2 = \sqrt{97}/2 = 4.924$

   H_o: median improvement ≤ 0
   H_1: median improvement > 0
   $\alpha = .05$

   C.R. $z < -z_{.05} = -1.645$ OR   C.R. $z > z_{.05} = 1.645$

   calculations: calculations:
   $x = 35$ $x = 62$
   $z_x = [(x+.5)-\mu_x]/\sigma_x$ $z_x = [(x-.5)-\mu_x]/\sigma_x$
   $\quad = [35.5 - 48.5]/4.924$ $\quad = [61.5 - 48.5]/4.924$
   $\quad = 13.0/4.924$ $\quad = 13/4.924$
   $\quad = -2.640$ $\quad = 2.640$

   conclusion:
   Reject H_o; there is sufficient evidence to conclude that median improvement > 0 (i.e., that the writing course is effective).

9. Let the first trial be group 1.
   claim: the populations have different distributions

emplo	A	B	C	D	E	F	G	H	I	J	K	L	M	N	O
F-S	0	13	5	15	15	126	28	-2	-5	31	3	51	3	14	37
R	-	6	4.5	8.5	8.5	14	10	-1	-4.5	11	2.5	13	2.5	7	12

   ```
   n = 14 non-zero ranks
   ΣR- =   5.5
   ΣR+ =  99.5
   ΣR  = 105.0
   ```

   check: $\Sigma R = n(n+1)/2 = 14(15)/2 = 105$

   H_o: the populations have the same distribution
   H_1: the populations have different distributions
   $\alpha = .01$

   C.R. $T \leq T_{L,14,.005} = 13$ OR   C.R. $T \leq T_{L,14,.005} = 13$
   $\qquad\qquad\qquad\qquad\qquad\qquad\qquad\qquad\qquad T \geq T_{U,14,.005} = 105-13 = 92$

   calculations: calculations:
   $T = 5.5$ $T = 99.5$

   conclusion:
   Reject H_o; there is sufficient evidence conclude that the populations have different distributions (in fact, that the times for the first trial are larger).

10. Below are the scores (in order) for each group. The group listed first is group 1, etc.

18-20	R	21-29	R	30+	R
97.1	2	97.9	6	97.0	1
97.7	5	98.2	8.5	97.3	3
98.0	7	98.2	8.5	97.5	4
98.4	10	99.0	14	98.6	12.5
98.5	11	99.6	15	98.6	12.5
	35		52.0		33.0

$n_1 = 5$ $R_1 = 35$
$n_2 = 5$ $R_2 = 52$
$n_3 = 5$ $R_3 = 33$
_____
$n = \Sigma n = 15$ $\Sigma R = 120$

check:
$\Sigma R = n(n+1)/2$
$\quad\quad = 15(16)/2$
$\quad\quad = 120$

H_o: the populations have the same distribution
H_1: the populations have different distributions
$\alpha = .05$
C.R. $H > \chi^2_{2,.05} = 5.991$
calculations:

$H = [12/n(n+1)] \cdot [\Sigma(R_j^2/n_j)] - 3(n+1)$
$\quad = [12/15(16)] \cdot [(35)^2/5 + (52)^2/5 + (33)^2/5] - 3(16)$
$\quad = [.05] \cdot [1003.6] - 48$
$\quad = 2.180$

conclusion:

Do not reject H_o; there is not sufficient evidence to conclude that the populations have different distributions.

NOTE: Compare these results to those of the parametric test in section 11-2, exercise #3.

11. Below are the scores (in order) for each group. The group listed first is considered group 1.
claim: the populations have the same distribution

Drug	R	Ctrl	R
508	1	549	4
512	2	563	5
516	3	565	6
603	8	598	7
621	9	652	12.5
624	10	668	14
627	11	674	15
652	12.5	676	16
711	18.5	703	17
729	20	711	18.5
747	21	772	22
777	23	789	25
787	24	800	26
	163.0	821	27.5
		821	27.5
		830	29
		837	30
			302.0

$n_1 = 13$ $\Sigma R_1 = 163$
$n_2 = 17$ $\Sigma R_2 = 302$
_____
$n = \Sigma n = 30$ $\Sigma R = 465$

check: $\Sigma R = n(n+1)/2$
$\quad\quad\quad\quad = 30(31)/2$
$\quad\quad\quad\quad = 465$

$R = \Sigma R_1 = 163$

$\mu_R = n_1(n+1)/2$
$\quad = 13(31)/2$
$\quad = 201.5$

$\sigma^2_R = n_1 n_2(n+1)/12$
$\quad = (13)(17)(31)/12$
$\quad = 570.9$

H_o: the populations have the same distribution
H_1: the populations have different distributions
$\alpha = .05$
C.R. $z < -z_{.025} = -1.960$
  $z > z_{.025} = 1.960$
calculations:
  $z_R = (R - \mu_R)/\sigma_R$
    $= (163 - 201.5)/\sqrt{570.9}$
    $= -38.5/23.894$
    $= -1.611$

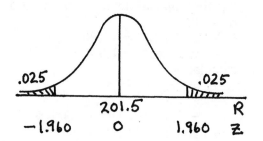

conclusion:
  Do not reject H_o; there is not sufficient evidence to reject the claim that the populations have the same distribution.

2. The following table summarizes the calculations.

x	R_x	y	R_y	d	d^2
91	5	4.56	2	3	9
92	6	6.48	6	0	0
82	2	5.99	5	-3	9
85	3	7.92	8	-5	25
87	4	5.36	4	0	0
80	1	3.32	1	0	0
94	7	7.32	7	0	0
97	8	5.27	3	5	25
	36		36	0	68

$r_s = 1 - [6(\Sigma d^2)] / [n(n^2-1)]$
  $= 1 - [6(68)] / [8(63)]$
  $= 1 - .810$
  $= .190$

H_o: $\rho_s = 0$
H_1: $\rho_s \neq 0$
$\alpha = .05$
C.R. $r_s < -.738$
  $r_s > .738$
calculations:
  $r_s = .190$
conclusion:
  Do not reject H_o; there is not sufficient evidence to conclude that $\rho_s \neq 0$.

3. Below are the scores (in order) for each group. The group listed first is considered group 1.

Beer	R	Liqr	R
.129	1	.182	9
.146	2	.185	10
.148	3	.190	12.5
.152	4	.205	15
.154	5	.220	17
.155	6	.224	18
.164	7	.225	19.5
.165	8	.226	20.5
.187	11	.227	21
.190	12.5	.234	22
.203	14	.241	23
.212	16	.247	24
	89.5	.253	25
		.257	26
			261.5

$n_1 = 12$ $\Sigma R_1 = 89.5$
$n_2 = 14$ $\Sigma R_2 = 261.5$
$n = \Sigma n = 26$ $\Sigma R = 351.0$

check: $\Sigma R = n(n+1)/2$
  $= 26(27)/2$
  $= 351$

$R = \Sigma R_1 = 89.5$

$\mu_R = n_1(n+1)/2$
  $= 12(27)/2$
  $= 162$

$\sigma_R^2 = n_1 n_2 (n+1)/12$
  $= (12)(14)(27)/12 = 378$

H_o: the populations have the same distribution
H_1: the populations have different distributions
$\alpha = .05$
C.R. $z < -z_{.025} = -1.96$
$z > z_{.025} = 1.96$
calculations:
$z_R = (R - \mu_R)/\sigma_R$
$= (89.5 - 162)/\sqrt{378}$
$= -72.5/19.442 = -3.729$

conclusion:
Reject H_o; there is sufficient evidence to reject the claim that the populations have the same distribution and to conclude that they have different distributions (in fact, population 1 has lower BAC levels).

Based on these results, it appears that the liquor drinkers have a higher BAC level and are (presumably) more dangerous.

14. The following table summarizes the calculations.

R_x	R_y	d	d^2
4	1	3	9
3	5	-2	4
5	6	-1	1
6	2	4	16
7	7	0	0
8	8	0	0
2	3	-1	1
1	4	-3	9
36	36	0	40

$r_s = 1 - [6(\Sigma d^2)]/[n(n^2-1)]$
$= 1 - [6(40)]/[8(63)]$
$= 1 - .476$
$= .524$

H_o: $\rho_s = 0$
H_1: $\rho_s \neq 0$
$\alpha = .05$
C.R. $r_s < -.738$
$r_s > .738$

calculations:
$r_s = .524$
conclusion:
Do not reject H_o; there is not sufficient evidence to reject the claim that $\rho_s = 0$.

15. Let the winners be group 1.
claim: the populations have the same distribution

pair	1	2	3	4	5	6	7	8
W-R	12	-5	-2	-2	6	0.5	3.5	0
R	7	-5	-2.5	-2.5	6	1	4	-

n = 7 non-zero ranks
$\Sigma R- = 10.0$
$\Sigma R+ = 18.0$
$\Sigma R = 28.0$ [check: $\Sigma R = n(n+1)/2 = 7(8)/2 = 28$]

H_o: the populations have the same distribution
H_1: the populations have different distributions
$\alpha = .05$
C.R. $T \leq T_{L,7,.025} = 2$ OR   C.R. $T \leq T_{L,7,.025} = 2$
$T \geq T_{U,7,.025} = 28\text{-}2 = 26$

calculations: calculations:
   $T = 10$ $T = 18$
conclusion:
   Do not reject H_o; there is not sufficient evidence to reject the claim that the populations have the same distribution.

Cumulative Review Exercises

1. a. Since $n_1 = 20$ and $n_2 = 10$, use Table A-10.
   claim: the sequence is random
   H_o: the sequence is random
   H_1: the sequence is not random
   $\alpha = .05$
   C.R. $G \leq 9$
   $G \geq 20$
   calculations:
   $G = 15$
   conclusion:
   Do not reject H_o; there is not sufficient evidence to reject the claim that the sequence is random.

   b. Let p be the proportion of women: $\hat{p} = x/n = 10/30 = .333$
   claim: $p \neq .50$
   H_o: $p = .50$
   H_1: $p \neq .50$
   $\alpha = .05$
   C.R. $z < -z_{.025} = -1.960$
   $z > z_{.025} = 1.960$
   calculations:
   $z_p = (\hat{p} - \mu_p)/\sigma_p$
   $= (.333 - .50)/\sqrt{(.5)(.5)/30}$
   $= -.167/.0913$
   $= -1.826$
   conclusion:
   Do not reject H_o; there is not enough evidence to conclude that $p \neq .50$.

   c. Let p be the proportion of women: $\hat{p} = x/n = 10/30 = .333$
   $\hat{p} \pm z_{.025} \cdot \sqrt{\hat{p}\hat{q}/n}$
   $.333 \pm 1.960 \cdot \sqrt{(.333)(.667)/30}$
   $.333 \pm .169$
   $.165 < p < .502$

   d. There is not sufficient evidence to conclude that there was a bias against either gender or that the sample was not obtained in a random sequence. Even if there had been significantly more of one gender than another, that might reflect the gender distribution of the passengers and not gender bias by the pollster. There do not appear to be any problems with the survey in these areas.

2. The histogram of Sunday rainfall amounts given below is similar to the one for Monday. The distribution is positively skewed and is not normal.

3. a. $\bar{x} = (\Sigma x)/n = 3.52/52 = .068$ inches
   b. $\tilde{x} = (x_{26} + x_{27})/2 = (0.00 + 0.00)/2 = 0.000$ inches
   c. $M = 0.000$ inches (occurs 37 times)
   d. $R = 1.28 - 0.00 = 0.00$ inches
   e. $s^2 = [n(\Sigma x^2) - (\Sigma x)^2]/[n(n-1)] = [52(2.2790) - (3.52)^2]/[52(51)] = .0400$
      $s = .200$ inches
   f. Outliers are scores more than 2 standard deviations from the mean.
      $\bar{x} \pm 2s$
      $.068 \pm 2(.200)$
      $-.332$ to $.468$

The only score beyond those limits is the 1.28 inches.

4. claim: the proportions of rainy days are not the same for all seven days

day of the week

		Mo	Tu	We	Th	Fr	Sa	Su	
rain?	Y	19	16	16	21	20	20	15	127
	N	33	36	36	31	32	32	37	237
		52	52	52	52	52	52	52	364

H_o: there is homogeneity of proportions
H_1: there is not homogeneity of proportions
$\alpha = .05$
C.R. $\chi > \chi^2_{6,.05} = 12.592$
calculations:

$\chi^2 = .0405 + .2531 + .2531 + .4499 + .1901 + .1901 + .5444$
$\quad\quad .0217 + .1356 + .1356 + .2411 + .1019 + .1019 + .2917$
$\quad = 2.951$

conclusion:
    Do not reject H_o; there is not enough evidence to say the proportions of rainy days are not the same for all seven days.

NOTE: A one-dimensional goodness-of-fit test is not appropriate since the 127 rainy days were constrained by a second dimension (e.g., there could be no more than 52 on any one day). In addition, the test does not require equal marginal column totals and it was not necessary to omit the 53rd Wednesday.

5. a. Let the After scores be group 1.
   claim: median difference = 0

pair	1	2	3	4	5	6	7	8	9	10
A-B	+	0	-	+	+	+	-	+	+	+

   n = 9 +'s and -'s

   H_o: median difference = 0
   H_1: median difference $\neq$ 0
   α = .05

   C.R. x ≤ $x_{L,9,.025}$ = 1   **OR**   C.R. x ≤ $x_{L,9,.025}$ = 1
   $\qquad\qquad\qquad\qquad\qquad\qquad$ x ≥ $x_{U,9,.025}$ = 9-1 = 8

   calculations:$\qquad\qquad\qquad$calculations:
    x = 2$\qquad\qquad\qquad\qquad\quad$x = 7

   conclusion:
      Do not reject H_o; there is not sufficient evidence to reject the claim that
      median difference = 0.

   b. Let the After scores be group 1.
   claim: the populations have the same distribution

pair	1	2	3	4	5	6	7	8	9	10
A-B	100	0	-10	120	140	110	-30	90	150	150
R	4	-	-1	6	7	5	-2	3	8.5	8.5

   n = 9 non-zero ranks
   ΣR- = 3
   ΣR+ = 42
   ΣR = 45
   check: ΣR = n(n+1)/2 = 9(10)/2 = 45

   H_o: the populations have the same distribution
   H_1: the populations have different distributions
   α = .05

   C.R. T ≤ $T_{L,9,.025}$ = 6   **OR**   C.R. T ≤ $T_{L,9,.025}$ = 6
   $\qquad\qquad\qquad\qquad\qquad\qquad$ T ≥ $T_{U,6,.025}$ = 45-6 = 39

   calculations:$\qquad\qquad\qquad$calculations:
    T = 3$\qquad\qquad\qquad\qquad\quad$T = 42

   conclusion:
      Reject H_o; there is sufficient evidence to reject the claim that the populations have the
      same distribution and to conclude that they have different distributions (in fact, the After
      scores tend to be larger).

c. Let the After scores be group 1.
   original claim: $\mu_d = 0$ [n $\leq$ 30 and σ_d unknown, use t]
   d = $x_1 - x_2$: 100 0 -10 120 140 110 -30 90 150 150
   $\quad$ n = 10
   $\quad \Sigma d = 820 \qquad\qquad \bar{d} = 82$
   $\quad \Sigma d^2 = 110200 \qquad s_d = 69.09$
   H_o: $\mu_d = 0$
   H_1: $\mu_d \neq 0$
   $\alpha = .05$
   C.R. t < $-t_{9,.025} = -2.262$
   $\qquad$ t > $t_{9,.025} = 2.262$
   calculations:

   $\quad t_{\bar{d}} = (\bar{d} - \mu_{\bar{d}})/s_{\bar{d}}$
   $\qquad = (82 - 0)/(69.09/\sqrt{10})$
   $\qquad = 82/21.847$
   $\qquad = 3.753$
   conclusion:
   $\quad$ Reject H_o; there is sufficient evidence to reject the claim that $\mu_d = 0$ and to conclude that
   $\quad \mu_d \neq 0$ (in fact, $\mu_d > 0$).

d. The test in part (a) failed to reject the hypothesis of no difference.
   The test in part (b) rejected the hypothesis of no difference in a "relatively close" decision.
   The test in part (c) "strongly" rejected the hypothesis of no difference.

   Each test used progressively more information from
   $\qquad$ (a) the sign of the difference
   $\quad$ to (b) the rank of the difference
   $\quad$ to (c) the actual value of the difference.

   In general, the more information a test uses, the better able it is to detect departures from
   the null hypothesis. These results also support the statement that non-parametric tests [like
   parts (a) and (b)] lack the sensitivity of parametric tests [like part (c)] and so require
   stronger evidence before a null hypothesis is rejected.

FINAL NOTE: Congratulations! You have completed statistics -- the course that everybody likes
to hate. I hope that this manual has helped to make the course a little more understandable for
you. I hope you leave the course with an appreciation of broad principles and not memories of
merely manipulating formulas. I wish you well in your continued studies, and that you achieve
your full potential wherever your journey of life may lead.